About Island Press

Since 1984, the nonprofit Island Press has been stimulating, shaping, and communicating the ideas that are essential for solving environmental problems worldwide. With more than 800 titles in print and some 40 new releases each year, we are the nation's leading publisher on environmental issues. We identify innovative thinkers and emerging trends in the environmental field. We work with world-renowned experts and authors to develop cross-disciplinary solutions to environmental challenges.

Island Press designs and implements coordinated book publication campaigns in order to communicate our critical messages in print, in person, and online using the latest technologies, programs, and the media. Our goal: to reach targeted audiences—scientists, policymakers, environmental advocates, the media, and concerned citizens—who can and will take action to protect the plants and animals that enrich our world, the ecosystems we need to survive, the water we drink, and the air we breathe.

Island Press gratefully acknowledges the support of its work by the Agua Fund, Inc., The Margaret A. Cargill Foundation, Betsy and Jesse Fink Foundation, The William and Flora Hewlett Foundation, The Kresge Foundation, The Forrest and Frances Lattner Foundation, The Andrew W. Mellon Foundation, The Curtis and Edith Munson Foundation, The Overbrook Foundation, The David and Lucile Packard Foundation, The Summit Foundation, Trust for Architectural Easements, The Winslow Foundation, and other generous donors.

The opinions expressed in this book are those of the author(s) and do not necessarily reflect the views of our donors.

Trophic Cascades

TROPHIC CASCADES:
PREDATORS, PREY, AND THE
CHANGING DYNAMICS OF NATURE

Edited by

John Terborgh and James A. Estes

Washington | *Covelo* | *London*

Library of Congress Cataloging-in-Publication Data

Trophic cascades: predators, prey, and the changing dynamics of nature / edited by John Terborgh and James A. Estes.
     p. cm.
  Includes bibliographical references and index.
  ISBN-13: 978-1-59726-486-0 (cloth : alk. paper)
  ISBN-10: 1-59726-486-5 (cloth : alk. paper)
  ISBN-13: 978-1-59726-487-7 (pbk. : alk. paper)
  ISBN-10: 1-59726-487-3 (pbk. : alk. paper)
1. Predation (Biology) 2. Predatory animals—Ecology. 3. Keystone species. I. Terborgh, John, 1936–
II. Estes, J. A. (James A.), 1945–
  QL758.T76 2010
  577′.16—dc22

                                                                                         2009042702

Printed using Bembo

Text design by Karen Wenk
Typesetting by Karen Wenk
Text font: Bembo

Printed on recycled, acid-free paper

Manufactured in the United States of America
10 9 8 7 6 5 4 3 2 1

Contents

Foreword

Predators often shape nature in ways that at first glance appear counterintuitive. More than 20 years ago, a colleague at the University of Washington showed me a photo of a sea otter floating on its back, holding a sea urchin with its two front paws, poised for a meal. I remember my surprise when he told me that without sea otters there wouldn't be kelp forests. "But wait," I said, "isn't that sea otter about to eat an urchin and not a piece of kelp?" He agreed that was the case. "So why should it matter to the kelp whether otters are around?" I asked. He then explained that it's because sea otters (as predators) keep urchin populations in check, and without otters, kelp-devouring urchins proliferate, essentially mow the kelp forests down, and leave an undersea desert in their wake.

I found this story fascinating and compelling: fascinating because it uncovered the hidden and intricate ways nature works, compelling because it demonstrated the broad and devastating effects that can follow the disappearance of a single species. The transformation of a majestic underwater kelp forest into an undersea ghost town remains a vivid and profoundly sad image.

It turns out that my colleague was making the final preparation for an article on the sea otter, kelp, and sea urchin saga that appeared in the journal *Science*. One important factor in the discovery of this tale is that sea otters, through reintroductions and natural population growth, were making a comeback years after they were hunted to local extinction. This provided scientists with the important and rare opportunity to observe the broader ecological consequences of a world with and without sea otters. This unusual series of events afforded an excellent proving ground for uncovering the indirect consequences of the loss of a predator on its food web, a phenomenon that has come to be known as a trophic cascade. This particular trophic cascade has become a classic example of the phenomenon, and further descriptions of it and many other trophic cascades appear in this volume.

When I was approached by the co-editors of this book to support this effort, I recognized the important and varied roles a volume on trophic cascades could play. First, although the ecological foundation for trophic cascades appeared in the scientific literature more than half a century ago, until now the tremendous scientific progress made in this field has remained scattered. In addition, over the years the extent and importance of trophic cascades in nature have been hotly debated.

This first comprehensive work on trophic cascades is excellent, thorough, and timely. It puts in one place the accumulated wisdom of an impressive array of ecological detectives who through careful long-term investigation have uncovered the complex dynamics of a wide range of ecotypes and regions. The reader will be treated to accounts of trophic cascades akin to adventure stories, full of intricate interactions, intrigue, unexpected consequences, and the discovery of pattern and order against a backdrop of noise and environmental degradation. The studies span the furthest reaches of the planet and include work in some of its most remote and pristine ecosystems. Some accounts go beyond consideration of trophic effects to examine, for example, behavioral responses of prey to predators, often called the ecology of fear. Together, these yarns form a rich tapestry. For those who are fascinated by the complex and mysterious ways nature works, the accounts in this book will not disappoint.

Human transformation of the natural world has been broad and durable and has accelerated in recent decades. With few exceptions the trajectory has been to a simplified and degraded natural world. Few if any pristine ecosystems remain, and the chance to understand them is passing by. Many of the studies recounted in this volume could not be duplicated today. Among the many transformations of nature wrought by humankind is a large and growing predator void. The diminution of top predators now extends to the ocean, the last great natural frontier on this planet, and to the furthest reaches of the deep sea. Forecasting the follow-on effects in ocean ecosystems is of critical importance. Is a great unraveling of ocean life taking place? Are once lush underwater paradises being transformed into wastelands? This volume may well contain important clues for answering these questions.

Most importantly, I believe this work will be of vital importance in addressing the urgent task of stemming the destruction of natural systems and working to restore them. Understanding the ways ecosystems work is critical to understanding and devising ways to repair them, and this comprehensive global portrait of trophic cascades will be invaluable in that effort. If the wisdom gathered into this book is harnessed for the restoration of degraded oceans, seas, rivers, and lands, in my view it will have served its highest and best purpose.

My hope is that this book attracts a broad audience far beyond the scientific community. To some the loss of predators may seem unimportant, irrelevant, perhaps even good. But what this book will teach all of us is that the loss of predators is just a starting point, with far-reaching ripple effects. The chances are good that somewhere in the vast chain of consequences of the great predator void there is something every one of us would sorely miss in a tangible, personal way. My hope is that this volume sparks concern and serves as a call to action for every reader.

In the end we will protect only what we care about. There is something in this volume that everyone can care about.

Ellen K. Pikitch, Executive Director,
Institute for Ocean Conservation Science

Preface

Since its origins in the early twentieth century, the science of ecology has emphasized physical processes to account for the structure and organization of ecosystems. Photosynthesis is the basic energy transduction process of life, and the primary productivity derived from it underlies all other biotic fluxes. Photosynthesis is supported by light, warmth, and minerals and nourished by carbon dioxide. It is therefore understandable that the early generations of ecologists found compelling links between these drivers of primary productivity and the structure of ecosystems (Lindeman 1942; Odum 1957; Walter 1964, 1968; Holdridge 1967; Rosenzweig 1973; Whittaker 1975). Biogeography has followed in a parallel tradition of seeking understanding through the analysis of physical variables and processes, such as area, degree of isolation, elevation, and vicariance (Wallace 1860; Croizat 1958; Walter 1964, 1968; MacArthur and Wilson 1967). Productivity in the sea has seemed even more tightly coupled to physical processes than vegetation on land, and oceanographers have accordingly devoted great attention to such processes as nutrient fluxes, temperature shifts, currents, upwelling, vertical mixing, and convergences (Parsons et al. 1983; Lalli and Parsons 1997).

Physical parameters of the environment are easy to measure and obviously correlate with major features of the biological world, so it is natural that they were emphasized first. A century of studying the relationships between the physical and biological worlds has yielded a powerful science, able to predict such fundamental attributes of ecosystems as the productivity of aquatic and terrestrial ecosystems, the geographic extent of major biomes, responses to gradients in rainfall and elevation, latitudinal diversity gradients, and much more (Whittaker 1975; Rosenzweig 1996; but see Chapter 16, this volume).

Coexisting with the bottom-up flow of productivity-driven phenomena is a top-down countercurrent. Products of photosynthesis are harvested by consumers, and these in turn are eaten by predators. Predators reign at the top of the food chain and prey on consumers such as grazers, browsers, frugivores, and, in the sea, the myriad life forms that feed on phytoplankton. But do predators eat *enough* consumers to make a difference at the producer level? Ecologists have been debating this question for decades. The question seems so simple, yet the answer has been exasperatingly elusive. But in the answer lies the key to how nature works, so the answer matters enormously, both as a matter of scientific understanding and as a point of departure for how to ensure the survival of biodiversity on Earth.

If predators limit consumers, then the plants that constitute the producer level at the bottom of the food chain can flourish. This is the famous "green world" hypothesis of Hairston, Smith, and Slobodkin (HSS 1960). But if predators are able only to catch the sick, the weak, and the old in prey populations, the prey (here referring to consumers) must be regulated from the bottom up by their principal resources. But the world is obviously green, so how can this be true? The retort of bottom-up advocates is that much of that which is green in the world around us is either toxic or so lacking in nutritional value that it is inedible to consumers. In short, the world is green because so much of it is inedible (Ehrlich and Raven 1964; Murdoch 1966).

A parallel line of debate extends to the ocean margins, where kelps and seagrasses are often the dominating features of shallow seascapes. Farther out to sea, across the vast surface waters of open oceans where photosynthesis occurs solely through phytoplankton, and in the dark abyssal realm where organisms obtain their energy through detrital fallout and chemosynthesis, alternatives to bottom-up control seem barely to have been considered. Despite the fact that most phytoplankton species are edible to a wide range of zooplankton, bottom-up regulation remains an article of faith for most biological oceanographers (Lalli and Parsons 1997; Chapters 6 and 19, this volume).

Those who hold the view that much vegetation is inedible or that rates of consumption are intrinsically exceeded by rates of production are skeptical about the capacity of top-down forces to structure ecosystems (Strong 1992; Polis et al. 2000). According to this point of view, strong top-down regulation is a curiosity or aberration, present in some circumstances but not in others, and overall, in the big picture, unimportant.

Predation and other types of top-down forces (e.g., herbivory, parasitism) have been relegated to secondary status for a simple reason: Top-down processes are difficult to study. The laws of thermodynamics require that predators be rare relative to their prey and that they range over much larger areas, rendering exclusion experiments difficult and expensive to implement. Consequently, most such experiments are conducted at the scale of a few square meters or less (Schmitz et al. 2000). Are such experiments going to convince anyone that orcas, great white sharks, wolves, tigers, and jaguars are important? At best, such inference requires a leap of faith (Carpenter and Kitchell 1988).

Acceptance by scientists of the simple logic of HSS has therefore lagged. In part, scientists have been held back by their own cultural baggage: institutionalized skepticism, the existence of alternative hypotheses that have not been definitively ruled out, and standards of evidence that exceed achievable limits. The native skepticism of science is most readily dispelled via rigorously controlled experiments. Yet experiments on the scale necessary to study the predator–prey dynamics of large vertebrates have been beyond the capacity and the means of the scientific establishment in the United States or anywhere else.

These scaling challenges have been hindrance enough on land, but in the sea they have been insurmountable. Large marine predators such as great white sharks, tuna, and toothed cetaceans routinely track prey over hundreds or thousands of kilometers of open ocean. No experiments are possible here. But the fact that appropriate experiments are beyond the reach of investigators does not mean that the large predators of the sea are unimportant or that they should be ignored by ecologists. They present a challenge to the ingenuity of the investigator, who must search the world for places, events, or special circumstances that create "natural experiments" that could not be done in a planned or replicated fashion. Perhaps the evidence to be derived from such "experiments" is not as airtight or repeatable as it would be from rigorously controlled and replicated treatments, but it is all we have or are likely to have in the foreseeable future as the basis for inferring how the world's largest ecosystem operates.

Biologists have recognized the central importance of predation since the mid-nineteenth century (Bates 1862; Müller 1879). But, in retrospect, it is now obvious that the green world hypothesis of HSS and an even earlier articulation of it by Charles Elton (1927) were ahead of their time. Many ecologists were convinced by the logic of the green world argument and conceded that HSS might be right, but there was no obvious way to put the idea to rigorous tests. Hairston, Smith, and Slobodkin earned a tip of the hat in textbooks, but in an era of increasing reductionism their thesis was put aside as unsubstantiated speculation.

Inspired by the insights of such distinguished forerunners as Charles Elton, Aldo Leopold, and HSS, a handful of ecologists sought to overcome the methodological handicaps inherent in the study of predator–prey systems through intensive investigations of systems that offered special tractability. Paine was the pioneer with his *Pisaster* removal experiment in 1966. The experiment was free of confounding variables, and the results were dramatic. The experiment was immediately hailed as a landmark and enshrined in textbooks. But skepticism remained. Perhaps the results were attributable to a particularly powerful keystone predator, perhaps targeting by the predator of the dominant competitor in the prey community was a lucky factor that would not pertain to other systems, perhaps the space-limited system of the rocky intertidal zone created special conditions that did not apply to open systems, perhaps the results depended in some unknown way on the multistage life cycles of most of the organisms involved, and so on. Such lingering doubts caused many scientists to regard Paine's results as an isolated example without far-reaching relevance to how biological systems work in general.

A GROWING CONSENSUS

Despite methodological handicaps, evidence corroborating key features of Paine's findings began to accumulate from carefully constructed case studies. Certainly the most dramatic and quotable of the early studies was the startling discovery by Estes and Palmisano (1974) of how Pacific kelp forests owe their existence to sea otters. Exuberant accounts of this signal breakthrough reverberated through the scientific and popular press, creating public awareness that biological interactions are important. Appearing in parallel with these early empirical investigations were seminal ideas that began to build a theoretical foundation under the concept of top-down regulation (Rosenzweig 1973; Fretwell 1977; Paine 1980; Okasanen et al. 1981). On the empirical side, Paine (1980) showed that food web dynamics were largely attributable to strong interactors, often called keystone species; the starfish *Pisaster* was the prime example. The next year, Oksanen and his colleagues confirmed the existence of predicted abrupt state shifts driven by herbivory and predation on arctic productivity gradients (Oksanen et al. 1981). Soon after that, Schoener and Toft (1983) and Pacala and Roughgarden (1984) demonstrated that *Anolis* lizards could exert strong top-down control on spiders and herbivorous insects, respectively, on small West Indian islands. Close on the heels of these meso-scale terrestrial studies came the remarkable whole ecosystem experiments of Carpenter and Kitchell (1988, 1993). Predicted dramatic consequences of the experimental removal or addition of the top trophic level from small lakes again stirred the popular (and scientific) imagination.

Similar narratives began to come in a rush from all types of ecosystems, from unbounded marine situations to freshwater streams and lakes, and from terrestrial systems from the Arctic to the tropics. Here are some of the better-known examples.

- In the absence of top carnivores, white-tailed deer irruptions occurred over large portions of the eastern United States, with consequent suppression of hemlock recruitment in northern forests and oak recruitment in mid-latitude forests and selective depletion of favored herbaceous plants (Alverson et al. 1988; McShea et al. 1997; Rooney et al. 2004).
- Cage experiments in tropical and temperate streams confirmed the structuring effects of top-down forcing and the importance of food chain length in determining whether autotrophs are enhanced or reduced by apex predators (Power 1990; Flecker 1992).
- Severe degradation of the vegetation of an oceanic island followed the reintroduction of a long-absent native herbivore (Campbell et al. 1991).
- The growth of balsam fir, as indicated by growth rings, is indirectly regulated by wolf predation on moose on Isle Royale (McLaren and Peterson 1994).
- Overharvest of fishes and invertebrates led to algal overgrowth of Jamaican coral reefs (Hughes 1994).
- Partial predator exclosures at a 1-square-kilometer scale in Yukon, Canada resulted in a surge in snowshoe hare densities (Krebs et al. 1995).
- "Mesopredator" release in coyote-free canyons of San Diego County, California had strongly negative consequences for bird populations (Crooks and Soulé 1999).
- Aspen stands and riparian thickets in Yellowstone National Park declined after wolf extirpation (Ripple and Larsen 2000; Beschta 2005) and have been recovering since wolf restoration (Ripple and Beschta 2007b).
- Woody plant recruitment was suppressed by hyperabundant herbivores on predator-free islets in tropical Lago Guri, Venezuela (Terborgh et al. 2001).

- Decimation of the cod fishery on Newfoundland's Grand Bank was succeeded by an outbreak of sea urchins, dogfish, skates, and lobsters (Worm and Myers 2003; Steneck et al. 2004; Frank et al. 2005).
- Vegetation was released on a remote oceanic island after local extirpation of a dominant herbivore (O'Dowd et al. 2003).
- Introduction of arctic foxes to islands in the Aleutian archipelago resulted in an ecosystem phase shift from grasslands to tundra (Croll et al. 2005).
- Decimation of great sharks in U.S. mid-Atlantic coastal waters preceded an outbreak of mollusk-eating cow-nosed rays and the resulting collapse of estuarine shellfisheries (Myers et al. 2007).
- Overfishing of cod in the Baltic Sea triggered a series of algal blooms via a multilevel trophic cascade (Casini et al. 2008).

As the number and variety of these case studies suggests, top-down forcing and its effects on lower trophic levels have not been overlooked by the scientific community, but they have been overshadowed by a dominant current of opinion holding that more was to be gained by investigating bottom-up forcing. A clear example of the secondary status afforded to top-down processes by funding agencies was the International Biological Program of 1964–1974, a large U.S. National Science Foundation–sponsored program of comparative research on ecosystems that emphasized measurements of climate, productivity, nutrient cycling, and other bottom-up processes. The current incarnations of big science programs in terrestrial ecology in the United States are the Long-Term Ecological Research program and National Ecological Observatory Network, designed primarily to monitor the progress and effects of climate change (Keller et al. 2008). In marine science, the counterpart of these programs is Global Ocean Ecosystem Dynamics (GLOBEC). "The aim of GLOBEC is to advance our understanding of the structure and functioning of the global ocean ecosystem, its major subsystems, and its response to physical forcing, as the empirical foundation to be used in forecasting responses of the marine ecosystem to global change" (see www.globec.org). In all these heavily funded programs, investigations of top-down processes have been relegated to a distant second place.

Single-minded focus on the role of bottom-up drivers has directed the flow of funding away from investigations of top-down processes, precluding even the most obvious experiments. For example, to date the U.S. National Science Foundation has not sponsored any large-scale experiments designed to investigate the effect of predator removal on terrestrial ecosystems. The only U.S. effort of which we are aware that controlled a mammalian predator on an ecologically significant scale was a coyote removal experiment, funded by a combination of state and private agencies, that used a 5-square-kilometer block design with replicates (Henke and Bryant 1999). The results demonstrated a powerful "Paine effect," as a six-member rodent community collapsed to one dominant species, *Dipodomys ordii*. Deer irruptions in the eastern United States motivated the U.S. Forest Service to install a series of 13- or 26-hectare exclosures in Pennsylvania (Tilghman 1989). Neither of these experiments is well known to ecologists because both efforts were conducted by wildlife managers and published in the literature of that field. We know of no other experiments on a scale greater than 1 hectare in the United States, although a partial predator exclusion experiment conducted on a 1-square-kilometer block design has been implemented in Yukon, Canada (Krebs et al. 1995). These facts provide unequivocal testimony that top-down forcing has been systematically neglected by the agencies that fund basic science in the United States.

We believe that the lack of recognition and acceptance of the top-down countercurrent as a ubiquitous and fundamental structuring force in nature is retarding progress in ecological science at a crucial time when the stability of ecosystems all over the world is being threatened by multiple human interventions, not the least of which is the systematic elimination of top predators. If disrupting the trophic cascade were to lead to the dire consequences predicted by theory and anticipated by empirical studies (including those cited earlier), then many of the world's ecosystems would already be in serious trouble. Our deep anxiety at this prospect inspired us to organize a conference that was held at the White Oak Plantation in Yulee, Florida (USA) on February 7–10, 2008. For making our stay at the plantation a memorable and pleasurable experience, we are most grateful to Tom Galligher, Troy Miller, and the rest of the unfailingly friendly and helpful White Oak staff. For sponsoring our stay at the plantation, we express our deep gratitude to the Howard T. Gilman Foundation. We warmly thank Patricia Alvarez and Stacey Reese for innumerable forms of assistance with production of the book. Defenders of Wildlife and the Pew Charitable Trusts provided critical financial support, the latter through a grant from the Institute for Ocean Conservation Science.

Our purpose in organizing this book was to elevate biotic forcing in ecology to what we believe is its proper place as coequal with physical forcing. The time for a synthesis is right. Evidence of top-down trophic cascades has been mounting, as indicated earlier, and has lately begun to snowball. The accumulated weight of this evidence has become overwhelming, as we shall demonstrate in the ensuing chapters. We hope that this book will be read by doubters and skeptics as well as the convinced, because we are confident that the theory, evidence, arguments, and interpretations presented herein will open a new era in ecology.

We shall demonstrate that top-down forces interact with bottom-up forces through a dynamic balance and that this balance confers structure on ecosystems and ultimately regulates their species composition and diversity. Because of the huge historical disparity in funding and, consequently, scientific attention given to investigating bottom-up and top-down processes, understanding of how top-down forces operate via all their myriad and intricate pathways lags far behind. Temperature, moisture, production potential, and the serendipitous constraints of history set broad limits on the distribution and abundance of species. Yet beyond these rather obvious macro-scale agents of physiological tolerance, it is the operation of top-down forces that regulates and sustains biodiversity on our planet. For this reason alone, the scientific community should focus much greater attention on top-down processes and their stabilizing effects on natural ecosystems.

John Terborgh and
James A. Estes

CHAPTER 1

Trophic Cascades: What They Are, How They Work, and Why They Matter

John Terborgh, Robert D. Holt, and James A. Estes

Humans have been waging war against predators since the dawn of history. Lion slayers were heroes of Greek mythology. Shepherds bred large, aggressive dogs to fend off wolves and bears in the Pyrenees, Carpathians, and elsewhere. Gamekeepers were hired by the great estates of Britain to eradicate foxes, goshawks, and badgers. In the United States, an agency of the federal government, the Biological Survey (later the U.S. Department of the Interior), hired hundreds of predator control agents to shoot, trap, and poison wolves, cougars, coyotes, eagles, and a host of lesser predators. Bounties and culls have been used in Alaska and Canada to control seals and sea lions in the name of fishery management. An almost endless list of such measures could be compiled.

Humans have been so effective at decimating or entirely eliminating predators over most of the land and sea that the effects of these persecutions are becoming apparent to the ordinary citizen. There is hardly a resident of suburban America today whose efforts to grow flowers or vegetables isn't thwarted by ubiquitous deer. Ask any gardener why there are so many deer, and the answer is a consistent refrain: "Because they don't have any predators." The line of reasoning from cause to effect is simple and linear. It is precisely what Hairston, Smith, and Slobodkin (HSS 1960) posited nearly 50 years ago.

Recognition on the part of official agencies that predators play important roles in nature has been belated but is now spreading in the United States, Canada, Europe, and the industrialized countries generally, where legal structures protect wildlife and managers intervene to mitigate human–wildlife conflicts. In these countries, the tide of opinion is changing. Whereas predators were actively persecuted a generation ago, they are now being restored. Examples from the United States include reintroduction of the gray wolf in Wyoming, Mexican wolf in Arizona and New Mexico, red wolf in North Carolina, lynx in Colorado, black-footed ferret in Wyoming, South Dakota, and Chihuahua, Mexico, and sea otters in southeast Alaska, Washington, and southern California. And after an absence of more than 100 years, jaguars are returning to the borderlands between the United States and Mexico.

Why should we celebrate this development and encourage its expansion to additional places and predators in other parts of the world? The answer is a complicated one, blending philosophical, aesthetic, practical, and scientific reasoning. In this book we shall be concerned primarily with the scientific reasons for sustaining predators while recognizing that more philosophical approaches are also valid.

Predators are important because they occupy the top rung of the trophic ladder and from that position regulate the food web below them. Top vertebrate predators are large bodied and can move over large areas, thus coupling the dynamics of seemingly distinct communities and ecosystems. Recently, the ability of predators to move flexibly between communities, responding opportunistically to shifts in prey abundance, has been suggested to be an important governor of food web stability (McCann et al. 2005; Holt 2009). Eliminating predators destabilizes ecosystems, setting off chain reactions that eventually cascade down the trophic ladder to the lowest rung. In 1980, Robert Paine coined the term *trophic cascade* to describe this process. The altered state that develops after the loss of apex predators is invariably simpler than the initial state, supporting less biodiversity. Thus, predators hold an important key to retaining the high levels of biodiversity we associate with primordial nature.

Three-level cascades are the simplest and most familiar case (e.g., wolf–deer–vegetation). Wolves eat deer and thereby indirectly benefit vegetation, depending on their efficiency in maintaining deer numbers at low levels. If wolves are efficient deer predators, deer populations remain low and the vegetation experiences only light herbivory; if they are inefficient, deer populations are higher and herbivory is heavier. This efficiency is analogous to Paine's (1980) interaction strength. However, unlike the case of Paine's *Pisaster*, it is generally not possible to use rigorous, controlled experiments to determine the impact of a large predator such as a wolf because such demonstrations require wolf removal and subsequent assessment of the demographic response of deer. Responses of deer and their allies to the local extirpation of wolves or other top predators have typically lagged by decades, during which time other factors, such as plant succession, hunting, land use changes, and other human activities, can intervene to complicate the picture (McShea et al. 1997).

The simple example of the wolf–deer–vegetation interaction introduces some key features of trophic cascades. First, predators harvest prey with a certain efficiency that can vary with the topography, vegetation, density, and evasive behavior of the prey and perhaps other factors such as other species of prey and predators (Berger 2008). Thus, the strength of the top-down interaction is not a simple property of predator and prey alone, but it depends on the context in which the interaction takes place. Second, we note that the wolf population interacts only with deer; it is at the top of the pyramid and regulated from the bottom up via the deer population that supports it. However, the deer population is in the middle of a bidirectional flow of resources. It depends on forage (a bottom-up process) and is preyed on by wolves (a top-down process). The density of the deer population thus depends on the balance of these two counter-current forces. Finally, the vegetation is also regulated both by bottom-up (water, sunlight, nutrients) and by top-down (herbivory) processes. In essence, this simple bidirectional interaction scheme is what Hairston, Smith, and Slobodkin proposed in 1960 in their famous "green world" hypothesis.

A common misimpression is that there is an either–or dichotomy between systems driven from the bottom up and those driven from the top down. Bottom-up and top-down processes are not in any way exclusive; they are complementary countercurrent flows, inextricably bound together. Bottom-up processes are fundamental and inescapable, driving photosynthesis and being supported by it. If photosynthesis increases, as along a climatic gradient for example, the responses are quantitative: more productivity, more herbivores, more predators being supported by those herbivores, and occasionally an increase in food chain length (Crête 1999). Only at the very lowest productivity levels, such as near the limits of vegetation in deserts or the high Arctic, does one find ecosystems with fewer than the three standard levels (Oksanen et al. 1981).

With the exception of microbial ecosystems supported by chemoautotrophs, photosynthetic productivity determines the availability of resources to higher levels, either directly (via chains starting with herbivory) or indirectly via detritivores (e.g., the deep sea, caves). How primary productivity is allocated among higher levels is determined not only by the efficiency of material and energy flux upward through the food chain but also from the top down through the trophic cascade. In the absence of herbivores or predators, the entire annual net productivity of a patch of vegetation must pass into the detrital food web. If herbivores are present, some of the productivity will accrue to them and less will recycle through detritus. If predators are added, the flow of resources will ascend one level further, and less may accrue to herbivores because their numbers are kept in check by the predators (although they will probably turn over faster). Thus, bottom-up processes determine the flow of resources into the system, whereas top-down processes influence how the resources are distributed among trophic levels.

Another issue that has led to confusion is whether trophic cascades are inherently static or dynamic. In fact, a trophic cascade is always dynamic, but the dynamism is not always manifest. When a trophic system is at a stable point (i.e., equilibrium), its component levels remain fairly constant. However, the appearance of stasis is illusory, suggesting the absence of any dynamic process. But under the surface, the interactions between levels are not static but rather highly dynamic. Predators are eating prey, prey are eating plants, plants are growing, and so on. A lot is happening, but the various interacting forces and flows are in balance, so the underlying dynamism is not apparent. Call it cryptic dynamism. The stable, equilibrial condition is quite properly called a trophic cascade because the term refers to the whole interacting system, not just to one or another of the states it can assume.

A final point that warrants clarification in these preliminary comments concerns the role of keystone species in trophic cascades. Paine's (1966) founding example of the starfish *Pisaster* has left an indelible mark on the literature. *Pisaster* is an unequivocal keystone species, defined by Mary Power and colleagues (1996) as one having effects on other elements of an ecosystem that are large relative to its numbers or biomass. The wolves of Yellowstone are another keystone species. Dramatic responses can be obtained by perturbing a keystone species, but are keystone species necessary as mediators of strong trophic cascades? The *Pisaster* example, and some other equally dramatic ones, has led some authors to conclude that keystone species are necessary ingredients of strong trophic cascades (Polis et al. 2000). We shall see definitively that this is not the case. Keystone species are notable because they concentrate much of the interaction strength of an entire trophic level in a single species, but across nature more generally, keystone species possessing such concentrated interaction strength are probably the exception rather than the rule.

Because of controversy over why the world is green, there has been a focus in much of the trophic cascade literature on indirect carnivore impacts on plants (or space occupiers in marine systems), via shifts in herbivore abundance and activity. The concept of a trophic cascade actually has a much broader scope than just indirect mutualisms between predators and plants; the basal species might be space occupiers in marine systems or detritivores and decomposers in soil food webs, for instance. There can also be trophic cascades between species, all of which are predators (e.g., in the Bahamas, lizard cuckoos may eat *Anolis* lizards or force them into hiding and so reduce predation on spiders). But the main heat in the literature on trophic cascades seems to arise from efforts to understand patterns in plant communities.

The basic question posed by HSS (and, they believed, answered by them) is this: To understand plant ecology (e.g., distributions of life forms within a community, or the distribution of plant species along environmental gradients), must one pay attention to the food webs supported by those plant communities? It is fair to say that plant ecology has traditionally focused on how plant form, life history, and species composition reflect the outcome of competitive interactions and population dynamics playing out in the context of various factors in the physical environment (e.g., climate, soil, disturbance regimes). This unilaterally bottom-up view of plant ecology essentially ignores herbivory as a deterministic force in structuring vegetation. But if trophic cascades are ubiquitous and large, as we are convinced, bottom-up forcing is only half the picture.

Some authors argue that trophic cascades are idiosyncratic in occurrence and not all that important. For instance, Polis et al. (2000) suggest that in contrast to aquatic systems, "community cascades . . . are absent or rare in terrestrial habitats" (473); furthermore, they claim that "support for even species-level cascades is limited in terrestrial systems" (474). The reasons they give for these assertions are that most food webs have a reticulate and heterogeneous structure and that many prey, plants in particular, are inedible.

Biological control of agricultural pests is a widespread application of trophic cascades to solve practical problems in applied terrestrial ecology, so the second quote from Polis et al., taken literally, is false. But agricultural systems, by design, tend to be low in species diversity and other kinds of heterogeneity, and crop plants have been bred to be edible (at least to us) at the expense of spines, secondary compounds, low-quality tissues, and other antiherbivore devices. So maybe biological control is the exception that proves the rule: Trophic cascades may be ubiquitous in the artificial landscapes of agroecosystems without being significant drivers of plant community structure and dynamics in natural ecosystems.

Our reading of the literature suggests that the claims of Polis et al. (2000) are greatly overstated. This volume is replete with convincing evidence of cascades in terrestrial and aquatic systems. Moreover, it should be noted that there are limitations in most experimental studies of trophic cascades in terrestrial biomes (Holt 2000). For understandable reasons, most manipulative studies are short term. For instance, in the review by Schmitz et al. (2000), 80 percent of the studies involved measurements over a single growing season, even though many of the target species were long-lived shrubs, trees, perennial herbs, and graminoids. The time scales of transient dynamics in trophic cascades in terrestrial systems are likely to be much longer than in many aquatic systems, where the basal producers (e.g., phytoplankton) have short generation times and so can respond very rapidly to shifts in herbivory. A small quantitative impact of herbivory observed in a single growing season in a terrestrial system that seems quite subtle, assessed, say, in terms of individual growth rates or tissue damage, could be magnified over time, for instance, by altering competitive ability. Shifts in community composition caused by altered herbivory regimes could necessitate colonization from a regional source pool, followed by shifts in local abundances, both of which could be very slow processes, especially if the vegetation is woody. Patterns in abundance as a function of trophic level along gradients reveal the importance of such within-level species sorting for elucidating natural patterns (Leibold 1996; Leibold et al. 1997). Finally, manipulative studies never remove *all* the natural enemies of herbivores. These include not just predators but parasitoids, pathogens, entomophagous nematodes, and so on. Experiments rarely run long enough so that the regional species pool of potentially important herbivores is sampled at an experimental site.

If one accepts that trophic cascades are important in natural ecosystems, applied ecologists should be deeply concerned because humans disrupt natural predator–prey systems in many ways. Generalist top vertebrate predators (e.g., the Florida panther [*Felis concolor floridiana*]) are at particular risk because of a perfect storm of multiple, correlated vulnerabilities. Top vertebrate predators tend to have low population densities and large home ranges, making them particularly vulnerable to habitat fragmentation. Moreover, low intrinsic growth rates imply weak demographic responses to increased mortality. Thus, small but sustained increases in mortality can inexorably drive such species to extinction. Because of their opportunistic diets and spatial mobility, they often come into direct contact with humans or our commensals (e.g., livestock), prompting humans to persecute them. Putting all these factors together, it is not surprising that among the species most at risk around the world are top predators such as tigers and the great sharks.

What does theoretical ecology have to say about trophic cascades? By *theory* we mean formal mathematical models that lay out explicit assumptions about the dynamic forces in ecological systems and draw out the logical consequences of those assumptions. Such models are often motivated by fine conceptual theory presented verbally, as in the stimulating papers by Fretwell (1977, 1987) and various chapters in this volume. There is a huge body of theory on predator–prey and food web dynamics that in a broad sense is relevant to trophic cascades. However, we shall not attempt an exhaustive review but instead shall re-examine the main thrust of some key older papers to provide a convenient summary of historical perspectives that are the conceptual foundation of many empirical studies of trophic cascades. Other chapters in this volume (Chapters 4, 17, and 18) deal with current theoretical issues.

HISTORICAL PERSPECTIVES ON THE THEORY OF TROPHIC CASCADES

All ecologists know that the world is complex. Some of us revel in that recognition (Polis 1991) and are deeply skeptical of theories based on simplifying assumptions. Others of us hope that simple models can be used like a surgeon's knife, cutting deftly through the cloying fat of complicating detail to get at the essential sinews of ecological reality. One complication that immediately arises when we contemplate theoretical studies of trophic cascades is that most food webs are highly complex (e.g., contemplate Figure 6 in Winemiller 1990), with many species locked in tangled webs of interactions. For both practical and analytical reasons, theoretical models in ecology must greatly simplify known complexities. For trophic cascades, the natural and admittedly grossly oversimplified starting point is the community module (sensu Holt 1997), represented by an unlinked food chain capped by a top predator that feeds on a herbivore population, which in turn is sustained by a basal plant population. (These can be viewed as single species at each level or as aggregate functional groups comprising several functionally equivalent species.)

The simplest models for unlinked food chains are based on Lotka–Volterra models, where all the per capita relationships, within and between species, are expressed by linear functional forms. May (1973a) once compared such simple models in ecology to the models of perfect crystals in physics. Perfect crystals do not exist, but developing a theory for such crystals nonetheless provides yardsticks, which can be used to gauge the consequences of various sorts of imperfections in crystal structures. In like manner, no ecologist, not even the woolliest theoretician, believes that a Lotka–Volterra model literally describes all the rich behavior of any actual ecological system, but such models may nonetheless capture some essential features that carry over to much more complex, realistic—and analytically opaque—models. Simple models give us an accessible and tractable starting point that serves as a springboard for tackling more complex and realistic models.

We shall begin with a continuous-time, differential equation model, where each trophic level is represented by a single equation, as follows:

$$\frac{dP}{dt} = P(b'a'N - m')$$

$$\frac{dN}{dt} = N(abR - a'P - m) \tag{1.1}$$

$$\frac{dR}{dt} = R(r - dR - aN)$$

Here P, N, and R are the abundances of the predator, the herbivore, and the plant, respectively; r is the intrinsic growth rate of the plant; d is a measure of its direct density dependence (e.g., competition for resources and space); a and a' are per capita attack rates; b and b' are conversion factors (relating consumption to births); and m and m' are density-independent mortality rates for the herbivore and predator, respectively.

Analyses by Stuart Pimm (1979) and various mathematicians (e.g., Hallam 1986; Freedman and Waltman 1977) in the 1970s and 1980s of the Lotka–Volterra model described by Equation 1.1 led to a number of conclusions:

- As one ascends the food chain, the conditions for persistence of each consecutive level become more stringent.
- Analysis of these persistence conditions shows that they are more likely to be met for the predator as the productivity of the plant increases (via higher r or lower d).

In other words, this model leads to the prediction that food chain length should increase with the productivity of the basal trophic level. Basically, if primary productivity is too low, too little energy will pass through the intermediate trophic level to sustain the top level as a viable population. The prediction that food chains should tend to increase in length with increasing productivity is a general feature of many models and has been demonstrated in laboratory microcosms (Kaunzinger and Morin 1998) and at the very low productivity end of natural variation in primary production (Aunapuu et al. 2008). However, even some unproductive ecosystems seem to be able to sustain a top predator, which may persist because of factors left out of traditional models (e.g., mobility, long generation lengths, adaptations to cope with resource scarcity). It is an open question whether, in general, natural variation in food chain length between communities is explained principally by variation in primary production or by the interplay of many distinct factors; the latter at present seems most likely (Post 2002; Holt in press).

- For this specific model, there are no alternative stable, noninvasible equilibria, so historical idiosyncrasies will not affect the ultimate community found at a site.
- Given that an equilibrium exists, it is locally and globally stable, so there is no limit cycle or chaotic dynamics.
- However, the resilience to perturbations—the time for the system to recover to its initial equilibrium after a disturbance (the shorter the recovery time, the greater the resilience)—decreases with increasing food chain length. Basically, there can be a compounding of perturbations up the food chain (Pimm 1979). So, in a certain sense, longer food chains are predicted to be dynamically more delicate in this simple model.

With these theoretical results in hand, the next step is to discern the degree to which they are general or instead reflect the many simplifying assumptions built into the Lotka–Volterra model. An important step toward generality was provided by Rosenzweig (1973), who developed a general, graphical, three-species food chain model, which in effect included nonlinear density dependence in the plant and nonlinear functional and numerical prey-dependent responses by the herbivore. He also carried out a formal local stability analysis, which led to a number of important theoretical conclusions:

- Stability requires the existence of direct density dependence (e.g., interference or competition for space) at one or more trophic levels (see also Wollkind 1976). This is a generalization that holds for all ecological models, regardless of their details.
- If higher trophic levels have weak direct density dependence, the basal level must have strong density dependence for the system to be stable.
- Intense predation can destabilize a food chain if the top predator has weak direct density dependence, as do herbivores, and the predator has a saturating functional response to its prey. This is a generalization of an insight that emerges from simple two-link predator–prey interactions when predation is effective at limiting prey numbers well below (the prey) carrying capacity (Rosenzweig 1971).
- One can construct examples in which moderate predation stabilizes an

otherwise strongly unstable plant–herbivore interaction if the top predator has direct density dependence and the herbivore only has weak direct
density dependence. It is noteworthy that nearly all vertebrate top predators in terrestrial ecosystems, except possibly alligators, snakes, and a few
other reptiles, have strong intraspecific density dependence mediated
by direct aggression or territoriality. The very trophic apparatuses that
permit vertebrate predation in the first place—sharp teeth, claws, and
talons—also provide arms for intraspecific conflict or necessitate the
avoidance of such conflict by spacing mechanisms such as territoriality.
This permits direct density dependence in predators to buffer them from
changes in their food supply, which as a byproduct can help stabilize the
entire system.

A theoretical example of predation stabilizing a plant–herbivore interaction
was sketched by May (1973a, 1973b), who showed using a nonlinear model
that a top predator with direct density dependence could persist stably atop a
three-species food chain, and the system would return to its equilibrium with
damped oscillations after disturbance. When the predator is removed, the inherent instability of the plant–herbivore interaction was unleashed, leading to oscillations of such large amplitude that extinction during the population troughs
would be likely.

The basic model of Rosenzweig (1973) was built upon in the celebrated
exploitation ecosystem article of Oksanen et al. (1981). One of the main conclusions of that article was that the prediction from Lotka–Volterra theory relating food chain length to primary productivity was more general, and it provided a scaffolding for understanding shifts in the relative importance of
top-down and bottom-up forces along environmental gradients in production.

It was not recognized until much later that the unstable dynamics in food
chains noted by Rosenzweig could lead not only to cycles but also to chaotic
dynamics (Hastings and Powell 1991). Chaotic dynamics can arise when each
trophic link has a saturating response, so that each consumer–resource interaction on its own tends toward unstable limit cycle behavior. In a sense, the species linked in a food chain act like coupled oscillators, which reveal much more
complex dynamics than do single oscillators on their own. The recognition
of the potential for chaotic dynamics led to a small cottage industry of work
by mathematicians on food chain models, full of recondite terms such as
"codimension-two Belyakov homoclinic bifurcations" (Kuznetsov et al. 2001).
Much of the ornate phenomenon analyzed in this literature has its mathematical charm, indeed elegance, but it is not immediately clear that these mathe-

matical details are all that relevant to natural systems. But leaving this quibble aside, some key qualitative messages do emerge from this body of mathematical work on unbranched food chain models and their exploration of instabilities that could be quite important for empirical studies, if unstable dynamics caused by coupled trophic interactions are pervasive in food webs.

First, in a wide range of circumstances, populations in a food chain may experience low densities, as the coupled system tracks a trajectory wandering over a dynamic attractor (for examples, see figures in Rinaldi et al. 2004). This means that there is often heightened extinction risk. Rinaldi et al. (2004) explored various mathematical aspects of the chaotic dynamics of the system, but for our purposes a biologically significant effect of this study (which the authors do not discuss) is that populations can plunge to very low densities, and so the food chain is likely to collapse.

A broad implication of these (and related) theoretical results for empirical studies is that trophic cascades may be manifest not only in changes in average abundance but in shifts in system stability and hence, potentially, extinction rates. However, a cautionary note is that although models suggest a range of scenarios that describe unstable dynamics, such dynamics are not as often observed in nature. The rarity of unstable dynamics in nature perhaps has multiple explanations, at least one of which is the interplay of spatial heterogeneity, mobility, and the presence in most landscapes of refugia (though perhaps different ones) for predator, prey, and producer. It is likely that the spatial mobility and behavioral flexibility of large vertebrate top predators in particular can provide important buffers moderating the inherent tendencies of nonlinear food chains to exhibit extremely unstable dynamics.

Second, when the dynamics of ecological systems are unstable, their responses to perturbations or systematic shifts in environmental conditions may often be surprising, basically because unstable dynamics magnify the impact of nonlinearities in the system (Abrams 2002; see Rinaldi et al. 2004 for a food chain dynamic). Our ecological intuition is not very good at predicting what happens in nonlinear systems with unstable dynamics with multiple feedbacks playing out over different time scales, and this surprising result could be viewed as a specific example of this general truism. Counterintuitive effects emerge in many ecological models that have unstable dynamics arising from nonlinear feedbacks; the response of abundances, averaged over the trajectory of the system, to a change in a system parameter may go in exactly the opposite direction to what is expected from an examination of the system's equilibria (Abrams 2002). It is difficult to assess these predictions in field systems because in nonlinear dynamics, populations fluctuate around some kind of potential equilibrium, but their long-term average is not in general the same as the numerical value of that equilibrium. Moreover, assessing trends in fluctuating time series in response to environmental change poses deep statistical challenges.

The bottom line of this body of theoretical work is that trophic cascades involving changes in average abundance of species at different trophic levels can also entail shifts in the dynamical behaviors of populations, such as the tendency to oscillate or the magnitude and time course of oscillations. There are some excellent examples of plant–herbivore interactions being strongly unstable in the absence of top predators. For instance, McCauley et al. (1999) showed in aquatic mesocosms that strongly unstable oscillations arose for *Daphnia* consuming algae. In a terrestrial example, voles explode on predator-free islands in the Baltic and overexploit their plant food resources. But where predators (mustelids) are present, vole populations remain bounded in their numerical fluctuations (Banks et al. 2004).

Today one seeks to understand the dynamics of two-level trophic systems (producer and consumer) through contrived experiments at micro and meso scales and field studies in such exotic places as arctic islands that lack natural predators (Aunapuu et al. 2008). Our short cultural memory, as encapsulated in Pauly's (1995) notion of the shifting baseline, has blinded many modern ecologists to the fact that food-limited herbivores dominated terrestrial ecosystems over much of the world until recent times. Until humans drove them extinct, proboscideans (elephants and their relatives) and other megaherbivores ranged over all the continents except Australia and Antarctica, from the Arctic Ocean to the Southern Seas (Burney and Flannery 2005). Such large animals are immune to predation as adults and are able to increase until limited by the food supply (Owen-Smith 1988). Herd-forming migratory ungulates constitute a second class of major herbivores that are largely free of predation and consequently regulated from the bottom up (see Chapters 15 and 16, this volume). Loss of these major classes of herbivores over most of the terrestrial realm appears to have altered much of the earth's terrestrial vegetation, so the consequences have been momentous (Bond 2005).

As noted earlier, and discussed further in Chapters 17 and 18, the theory of trophic cascades has been refined and greatly advanced since publication of foundation papers by HSS, Paine, Rosenzweig, Fretwell, and Oksanen et al. One particular focus of empirical and theoretical research at present is elucidating the role of diversity at different trophic levels in modulating the strength of trophic cascades and patterns of abundance along productivity gradients. Diversity in the prey trophic level can at times moderate top-down control (Leibold et al. 1997; Stachowicz et al. 2007). Diversity in the predator level can either increase top-down effects (Straub and Snyder 2008) or weaken them (Stachowicz et al. 2007). The former often reflects niche complementarity (e.g., different predators feed in different microhabitats or at different times of day). The latter is particularly likely when intraguild predation and interference between predator species is strong. This is frequently observed in biological control of agricultural pests (where it can lead to a conflict between the conservation of predator diversity and the efficacy of control; Straub et al. 2008), but is also ubiquitous in vertebrate carnivore guilds (Sergio and Hiraldo 2008; Hunter and Caro 2008). Variation in predator species diversity across time or space can thus lead to complex mosaic patterns in the strength of trophic cascades.

The basic processes of top-down control are understood and have received ample empirical support from a global array of ecosystems, as will be documented in this book. The field now stands at a new threshold, one that promises enormous dividends in enhanced understanding of the way in which ecosystems work. The new plateau of understanding rests on the concepts of alternative states, positive feedback loops, catastrophic regime shifts, and hysteresis (Scheffer et al. 2001).

Alternative stable states, as the name implies, are alternative configurations of a given ecosystem and correspond to the existence of alternative attractors in a dynamic system. Alternative states were first imagined as a theoretical possibility by Richard Lewontin in 1969 and predicted in the formulation of Oksanen et al. (1981), but the concept has been strengthened through further theoretical insights and widespread empirical support. As Lewontin noted, the possibility of alternative stable states raises the specter that history leaves an indelible footprint on current community configurations. If we consider the species composition of a community, alternative equilibria may be generated from a regional species pool because of strong interspecific interactions. If the Lotka–Volterra equations describe community interactions between n species, then only a single equilibrium with all species present can possibly exist. But this set of n species may contain a number of subsets that, when present and established, are able to prevent invasion by species left out of this subset. When alternative stable states exist, the sequence of colonization events matters greatly in determining the final community configuration. But if interactions within or between species are strongly nonlinear (i.e., non-Lotka–Volterra models), then alternative stable states can exist, even with all the same players being present.

A fundamental objective of ecological science is to understand how ecosystem properties and community structure track major external variables and how ecosystems respond to different kinds of perturbations. Community ecologists distinguish between two kinds of perturbations: pulse perturbations and press (Bender et al. 1984). In the former, one imagines that there is a single, sharp perturbation to the system, such as a quick culling of a dominant species (without driving it to extinction), whereas in the latter the perturbation is sustained over an indefinite time horizon. A press perturbation in effect takes an original system and transforms it into a new system. The question is how much of the structure of the original system carries over to the new system.

Figure 1.1 (modified from Scheffer et al. 2001) helps clarify two kinds of responses one can observe to either press or pulse perturbations. We imagine there is a state variable we are interested in (e.g., abundance of a focal species), and there is a parameter of the system that can be directly perturbed (e.g., by experimental manipulation). Systems labeled "A" and "B" in Figure 1.1 both have nonlinear positive and negative feedbacks defining how they respond to press and pulse perturbations and how they settle into an equilibrium (or more than one) for any given fixed value of the key parameter. If we abruptly increase

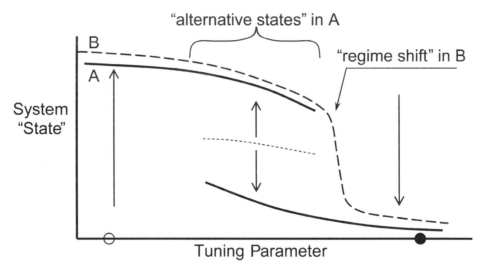

Figure 1.1. A schematic depiction of the concepts of regime shifts, alternative states, and hysteresis. Two ecosystems, A and B, are shown, where the equilibrial states of those systems vary with a tuning parameter, an independent variable extrinsic to the system (e.g., nitrogen loading in a lake, or *Pisaster* abundance in a small patch of the intertidal zone trod by Bob Paine). The equilibria are locally stable; small perturbations tend to return to the equilibrium, where the system started. In both cases, a large change (from the open to closed circles) in the tuning parameter implies a large change in state. The nonlinear forces in B are such that the system stays largely in one state or another, with an abrupt transition between them. But for every value of the tuning parameter, the system settles into just one state. For A, there is a zone of alternative stable states, and where the system ends up depends on where it started out.

the value of the parameter from low (open circle) to high (closed circle) values and leave it there (a pulse perturbation), we expect equal changes in both systems. But if we change the system between these values and do so slowly (so each system stays near equilibrium), we see different responses in the two systems. In B, there is a sharp shift in state. Such shifts have been called regime shifts in the literature. But note that at any given value of the parameter, there is a single resulting system state.

In system A, there also can be sharp transitions. But in contrast to system B, the value at which the system changes depends on the direction of change in the parameter. Over a certain range of parameter values, it is necessary to know where the system started to know where it is. This dynamic structure is called hysteresis. Moreover, in this parameter range, pulse perturbations that change the state of the system can lead to abrupt shifts rather than a tendency to return back to where it started.

The state of any ecological system is not really a mathematical point, a hypothetical equilibrium of fixed and unchanging abundances, but rather (in the jargon of dynamic systems) an attractor, a bounded regime within which fluctuations—which are always present in the natural world, albeit to differing degrees—are contained. Dynamic systems driven by nonlinear interactions can have multiple attractors, alternative states toward which their trajectories tend, depending on where they start. A very important issue in ecology is understanding when systems have alternative states and when they do not. We believe this is of primary significance when we reflect on the potential importance of trophic cascades, particularly in the context of conservation and restoration.

Trophic cascades can involve regime shifts and potentially even alternative stable states. When a stable trophic system is powerfully perturbed in a press fashion, say by addition or removal of a top level, the remaining system can be destabilized and enters a transient state. Adding a top predator (e.g., largemouth bass) to a pond containing only planktivorous fish, zooplankton, and phytoplankton results in major shifts at lower levels. Bass reduce the density of planktivorous fish, releasing zooplankton to increase, whereupon abundant zooplankton filter out phytoplankton, a process that clarifies the water column. All this takes place over a period of months to a year. During the transition, the system is rapidly changing.

In other systems, the transitions arising from such perturbations may play out over centuries or even millennia. An upsurge in herbivory on seedlings in a temperate forest may have no obvious impact on the structure of the forest until several centuries have passed, as recruitment of a different suite of species replaces the occasional but inevitable death of canopy dominants. It can be very difficult to distinguish between true alternative stable states and sluggish transitions between reversible states.

The simplest way for nonlinearities to be introduced is for species to experience strong positive intraspecific density dependence at low densities (Allee effects; for a recent review, see Courchamp et al. 2008). This in turn can arise from intrinsic factors. For instance, sexual outcrossing species may find it difficult to find mates at low densities. Well-established species in communities may not be able to recover if their numbers plummet to very low levels. So all communities comprising sexual species thus can contain subsets of alternative, more depauperate communities, where reestablishment of species that have gone locally extinct may be difficult.

Allee effects can also arise from trophic interactions, and this theoretically plausible effect may be more pertinent to the theme of trophic cascades. Type III functional responses by generalist predators are accelerating at low prey numbers (which can be stabilizing if prey are kept in check at low levels) but saturating at high levels, which means that high densities of prey can escape predator control. This process has been suggested as an explanation for outbreaks of rodent populations (Sinclair et al. 1990), shifting between a stable low-density state and another stable high-density state, permitted by transient pulses in food availability for the rodents. Although a number of examples of this effect have been suggested, it has been difficult to definitively show that this plausible process actually occurs (Sinclair 2003). Sinclair and Metzger (2009) note one example in Kruger National Park, where initially high numbers of wildebeest were reduced by culling. After culling, their numbers continued to decrease because of lion predation. This pattern suggests that two alternative states describe the nonlinear dynamics of the full system. However, it is difficult to convince skeptics on this point, in part because it is difficult to perform experiments at the appropriate spatial and temporal scales.

Transitions between alternative states or regime shifts may be fast or very slow and may be triggered by a variety of abiotic or biotic perturbations. A particularly familiar type of press perturbation is the addition or deletion of a top trophic level. Examples involving keystone predators, the starfish *Pisaster* and the largemouth bass, were discussed earlier. Other examples come from the marine realm, where overfishing has led to the collapse of many fish stocks to the point of commercial extinction. Van Leeuwen et al. (2008) point out that even though fishing has been totally banned for many of these stocks, the fish populations show no sign of recovering. They note that the literature contains many plausible mechanisms that can stabilize a fish population, once abundant, to low levels after a crash. For instance, the decrease in cod numbers may have led to an upsurge in planktivore abundance. Because these planktivores can prey on cod eggs and can compete with cod larvae for food resources, this can check population growth in the cod.

For a terrestrial example, one can point to ecosystems dependent on fire. Fire is a physical driver of alternative states that operates with positive feedbacks and hysteresis. Fire-adapted vegetation typically includes plants that dry out aboveground, creating fuel, while remaining viable belowground. If fires are frequent enough, such plants tend to dominate the vegetation of coarse, porous soils or regions subject to long, hot dry seasons. For example, vast portions of the southeastern United States were once occupied by the fire-dependent longleaf pine (*Pinus palustris*)–wiregrass (*Aristida stricta*) association. Wiregrass provided the fuel. Modern Americans have imposed fire suppression over much of this region, allowing forests of oak and other pine species to grow up in place of longleaf pine. Moist, shady oak forests, lacking a grassy ground layer, burn reluctantly and infrequently and are stable to occasional cool ground fires. Restoration of the longleaf pine system entails opening up the oak canopy and burning frequently, typically every 2–5 years. Thus, a high fire frequency is needed for the oak–longleaf pine transition, whereas a much lower frequency is needed for the reverse, pine–oak transition.

The knowledge that ecosystems can assume alternative states and that these alternative states are, in principle, reversible presents managers with a powerful tool for ecological restoration. With respect to fire ecology, the reversibility of alternative states and the conditions needed to stabilize them are quite well understood. In contrast, reversing the eutrophication of ponds and streams is a challenge fraught with difficulties (Chapter 4, this volume). As for trophic cascades mediated by biotic forcing, there are some encouraging initial signs in the recovery of kelp forests following the return of sea otters to various parts of their historical range in the North Pacific (Estes and Duggins 1995) and in the recovery of willows, cottonwoods, and aspens in Yellowstone and Jasper national parks following wolf restoration (Beschta 2003; Beschta and Ripple 2007a, 2007b). We believe an important future dimension of restoration ecology lies in the manipulation of trophic cascades, a prospect that will require managers to reexamine existing methods.

The intent of this chapter has been to introduce the reader to the basic concepts of what trophic cascades are and how they operate and to provide a thumbnail history of some of the basic approaches that have been taken to develop a theory of trophic cascades. Additional layers of complexity and detail will be introduced and discussed in later chapters.

The broad purpose of this volume is to provide an overview of the importance of large apex predators in maintaining trophic cascades across global ecosystems. The book is organized into four parts. The first consists of Chapters 2–

6, covering aquatic ecosystems. Terrestrial ecosystems from the Arctic to the tropics are covered in the second part (Chapters 7–12). The five chapters of the third part (Chapters 13–17) cover topics that cut across the divide between aquatic and terrestrial systems. And the final part (Chapters 18–21) presents a synthesis of concepts and evidence to make the case that trophic cascades regulate the organization, dynamics, and diversity of all natural ecosystems.

In the final synthetic chapter, we conclude that trophic cascades are the key to understanding how ecosystems function. And if this should prove true, ecology will finally have found its holy grail: the power to predict the responses of ecosystems to many kinds of abiotic and biotic perturbations.

CHAPTER 2

Food Chain Dynamics and Trophic Cascades in Intertidal Habitats

Robert T. Paine

This chapter reviews the origins of two fundamental food web concepts: keystone species, those whose impacts are disproportionately significant as determinants of community structure and organization, and trophic cascades, the dynamic processes unleashed whenever numbers of apex predators are decreased by exploitation, increased through restoration, or experimentally manipulated.

The conceptual evolution underlying the development of these concepts arose out of the experimental tractability of simple ecosystems such as those on exposed rocky shorelines. Marine intertidal shores provide exceptional opportunities for understanding the functional role of species and thereby understanding how biologically complex natural systems are organized, whether these systems can resist imposed stresses, and how well they can respond and recover. A deeper understanding of such issues is central to the topics explored in this book.

Three attributes of rocky intertidal ecosystems converge to facilitate experimental manipulations. First, many of the resident species are readily observable, providing ease of identification and accumulation of essential and often quirky natural historical detail. Second, the exposed surface allows experimental contraptions, from barriers to cages, and a host of artificial habitats to be fastened securely, endowing the sundry research enterprises with the enviable trait of experimental tractability. Third, the life cycles of the currently appreciated, functionally important species are neither too long (centuries) nor too short (days), and thus, on average, the basic demography matches that arbitrary time span acceptable to funding agencies.

KEYSTONE SPECIES AND TROPHIC CASCADES

The sweeping generalizations of Hairston, Smith, and Slobodkin (HSS 1960) ushered in a dynamic vision of how natural communities are structured by trophic level interactions, although interest in food webs or cycles (Elton 1927) and energy transfer between species (Lindeman 1942) began much earlier. The HSS model required that whole trophic levels act "as a single exploitative population" (Oksanen et al. 1981: 258) in their influence, or lack of it, on the immediately lower one. The notion of discrete, coherent, and readily identifiable trophic levels is an acknowledged oversimplification. Rampant omnivory is a challenging embarrassment (Thompson et al. 2007), and Darnell's (1961) term *trophic spectrum* is certainly a more appropriate descriptor of how varied food choice can be. I accept the blurring but proceed without apology as though species can be unambiguously assigned to a level. At the time of my initial research, the terms *bottom-up* and *top-down* and the related controversies had not yet arrived. It should be obvious that bottom-up influences must be present, as identified in Power's (1992) ecumenical review. I concur, and despite my bias toward a top-down perspective, I have often joked that turning off the sun would be terrible news for all earthly life systems. An early demonstration that higher trophic levels, even a single consumer species, could alter radically the species composition of the next lower level and that these functional changes could trickle still lower was Brooks and Dodson's (1965) landmark study of predation and the composition of freshwater plankton. It was followed immediately by experimental intertidal studies focused on community-level phenomena rather than population dynamics. I demonstrated that removal of the starfish *Pisaster ochraceus* led to dramatic shifts in local species composition on a rocky shore (Paine 1966). A similar experiment involving the sea urchin *Strongylocentrotus purpuratus* was begun in 1964 (Paine and Vadas 1969). By 1966, Dayton's (1971) classic research was well under way. These intertidal studies all conveyed a similar message: A single species at higher trophic levels could control species composition at the level of their prey. The term *keystone* (Paine 1969a) was applied to convey the sense that a single consumer, when removed, would initiate a collapse of the local prey assemblage from a mix of species toward a monodominant stand of a competitively superior species. The mechanisms underlying the pairwise interactions were clarified almost simultaneously: "Some primary consumer (1) is capable of monopolizing a basic resource and outcompeting or at least excluding other species and (2) is itself preferentially consumed by the keystone species" (Paine 1969b: 950). Food preference and superior competitive prowess are the critical ingredients; both can be measured experimentally. Such pairwise interactions have been extensively demonstrated on some rocky shores: by Menge et al. (1994) and Robles et al. (1995) for *Pisaster* along eastern Pacific shores; by Paine (1971) and Menge et al. (1999) with another starfish in New Zealand; by Robles (1987) in Southern California, where the keystone is a large crustacean; and by Castilla and Duran (1985) in Chile, where a carnivorous gastropod assumes the keystone role. It is instructive to note the polyphyletic membership in the keystone species club, suggesting that behavior and ecology rather than morphology govern the relationship.

I note here that field ecologists have not been reluctant to designate with special terms species they believe to be important functionally or to play significant roles. The pioneering plant ecologist Clements was especially adept, leading to the claim that ecology was the discipline abandoned to terminology. Many of those papers are essentially uninterpretable. Elton's (1927) concept of "key industry species" (the copepod *Calanus*) seems to have fallen out of favor. Keystones, "foundation species" (Dayton 1972), "ecological engineers" (Jones et al. 1994), and a variety of terms all designating species of importance to conservation have followed. Such jargon plays a role in the discussion as long as its meaning is precisely defined. When the term is embraced too enthusiastically and extended beyond its original intent, it loses its utility. *Keystone species* suffered that fate and was justifiably criticized (Mills et al. 1993; Hurlbert 1997). By refining and extending the original definition and providing numerous examples, Power et al. (1996) resurrected it to close to its original, mechanistic, species-specific intent.

The term *trophic cascade* (Paine 1980) was born out of desperation when I was asked to give the 1979 Tansley Lecture by the British Ecological Society. By then, lots of other strong top-down effects had accumulated from both observations and experimental manipulation of intertidal and other assemblages. Furthermore, what remains the seminal work on marine cascades, Estes and Palmisano (1974), had been published. All that was necessary was to bundle that work with the numerous pairwise studies showing strong top-down effects and generate a term. *Trophic cascade* was it. Like keystones, not all ecosystems have them, context is known to be important, but meta-analyses continue to suggest that these prototypical top-down relationships characterize many natural systems (Pace et al. 1999; Pinnegar et al. 2000; Shurin et al. 2002).

These terms—*keystones* and *trophic cascades*—are serially and conceptually related to food web architecture and dynamics. A keystone regardless of its phyletic affiliation can be recognized by an effect on its community disproportionate to its abundance or mass (Power et al. 1996). Basically a keystone controls a competitively superior prey and thus facilitates indirectly multispecies coexistence at that level. A cascade occurs when two or more pairwise interactions, each individually characterized by top-down control, become linked. In this sense there are alternating effects on adjacent trophic levels; all species in the system become linked either directly or indirectly and either benefit or suffer depending on trophic level placement. The clearest distinctions between systems with or without trophic cascades can be drawn by looking at how large the change in system primary productivity is and how taxonomically distinctive the derived alternative states might be.

APPLYING THESE TERMS TO NATURAL SYSTEMS

Figure 2.1 portrays two versions of an imaginary food web topology. The upper panel (A) treats all linkages as equivalent with the intent of identifying the magnitude of connectedness (Paine 1980). Such illustrations can become endlessly complex as species are added and even minor trophic nuances included. The lower panel (B) treats species as dissimilar in the consequences of pairwise interactions. I dismiss the Figure 2.1A caricature because, while reflecting complexity, it ignores dynamics. Figure 2.1B identifies three variations on the theme that interactions can matter. On the right is a food chain in which species are dynamically linked. Many pairwise studies fall into this category when extracted from a broader community context. These and the middle chain constitute what I call a pincushion configuration; examples are often rare prey and their still rarer and trophically specialized consumers. Sponges, hydroids, and bryozoa and their nudibranch (gastropod) predators, or nemertean worms, a phylum of trophic specialists on annelid worms, microcrustaceans, and even barnacles provide marine examples. The dynamics can vary from donor-controlled to tri-trophic Lotka–Volterra interactions. Pincushion patterns add complexity and beauty. Does their presence contribute meaningfully to assemblage function? We don't know, but I doubt it. Figure 2.1B further suggests that the presence of pincushions is indirectly linked to that of the keystone predator. A more detailed relationship, shown on the left side, illustrates the elements of a cascade, much as in Paine (1980), with controlling top-down influences, more

A. "All animals are equal."

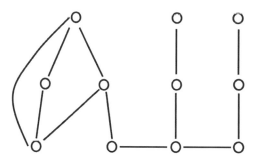

B. "But some animals are more equal than others."

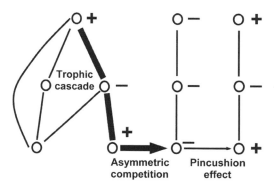

Figure 2.1. Two structurally different approaches to food web topology, expressed as a quote from George Orwell's *Animal Farm*. Panel A is simply descriptive and can be made endlessly complex as links are added. Panel B weights the strengths of interaction, as indicated by thicker lines. Thick lines characterize a cascade; lateral arrows identify the intensity of asymmetric competition and point from winner to loser. Signs (+,−) suggest changes in trophic-level population density or mass. See text and Paine (1980).

than two trophic levels, and effects translated laterally when competition is strongly asymmetric.

A keystone becomes recognized not so much because of a demographic influence on its prey but because of that prey's potential to outcompete other resource-limited neighbors. Trophic cascades extend the relationship to three or more trophic levels, are characterized by rampant indirect effects, and alter the assemblage's structural appearance and production, especially of the basal state. This stipulation is necessary because in many intertidal communities sessile animals dominate the limiting resource, space, but can have little or no

influence on the productivity of their planktonic prey. When macroalgae occupy that space, top-down effects that either increase or decrease productivity
can be anticipated, a phenomenon dependent on whether there are odd or
even numbers of the imposed trophic levels.

A SYSTEM SUBJECT TO A FEW STRONG INTERACTIONS: TATOOSH ISLAND

The venue for the development of these descriptors and the supporting observations and research was the Cape Flattery region, Washington State, especially
Tatoosh Island. Many details can be found in Paine and Levin (1981), Leigh et
al. (1987), and Wootton (1993, 1997). Tatoosh is a 16- to 18-hectare island immediately adjacent to deep water. Historically, it was occupied by Native Americans, the Makahs. In the mid-1800s a light station was constructed, and by the
early twentieth century all access was restricted. The U.S. Coast Guard station
was automated in 1976. I believe that the intertidal biota has been free of major
human influence for about 100 years, perhaps even 150. Sea otters were successfully reintroduced to the region in 1969–1970. The biota seems intact, with
the possible exception of northern fur seals; invasives (the algae *Sargassum mu-
ticum* and *Codium fragile*) are uncommon. Although the site is certainly not pristine (no modern site can be), it does not appear to have been biologically rav-
aged like the Mediterranean, Caribbean, or all mainland estuaries (Jackson et al.
2001). I next discuss two derived properties of the Tatoosh community, important ecological properties that are possibly best, and usually only, probed in systems characterized by strong top-down control and the resultant alternative
states.

Primary Productivity

Collations of global patterns of primary production clearly need refinement
and updating; nonetheless, intertidal assemblages and estuaries are remarkably
productive in comparison with most other ecosystems on a grams of carbon
per square meter per year basis (Valiela 1995). I suspect they also show greater
variation within a community type and that this variation can be unambiguously attributed to controlling top-down interactions exercised at two or more
trophic levels, much in the spirit of HSS (1960). For instance, Duggins (1980)
suggested that the productivity of urchin-dominated coralline algal barrens
(barrens are grazer-dominated space) was so small as to be essentially unmea-
surable. Reinvasion of his Alaskan sites by sea otters led to the virtual disappear-

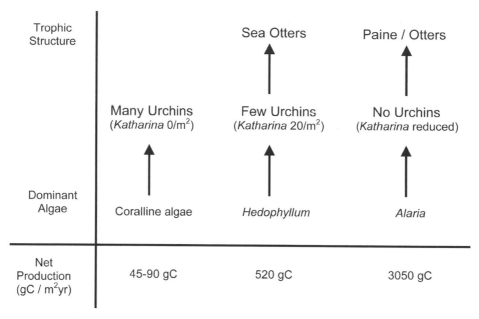

Figure 2.2. The linkage between trophic structure and estimated primary productivity at Tatoosh Island. Shifts in the identity of two invertebrate herbivores and their "predators" translate into a factor of about 50 in net primary productivity.

ance of urchins and initiated the development of an alternative macroalgal state that, depending on its recovery status, varied in productivity from 1,774 to 4,806 grams of carbon per square meter per year.

Figure 2.2 portrays three algal productivity scenarios at Tatoosh, all minimal estimates. Before otter reintroduction, sea urchin barrens were conspicuous around the island wherever physical conditions permitted their establishment. In barrens, the algal moiety consists of slow-growing, grazer-resistant coralline algae, and fleshy macrophytes are essentially absent. I have generously estimated net coralline productivity as ranging from 45 to 90 grams of carbon per square meter per year. Intertidal and subtidal urchin populations were decimated shortly after the otter's return and the barrens disappeared, being replaced by a canopy of perennial kelp, *Hedophyllum sessile*. A minimal estimate of its annual productivity is about 10.4 kilograms wet mass or about 520 grams of carbon per square meter per year (Paine 2002). The generally dense *Hedophyllum* canopy protects a large-bodied (8–10 centimeter) and dense (ca. twenty per square meter) grazer, the chiton *Katharina tunicata*, from gulls and otters. However, when I play chiton consumer as Native Americans do (Salomon et al. 2007; Paine, personal observation), another state develops. Fast-growing annual kelps, especially *Alaria marginata*, dominate the canopy, suppress *Hedophyllum*, but magnify estimates of net production to about 3,100 grams of carbon per square meter per year (7-year range, 1,300–5,800 grams of carbon per square meter per year; Paine 2002). A factor of approximately 50 separates the productivity extremes characterizing these taxonomically distinctive alternative states (Figure 2.2). Potentially, such differences provide a metric by which to evaluate the ecological legitimacy of whether an alternative state has formed.

Resilience

All natural assemblages vary over time in biological character and interaction intensity. Most often, such population variation is unremarkable, unnoticed, and unrecorded. Occasionally, however, forces generated either external to the system (e.g., a severe El Niño) or internally (e.g., a plague) can force the system into a new or even unanticipated alternative state. Resilience (Holling 1973) basically reflects the rate and trajectory with which a system bounces between one alternative state and another. Ives and Carpenter (2007) have reviewed the subject.

There is little doubt that strong top-down influences radically alter the next lower level: Mussels can replace a complex of other species in the absence of *Pisaster* (Paine 1966; Menge et al. 1994), coralline algae–dominated urchin barrens can become forests of kelp (Paine and Vadas 1969; Estes and Palmisano 1974; Estes and Duggins 1995; Chapter 3, this volume), recumbent bryozoa can be replaced by statuesque tunicates (Sutherland 1974), and estuaries dominated by *Spartina* can be transformed into mudflats (Silliman et al. 2005). Numerous examples of such top-down forcing, generating a striking shift in community composition, can be found in work in intertidal or shallow water marine ecosystems. Are or can they be stable? Most studies suggest the answer to be "no"; relax the forcing, and the system recovers at a rate roughly calibrated to the generation time of its major constituents. An example from Tatoosh follows.

The 1997–1998 El Niño represented a major physical perturbation along the northeastern Pacific shoreline. It is not known whether some species benefited (e.g., enhanced growth rates); many suffered, including the intertidal kelp *Hedophyllum*. The species is perennial, tough, or at least less palatable than its primary algal competitor (*Alaria marginata*), and it coexists with its consumers. Figure 2.3 illustrates its dominance of the mid-intertidal, as determined by multiple percentage cover estimates, from 1978 to 1997. The El Niño changed that. By early spring 1998, *Hedophyllum* had been nearly eliminated and replaced by barnacles in the upper half of its natural distribution. Continued sampling, 1998 through 2007, showed that mussels recruited successfully to these barnacles, transforming intertidal benches previously dominated by a kelp to a developing mussel bed. Paine and Trimble (2004) have discussed the initial

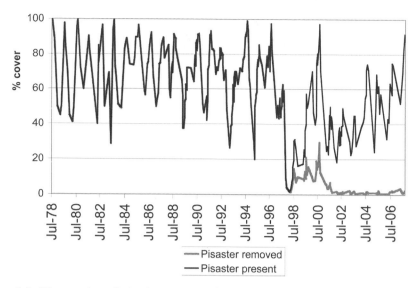

Figure 2.3. Time series of visual estimates of percentage cover of the kelp *Hedophyllum*. Means from 6 Tatoosh sites based on arc-sine, square root transformations are shown for the interval 1978 to 2007. The 1997–1998 El Niño substantially depressed the kelp population. Early recovery is illusory and based on the sampling technique; it takes a few years for the surviving mussels to outcompete *Hedophyllum*. From about 2001 through 2007 kelp recovery has been sustained at 3 sites characterized by the presence of *Pisaster*, whereas mussels persist at 3 other sites at which the predator's presence has been systematically reduced.

stages. Here I extend the analysis from 2003 through 2007. Specifically, at four sites where *Pisaster* has been allowed to reenter, mussels have been progressively eliminated; at two of these four sites, mussels are gone and *Hedophyllum* is recovering. Where *Pisaster* has been removed at every opportunity, the mussel bed has persisted and even modestly expanded. More to the point, mussels continue to grow, and as of August 2007, the largest of the matrix-forming individuals were 15–16 centimeters in shell length, close to the upper size limit that *Pisaster* can successfully attack in the intertidal. Table 2.1 illustrates that the mussel beds and kelp-dominated surfaces are biologically distinctive and thus alternative. That is, though sharing many species or taxa, the assemblage dominants are conspicuously different.

Lessons

I draw five lessons from these data. First, alteration of primary productivity by top-down interactions probably provides the most convincing evidence that a site's biological character has changed. The decision point is entirely arbitrary and could be disguised by compensating species replacements. That process appears not to operate on intertidal shores. Second, *alternative* should be defined

Table 2.1. Differing species compositions in the mussel-dominated (*Mytilus californianus*) and kelp-dominated (*Hedophyllum sessile*) community states at Tatoosh Island.

Mussel Bed	Sessile Species (% Cover)	Algal Stand
72	*Mytilus californianus*	0
2	*Hedophyllum sessile*	60
12	Barnacles (3 species)	0
>0.1	Sponge	0
0	Crustose coralline algae	26
5	Erect coralline algae	12
9	Bare space	0

Mussel Bed	Mobile Species (N/m²)	Algal Stand
18	Carnivorous gastropods	1
1	Sea urchins	6.1
15	*Lottia pelta*	15.4
0	*Acmaea mitra*	3.9
0	*Tonicella lineata*	10.6
0	*Katharina tunicata*	20
0.5	*Anthopleura* spp.	0

by criteria other than just statistically significant compositional differences. Third, perturbations of approximately equivalent magnitude may have dissimilar ecological signatures. As a major physical forcing example, the 1982–1983 El Niño had little effect on *Hedophyllum*, in contrast to its 1997–1998 cousin. I attribute the differences in response to contingency: Mussels recruited extensively during the latter event and not during the earlier one. Fourth, discussion of stability or resilience requires dissection of often subtle differences in time series of recovery. Kelp and mussels are clearly alternative, but are they stable? When *Pisaster* is present and can become abundant, the answer is "no": A mussel stage is entirely transitory, and the system reverts to its original condition within a few years. But, finally, if *Pisaster* is removed or fails to respond numerically, mussels probably have the capacity to grow to an invulnerable size (Paine 1976) and therefore persist even in the presence of their major consumer. An analogy with tree survival in terrestrial forests is tempting.

A SAMPLER OF INTERTIDAL SYSTEMS

We were implicitly challenged with an impossible mission: to examine and summarize whether cascades existed in our allotted ecosystem and, if so, under what conditions. Here I briefly interpret four high-quality studies that capture the sources and nature of the documented variation. Intertidal surfaces have proven to be magnets for attracting experimental manipulations, and there is no dearth of investigations; direct and indirect effects have been reviewed by Lubchenco and Gaines (1981), Hawkins and Hartnoll (1983), Menge and Farrell (1989), Menge (1995), and Menge and Branch (2001). *Marine Community Ecology* (Bertness et al. 2001) should be consulted for comprehensive reviews. These sources identify a realistic complexity to the trophic simplifications characteristic of all interaction webs (e.g., Figure 2.1B and many others in this book). For instance, Menge and Branch (2001, Table 9.1) list thirteen studies exploring the consequences of pairwise interactions. Menge (1995) has examined twenty-three food webs in South Africa, Chile, the eastern Pacific, New England, and Central America. Forty percent of the change in community structure could be attributed to indirect effects; 35 percent had "keystone predation." Collectively, these interactions do not identify a cascade, unless one invokes a third level, usually a human one. The four systems described next were chosen to identify the kinds of ecological variation that must be confronted before any mechanistic synthesis should be attempted.

Ythan Estuary

The Ythan is an entirely soft sediment, intertidal, 185-hectare site supporting a rich cast of invertebrates and their vertebrate (primarily fish, birds) predators (Hall and Raffaelli 1991; Raffaelli and Hall 1992). Initial manipulation involved exclusions (birds) and enclosures (fishes and predatory crustaceans). At least ninety-two taxa were identified, most to species at higher trophic levels. Taxonomic resolution, a universal problem in all community studies, suffered for lower levels; for instance, the taxon "harpacticoid copepods" contained at least 100 species. These studies discovered no strong linkages between species, no important top-down influences, no evidence of rampant indirect effects, and thus no cascades. Whether this conclusion supports my 1980 contention that strong interactions are a requisite for cascades can be argued; however, their research provides some evidence that multilevel cascades need not be universal and should not be found whenever strong, single-species, top-down influences are lacking. The Ythan research suggests that such interactions may be difficult to identify in systems that are decomposer or nutrient (bottom-up) based. Thus, weakly interacting species initially appeared to predominate in the Ythan, and "none of the manipulations produced any large changes in prey abundance" (Raffaelli and Hall 1992: 556). However, when Emmerson and Raffaelli (2004) subsequently experimented with certain pairwise interactions in the Ythan, strong controlling interactions emerged.

Because the Ythan despite manipulation provides no evidence of a community-wide cascade, it might be viewed as a counterpart to an intact tropical forest in which numerous intertwined food chains dilute a top-down signal, apparently weak interactions predominate, and there are few hints of an alternative state. Certainly, estimating minimally altered primary production in the Ythan would prove futile and provide limited support for a change of state. However, an important caveat exists. In highly diverse systems with rampant redundancy within functionally similar groups, their collective impact could be comparable to that of a single strongly interacting species (Chapter 8, this volume). The flooding of Lago Guri decomposed the original dry forest into simpler ecological units, which immediately revealed dramatic food web effects (Chapter 8, this volume). Whether the Ythan estuary ecosystem is truly immune to dominating top-down influences remains a challenge to be resolved at some future date.

Eastern United States *Spartina* Salt Marshes

Salt marshes are a worldwide intertidal feature wherever conditions permit. They are prominent along the eastern and southeastern U.S. seaboard. They continue to attract attention because of high productivity (Valiela 1995) and experimental tractability (Silliman and Bertness 2002). Relative to rocky shores and coral reefs, they are biologically simple, less diverse ecosystems delicately balanced between marine and terrestrial stressors. Furthermore, estuaries and their salt marshes are subject to numerous possible anthropogenic forcings ranging from chemical poisoning and sea level rise to biological alteration. Silliman and coworkers (Silliman and Bertness 2002; Silliman et al. 2005) nonetheless have proven *Spartina*-dominated marshes to be remarkably responsive to top-down influences, by showing beyond reasonable doubt that the periwinkle *Littoraria* can transform *Spartina* marshes into mud flats if snail densities about twice normal or higher are attained. Results of caging experiments seem unambiguous: Snails at high densities force a shift from grass to mud. But what factors release the snails?

A convincing argument can be made that overfishing of blue crabs, themselves major consumers of the snails, drives the shift. However, establishing that linkage as biological reality is fraught with the usual challenges of linking effect to cause, especially at extended spatial scales. Furthermore, crab effects are confounded by shifts in abundance in other snail predators: Raccoons are surely up, terrapins down. Thus the fog generated by ecological uncertainty, inadequate crab data, and industrial denial of any extended consequences of their activities renders a definitive assessment of their activities difficult. A four-level cascade, humans → blue crabs → snails → *Spartina*, seems highly likely. It contains all the hallmarks of a legitimate cascade: a substantial role for indirect, trophic facilitation and an unquestionable shift in community appearance and probably functioning. A capacity for the marsh to revert to grass at reduced snail densities implies that, although the states are alternative, neither is stable.

The Baltic Intertidal

These rocky sites provide a glimpse of how assemblages exposed to elevated nutrient loading from sustained coastal eutrophication are organized (Worm and Lotze 2006, and references therein). A Baltic "pristine" system would probably be dominated by brown algae, especially fucoids (rockweeds) and Laminariales (true kelps). Competitively superior, bloom-forming annuals are normally controlled by grazers, especially small gastropods and isopods. However, anthropogenic influences tend to reduce the herbivore load, favoring persistence of these annuals, which in turn negatively affect the perennial brown algae. The latter have been reduced 30–60 percent with eutrophication, implying that nutrient addition will be the ultimate driver of community structure on this rocky shore. Interestingly, these monodominant stands of weedy species need not persist; they need only be present during the restricted interval when fucoid zygotes are being released. A decrease in biodiversity is one expected consequence.

Bottom-up forcing (nutrient addition) surely generates an alternative state when judged by the identity of the dominant species. Would this state persist if the eutrophication influences were reduced? Almost certainly not, unless the grazing moiety had gone locally extinct, which seems unlikely. Can these dramatic shifts in community structure, supported by experimental manipulation of both grazer density and nutrient loads, be called a cascade? From appearances, the basic food web connectivity seems little changed. What has been altered are the rate processes; grazers are overwhelmed, and therefore competitively dominant annual algae flourish. Thus, despite potentially controlling top-down processes, an artificially elevated nutrient subsidy determines community structure.

Exposed Rocky Intertidal, Central Chile

J. C. Castilla and colleagues (Castilla and Duran 1985; Duran and Castilla 1989; Castilla 1999), by a negotiated and peaceful exclusion of humans from a 0.8-kilometer stretch of previously overexploited shoreline, have identified the magnitude of human influence. Simply excluding the local artisanal and recreational exploiters of intertidal algae, mollusks, crustaceans, and fishes generated ecologically instantaneous and sweeping results. Grazers increased, algae de-creased, the mussel–barnacle competitive interaction was altered, and a gastro-pod rather than a locally abundant carnivorous starfish was rapidly identified as a keystone species. Such altered food chain dynamics should be expected when the alpha consumer (human) impact is reduced from substantial to zero. There is no doubt that three trophic levels are involved; a novice observer would no-tice little difference in assemblage appearance when compared with adjacent, exploited sites (local fishers wouldn't be as naive), and systematists would as-semble identical species lists from the protected and exploited areas. Yet ram-pant indirect consequences are apparent. Do these changes constitute a cascade in the sense of the Estes and Duggins (1995) or Silliman and Bertness (2002) studies in which such forcing led to major structural, and probably productivity, shifts of the benthic algal assemblage?

Castilla's human exclusion experiment identified a potent human influ-ence, implying a global ubiquity to our influence. The challenge is to quantify its impact, overwhelming in Castilla's system and more subtle elsewhere. Once one accepts an undeniable human influence on probably all nearshore ecosys-tems (Jackson et al. 2001), adding this level should legitimize the transforma-tion of well-documented pairwise interactions to tri-trophic and thus potential cascade status.

CONCLUSIONS

The intertidal research summarized here identifies two issues that permeate all trophic ecology. First, many of the insights into how such systems work are based on controlled manipulations of both animal and plant components. The term *trophic cascade* was intended to capture linked and dynamic relationships. Second, in many of these studies there was a single focal species. This attribute shifts interpretations of how integrated ecosystems function from trophic levels, trophospecies, or guilds to species for which individual roles can be assigned. My preoccupation with such roles has an important, and honorable, ecological history. At the community level, its origins can be traced at least to Elton (1927), his insistence that ecologists consider status and what a species is doing, and his amusing example of understanding the functional roles of badgers and vicars in their own communities. On Chilean rocky shores, a giant tunicate plays a remarkably similar role (that of competitive dominant) to an outer coast *Mytilus californianus*, despite their phylogenetic remoteness (Paine and Suchanek 1983). In studies on per capita strengths of interaction (Paine 1992), sea urchins and a chiton were judged to be strong, whereas within chitons the influence on their algal prey varied from negative to positive. On other shores, numerous examples exist of algae and animals competing for the limiting resource, space (Dayton 1971; Buss and Jackson 1979). Furthermore, the aggregation of species into trophic collectives (e.g., trophospecies) introduces a suite of analytical challenges. For instance, Schall and Pianka (1978) described how higher taxa—ecologically similar xeric-adapted birds and lizards, in their example—showed an inverse relationship. Such complementarity on land or interkingdom competition on shores reduces substantially the utility of species aggregated for analytical convenience and certainly disguises asymmetric species roles, rendering analysis less realistic. Who gets included or excluded seems entirely arbitrary.

There is an intellectual chasm between approaches centered on an individual species' role and that of a whole trophic level; the former can characterize many intensely experimental, pairwise food chain interactions, the latter opportunistic and often observational approaches. Both provide numerous ecological success stories and deep insights into how natural communities are organized and function. Food chains and keystone species fit comfortably into the first category; trophic cascades, into the latter. What was emphasized in the system sampler presented in this chapter is the variation so typical of all ecology. Scale and context are important, the role of decomposers in directing community interactions is essentially unknown, and many rules govern the ecosystem structure game. Humans are certainly the overdominant keystones and will be the ultimate losers if the rules are not adequately understood and global ecosystems continue to deteriorate. Readily accessible and experimentally tractable intertidal systems provide an unequaled platform on which to test ideas, to distinguish the trivial from the consequential, and to relate these to the roles of species, thus invoking their individual evolutionary heritages.

ACKNOWLEDGMENTS

Intense, long-term single-site research requires a support system. I thank the National Science Foundation (Biological Oceanography) and the Andrew W. Mellon Foundation for underwriting my work on Tatoosh; Cathy Pfister, Tim Wootton, Alan Trimble, and a host of others for challenging and sharpening my views; and, especially, the Makah Tribal Council for granting permission to study on their lands.

CHAPTER 3

Some Effects of Apex Predators in Higher-Latitude Coastal Oceans

James A. Estes, Charles H. Peterson, and Robert S. Steneck

There have been numerous published accounts of predation and top-down forcing from temperate-latitude coastal oceans (Connell and Gillanders 2007; McClahanan and Branch 2008). These studies include diverse predators, various methodological approaches, and a broad range of geographic regions. This literature has become so extensive that we cannot possibly review it all in the space available here. Our goal is thus to provide a sampling of the details, a more superficial survey of the better-known or more persuasive studies, and a synthesis of the principles and generalizations that are emerging from this published literature.

Our chapter centers on three case studies: sea otters in the northeast Pacific Ocean, sharks in estuaries of the central U.S. Atlantic seaboard, and cod in the Gulf of Maine. We have chosen these particular examples because each was assembled around a progression of field studies conducted over many years, the resulting evidence for direct and indirect effects of predation is diverse and compelling, and collectively we have worked in each of these systems. We will follow these detailed accounts with a series of vignettes that further chronicle the ecological roles of predators in temperate marine systems around the world.

SEA OTTERS IN THE NORTH PACIFIC OCEAN

Sea otters ranged across the North Pacific rim for several million years before anatomically modern humans peopled the region. The discovery of abundant sea otter populations by the Bering Expedition in 1740–1741 initiated the Pacific maritime fur trade, which motivated Russian colonization of northwestern North America and led to the near extinction of sea otters.

The Fur Trade: A Fortuitous Natural Experiment

Only a handful of very small sea otter populations survived the fur trade (Kenyon 1969). These occurred in Russia, southwest Alaska, and central California. Additional populations were reestablished by translocations to southeast Alaska, British Columbia, Washington, and southern California (Jameson et al. 1982). The growth of these colonies, coupled with the species' sedentary nature, created a patchy distribution of habitats with and without sea otters. The sea otter's keystone role in kelp forest ecosystems was discovered in the early 1970s through comparisons of islands in the Aleutian archipelago where the species had recovered with those where it had not (Estes and Palmisano 1974; Estes et al. 1978) and through experimental studies of competitive interactions between kelp species at Amchitka Island, where otters were numerous at the time (Dayton 1975). The results of these early studies have been substantiated by similar studies in other areas (e.g., Breen et al. 1982; Kvitek et al. 1992, 1998). Elucidating the workings of the interaction chain has relied on fortuitous "experiments" such as documenting temporal changes at unoccupied sites as they were recolonized by expanding sea otter populations (Estes and Duggins 1995) and, more recently, by documenting temporal changes in the Aleutian archipelago as sea otter numbers collapsed because of increased killer whale predation (Estes et al. 1998).

Top-Down Forcing and Trophic Cascades

Sea otter predation reduces the size and density of numerous shellfish species in soft sediment and reef systems (Wendell et al. 1986; Kvitek et al. 1992; Estes and Duggins 1995). For example, sea urchin biomass density on shallow reefs is typically 10–100 times greater at locations lacking sea otters. Because sea urchins are herbivores, sea otters, sea urchins, and kelp interact via a trophic cascade (Paine 1980; Carpenter and Kitchell 1993), thus leading to rocky reef ecosystems that are either adorned with kelps (hereafter called kelp forests) or extensively deforested (hereafter called urchin barrens), depending on the presence or absence of sea otters. Intermediate configurations between kelp forests and urchin barrens are rarely observed, thus indicating that these states are highly unstable or transitory. Kelp forests and urchin barrens therefore are known as phase states, and the transitions between these phase states are called phase shifts (Lewontin 1969; Sutherland 1974; Done 1992; Hughes 1994).

The occurrence of kelp forests or urchin barrens is a predictable consequence of the presence or absence of sea otters in outer coast reef environments across much of the eastern North Pacific Ocean (Estes and Duggins 1995). By contrast, population and community structure varies substantially over time within these phase states, depending on invertebrate recruitment dynamics, physical disturbances, ocean temperature changes, and a host of other factors.

Complex Interactions

The sea otter's influence on reef systems extends well beyond the previously described trophic cascade (Figure 3.1). Kelp forests affect other species through increased production, the creation of three-dimensional habitat, and reductions in wave height and current velocity. These processes play out in numerous ways. For example, barnacles and mussels grow three to four times faster in otter-dominated kelp forests than in otter-free urchin barrens (Duggins et al. 1989); rock greenling (*Hexagrammos lagocephalus*, a common kelp forest fish in the Aleutian Islands) are roughly ten times more abundant in otter-dominated kelp forests than they are in otter-free urchin barrens (Reisewitz et al. 2005); the diets of glaucous winged gulls contain about 90 percent fish in otter-dominated kelp forests and about 90 percent intertidal invertebrates in otter-free urchin barrens (Irons et al. 1986); the loss of otters causes bald eagles to shift their diet from a roughly even mix of fish, marine mammals, and seabirds to one dominated by seabirds (about 80 percent by number of prey consumed; Anthony et al. 2008); sea otters compete with various benthic feeding sea ducks (eiders and scoters), thus limiting their populations (Irons, Byrd, and Estes, unpublished data); and predatory starfish are eaten by sea otters, thus reducing starfish size, abundance, and interaction strengths as predators of mussels and barnacles (Vicknair and Estes, unpublished manuscript).

Evolutionary Effects

The preceding summary outlines strong interactions between sea otters and numerous other species through direct and indirect food web linkages. How might these interactions have played out over evolutionary time scales? One intriguing possibility is that sea otters and their ancestors thwarted an evolutionary arms

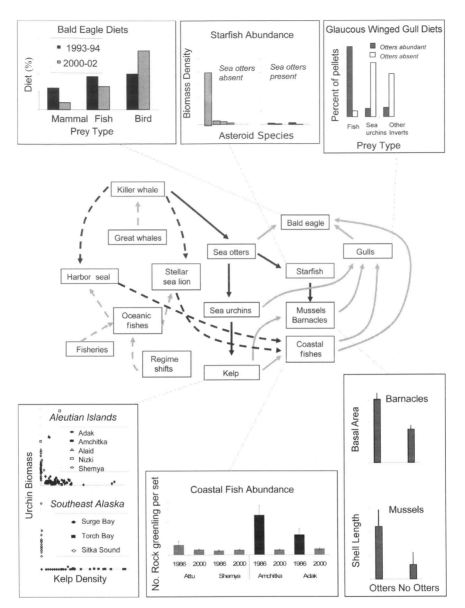

Figure 3.1. Selected food web showing key elements of North Pacific kelp forest system. Top-down forcing processes are indicated in black, bottom-up forcing processes in gray. Sold lines between species indicate linkages for which the evidence of a strong interaction is strong; dashed lines are linkages for which the evidence of a strong interaction is more speculative. The surrounding panels connected to particular species by dotted lines show the effects of sea otter predation on the indicated parameters.

race for defense and resistance between the kelps and their herbivores. Although predatory fishes and lobsters also drive trophic cascades in southern hemisphere kelp forests (Babcock et al. 1999), these do not appear to be as powerful or pervasive as the sea otter–induced trophic cascade in the North Pacific Ocean. A comparison of plant chemical defenses (tissue phlorotannin concentrations) and the resistance of herbivores to these putative defenses between western North America and Australasia (where the plants and herbivores evolved without sea otters or their ancestors) supports the arms race hypothesis. On average, phlorotannin concentrations in Australasian kelps and rockweeds are ten times higher than they are in northeast Pacific species, whereas Australasian sea urchins and gastropods are less deterred by these compounds (Steinberg et al. 1995). This coevolutionary model might explain why northern hemisphere kelp forests collapse so spectacularly in response to sea urchin population outbreaks, why the world's largest abalones (species for which growth rate is reduced by phlorotannins; Winter and Estes 1992) evolved in the North Pacific Ocean (Estes et al. 2005), and why the kelp-eating hydrodamaline sirenians (Steller sea cows and their ancestors) radiated from a pantropically distributed ancestor into the North Pacific Ocean but not elsewhere.

Ecosystem Collapse

After nearly a century of recovery from the ravages of the Pacific maritime fur trade, sea otter numbers in southwest Alaska began a precipitous decline in about 1990 (Estes et al. 1998; Doroff et al. 2003). Not surprisingly, the kelp forest ecosystem quickly followed suit by shifting to the urchin-dominated phase state (Estes et al. 2004). The sea otter collapse appears to have been caused by killer whale predation (Estes et al. 1998), and similar but somewhat earlier population declines by harbor seals and Steller sea lions may have been driven by this same process (Williams et al. 2004). The mystifying question is why killer whales changed their feeding habits. Springer et al. (2003) proposed an explanation that again involves anthropogenic disturbances to predator–prey interactions, in this case one initiated by post–World War II industrial whaling. Springer and colleagues proposed that the depletion of whale populations reduced prey availability for transient (marine mammal–eating) killer whales, thus causing these megapredators to expand their diets to include increased numbers of pinnipeds and sea otters, thereby driving their populations rapidly downward. Although this proposal remains hypothetical, it suggests an even more extensive and complex role for predation and top-down forcing in higher-latitude ocean ecosystems (Figure 3.1).

SHARKS AND EAST COAST ESTUARIES

Dating from the approximate advent of industrialized fishing by highly mecha-nized and efficient fleets, and in some cases much earlier (Rosenberg et al. 2005; Roberts 2007), abundances of large predatory fishes have declined dra-matically throughout much of the world's oceans (Myers and Worm 2003). This influence, which includes reductions in apex predatory sharks in both open ocean and coastal seas (Musick 1993; Baum et al. 2003), is reflected in the re-duction in average trophic level in fishery landings (Pauly et al. 1998). The great sharks represent a particular conservation challenge to fishery managers be-cause their generally low fecundity and slow maturation rates deprive them of the demographic resilience to respond readily to exploitation.

Consequences to Estuarine Ecosystems of Trophic Cascades Following Losses of Apex Sharks

A test of the hypothesis that removal of apex predatory sharks by overfishing can have important indirect impacts on estuarine ecosystems was conducted by Myers et al. (2007) along the Atlantic seaboard. These authors analyzed survey data on the great sharks and the smaller elasmobranchs that formed their prey. All eleven great sharks in this guild exhibited significant population declines over the past 35 years, ranging from 87 percent in sandbar sharks to 99 percent or more for bull, dusky, and smooth hammerhead sharks (Figure 3.2). The aver-age sizes of blacktip, bull, dusky, sandbar, and tiger sharks declined by 17–47 percent, consistent with intensified exploitation.

Over this same 35-year period, meta-analyses of survey data on the meso-predatory elasmobranch prey of these great sharks revealed that twelve of the fourteen species analyzed increased significantly in abundance (Myers et al. 2007). Among the largest of these population increases was an approximately twenty-fold increase in cownose ray (*Rhinoptera bonasus*) abundance (Figure 3.2). The eastern seaboard population of this species spends summers in the shallow waters of Raritan Bay, Delaware Bay, Chesapeake Bay, and Pamlico Sound. Myers et al. (2007) computed that the Chesapeake Bay population of cownose rays now totals more than 40 million.

Cownose rays consume shellfish of commercial and recreational value, in-cluding soft-shell clams (*Mya arenaria*), oysters (*Crassostrea virginica*), hard clams (*Mercenaria mercenaria*), and bay scallops (*Argopecten irradians*), as well as other clams not taken in fisheries, such as *Macoma balthica*. The projected consump-tion of bivalves by the current population of cownose rays over the 100 days of

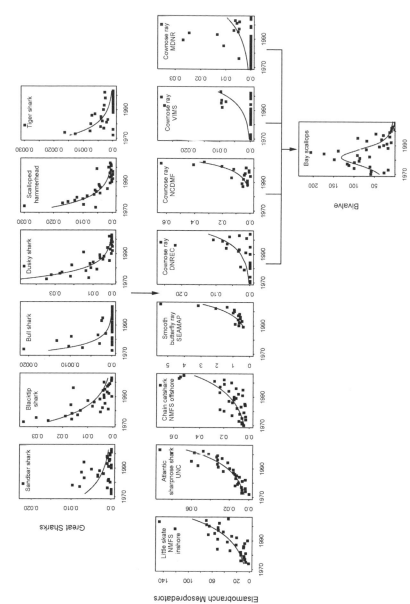

Figure 3.2. Patterns of temporal covariation among the declines of great sharks, the rise of small elasmobranchs, and the declines of bay scallops in East Coast estuaries and coastal oceans (from Myers et al. 2007). DNREC = Delaware Department of Natural Resources and Environmental Control, Division of Fish & Wildlife; MDNR = Maryland Department of Natural Resources, Fisheries Service; NCDMF = North Carolina Department of Environment and Natural Resources, Division of Marine Fisheries; NMFS = National Marine Fisheries Service; SEAMAP = Southeast Area Monitoring and Assessment Program, South Atlantic; UNC = University of North Carolina Institute of Marine Sciences longline shark monitoring survey; VIMS = Virginia Institute of Marine Science.

summer occupation of Chesapeake Bay totals 840,000 metric tons. In contrast, the total harvest of bivalves in Virginia and Maryland in 2003 was only 300 metric tons (Myers et al. 2007). Such intense demand for bivalves by an exploding population of rays suggests a potential for extending the impact of the losses of great sharks down another trophic level in a trophic cascade.

Peterson and colleagues (e.g., Peterson et al. 1989, 2001) had been studying the impacts of cownose ray feeding on bay scallops in North Carolina sounds during the late summer southward migration for nearly two decades. This work provides observational and experimental evidence that the impacts of cownose ray predation on bay scallops have grown along with ray abundances. Field sampling of bay scallops in several scalloping grounds immediately before and after the late summer migration of cownose rays showed no detectable change in scallop abundances in 1983 or in 1984, whereas repetition of the observations in 2002–2004 showed that bay scallops were virtually eliminated from all important scalloping grounds (Figure 3.2). Use of stockades constructed of closely spaced vertical poles that excluded rays demonstrated experimentally that bay scallop mortality during these recent years could be attributed to predation by the cownose rays. Thus, the loss of great sharks at the top of the food web led to a trophic cascade that indirectly eliminated a century-old shellfishery.

The further implications of the exploding cownose ray population after release from control by great sharks are profound but as yet untested. After cownose rays have consumed the visibly detectable epifaunal bivalves such as scallops and oysters, they may turn to the abundant bivalves in seagrass habitats (Orth 1975). Bivalves are much denser inside seagrass beds than on unvegetated bottom because the roots and rhizomes provide protection from typical predatory invertebrates such as crabs and whelks (Peterson 1982; Summerson and Peterson 1984). Seagrass habitat provides an important nursery for juvenile fishes and crustaceans (Heck et al. 2003), so the destruction of seagrass beds by foraging cownose rays implies a possible extension of the shark–ray cascade to additional species.

Another hypothesized effect of hyperabundant cownose rays on estuarine habitat involves their consumption of oysters. Oysters influence habitat type and quality in two important ways (Grabowski and Peterson 2007; Coen et al. 2007): by filtering enough particulates from the water column to reduce turbidity and enhance light penetration (Newell and Koch 2004) and by forming biogenic reefs that provide habitat for various other species (Lenihan et al. 2001). The explosion of cownose rays may well have contributed to the multidecade decline in oysters and is certainly inhibiting restoration efforts (National Research Council 2004).

Generality of Shark-Topped Trophic Cascades in Estuarine and Coastal Systems

Various evidence suggests that the great shark–ray–benthic mollusk trophic cascade is geographically widespread. First, the review by Libralato et al. (2006) indicates that great sharks are often keystone predators. Second, analyses of reef systems in the Hawaiian archipelago (Parrish and Boland 2004) and the Caribbean (Bascompte et al. 2005) further indicate that apex sharks initiate strong predatory interactions with their prey species. Third, rays, skates, and smaller sharks are rarely eaten by predators other than great sharks, implying little functional redundancy beyond members of the great shark guild. Fourth, many rays are well known worldwide as consumers of bivalves and other benthic mollusks. For example, the recent explosion of longheaded eagle rays has eliminated valuable wild shellfish and cultured shellfish stocks in Ariake Sound, Japan (Yamaguchi et al. 2005), probably a consequence of overfishing of great sharks. Fifth, although Bascompte et al. (2005) show that omnivory can reduce the likelihood and strength of trophic cascades, great sharks are piscivorous and rarely if ever include clams and benthic mollusks in their diets.

COD AND LOBSTERS IN THE GULF OF MAINE

Vast numbers of cod (*Gadus morhua*) that occasionally exceeded 90 kilograms in body mass once populated the North Atlantic Ocean (Steneck 1997). Archaeological evidence suggests this large-bodied, large-mouthed trophic generalist may have been the most important apex predator in shallow coastal ecosystems of the North Atlantic (Steneck et al. 2004). Like large predatory fishes elsewhere (Myers and Worm 2003), the abundance and size of cod were reduced greatly by fishing. Intensive cod fishing occurred during prehistoric periods (4,500 to 500 years before present) but expanded with European colonization and the establishment of small coastal fishing villages in the early 1600s. Coastal fish stocks first showed signs of localized nearshore depletion as early as 2,000 years ago (Bourque et al. 2007). Shore-based fishing continued into the eighteenth and nineteenth centuries but at ever-increasing distances from home ports (O'Leary 1996; Rosenberg et al. 2005). With expanded takes, body size and abundance have declined such that no cod exceeding 90 kilograms has been reported from the North Atlantic Ocean since the late 1800s (Collette and Klein-MacPhee 2002). Beginning in the 1930s, expanding zones of depletion radiated from coastal ports as industrial-scale fisheries and associated technology escalated. The most recent and most publicized collapses occurred on offshore banks in the United States and Canada since the 1990s (Steneck 1997).

Cod have remarkably local distributional affinities (Ames 2003). Tagging and genetic studies show that inshore stocks are demographically distinct from those offshore (Ruzzante et al. 1996), and these stocks consist of subpopulations with specific spawning, feeding, and nursery grounds. This spatial structure may explain the overall protracted and geographically asynchronous nature of the cod decline and explain why the final collapse of the fishery in the late 1980s and early 1990s was so abrupt.

Consequences of Overfishing

The depletion of cod and other predatory fishes has strongly affected the structure and function of coastal ecosystems in the northwest Atlantic Ocean. The key evidence comes from spatial contrasts detected in the 1980s between coastal areas where predatory fishes were ecologically extinct, and less intensively fished offshore seamounts. The offshore habitats supported fewer lobsters, crabs, and herbivorous sea urchins, had more abundant kelp (Vadas and Steneck 1995), and were characterized by higher predator attack rates on adults of all three invertebrate groups (Witman and Sebens 1992; Vadas and Steneck 1995; Steneck 1997).

The abundant lobster and sea urchin populations that developed in the coastal zone after the cod depletions became the primary target of local fisheries (Figure 3.3). By 1993, the value of sea urchins in Maine was second only to that of lobsters. As sea urchin populations were reduced in the fishery, so too was the rate of herbivory (Steneck 1997). In less than a decade, sea urchins became so rare that they could no longer be found over large areas of the coast (Andrew et al. 2002; Steneck et al. 2004) and as a result, kelp forests came to dominate the coastal ecosystem once again (Figure 3.3). This recent kelp forest recovery in the Gulf of Maine superficially resembles the initial phase state, although the present system is devoid of large vertebrate predators. Not surprisingly, the regulatory processes that maintain the kelp forest are now quite different from those in the earlier predator-dominated system. The combination of abundant algae and a lack of large predatory fishes has favored Jonah crabs (*Cancer borealis*), which have de facto assumed the role of apex predator. This was well illustrated when 36,000 adult urchins were relocated over a 2-year period to six widely spaced patches in an area that had been an urchin barren a decade earlier, only to be eaten by the now abundant Jonah crabs (Leland 2002).

Ecosystem release from predation played out differently elsewhere in the northwest Atlantic Ocean. Hyperabundant sea urchins in Nova Scotia proved vulnerable to a thermally triggered waterborne disease, leading to sea urchin mass mortality and kelp reforestation. Nova Scotian coastal reefs have fluctuated

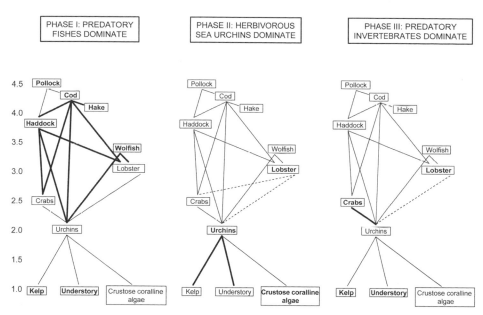

Figure 3.3. A schematic showing the progression of change in the structure and function of kelp forests in the Gulf of Maine. Bold lines and species names indicate comparatively great interaction strengths or great abundances, respectively (redrawn from Bourque et al. 2007).

between kelp forests and urchin barrens three times since 1965 (Steneck et al. 2002).

Overall, the loss of large predatory fishes from coastal zones of the western North Atlantic has caused a shift from strongly top-down to more bottom-up control. Today, larval settlement and available nursery habitat control the demography of lobsters (Steneck and Wilson 2001), crabs (Palma et al. 1999), and sea urchins (Vavrinec 2003). For all three species, nursery habitats have become the limiting resource (i.e., bottom-up) as opposed to predation on adults (top-down) that regulated abundances in the past.

Fishing as the Prime Ecosystem Driver

Although climate and physical oceanographic change have been invoked by some as an explanation for the extirpation of groundfish in the western North Atlantic, we think this unlikely. Evidence of asynchronous declines argues against such large-scale drivers. For example, the prehistoric cod decline at North Haven Island in Maine (Bourque et al. 2007) was not observed in other coastal midden sites (Lotze et al. 2006). Much better documentation of small-scale fishing operations in Maine in the 1600s reveals locally increased fishing efforts followed by local depletions that were not noted at other fishing stations elsewhere along the coast. Similarly, the coastal decline of cod in Maine in the 1930s was not evident at Georges Bank. Asynchronous cod declines in the Gulf of St. Lawrence, Scotian Shelf, Grand Banks, and Georges Bank all show patterns of decline related to fishing effort and do not accord with hypotheses based on a climate-driven event.

The loss of apex predators corresponds with increased abundances of mesopredators such as lobsters, shrimp, smaller fish, and herbivores. Again, these changes are synchronous with predator declines but are not regionally correlated. For example, increases in lobsters along the coast of Maine correspond with the extirpation of coastal cod stocks in the 1930s. More recent offshore declines of predators correspond to more recent increases in crabs (Frank et al. 2005) and shrimp (Worm and Myers 2003) off Canada's Scotian Shelf. Greene and Pershing (2007) have suggested that physical oceanographic changes in polar regions since 1990 may have contributed to changes in the abundance of mesopredatory fishes and crabs reported by Frank et al. (2005). Although the changes described by Frank and colleagues for the 1990s do generally correspond to the timing of the regime shift described by Greene and Pershing (2007) for the Scotian Shelf, there is no evidence that similarly "unique conditions" occurred at other locations where shrimp or crabs had increased in abundance earlier.

Ecological and Economic Simplification

Fishing down food webs in the Gulf of Maine has resulted in hundreds of kilometers of coast with low biological and economic diversity. Bloodworms used for bait are worth more to Maine's economy than cod. The trophic level dysfunction (sensu Steneck et al. 2004) of both apex predators and herbivores leaves a coastal zone suited for crabs and especially lobsters, the latter attaining population densities that exceed one per square meter along much of the coast of Maine (Steneck and Wilson 2001). Although the economic value of lobsters is high, this one species accounts for more than 80 percent of the total value of Maine's fisheries (the remaining 20 percent is contributed by forty-two harvested species). Thus, if a disease such as the one that decimated Rhode Island's lobster stocks (Castro and Angell 2000) were to infect lobsters in the Gulf of Maine, the result would be a socioeconomic disaster.

OTHER CASE STUDIES

Strong direct and indirect effects from top-down forcing and trophic cascades are known for various other temperate ecosystems around the world (Table 3.1).

Table 3.1. Synopsis of case study accounts of trophic cascades in temperate latitude coastal marine ecosystems.

Location	Species	Effects	Density or Trait	Source
Aleutian Islands	Sea otters Sea urchins Kelp	Trophic cascade; indirect influences on other species and food web processes; coevolutionary impacts	Density	Estes and Palmisano (1974)
Eastern U.S. estuaries	Great sharks Small elasmobranchs Filter-feeding bivalves	Trophic cascade	Density	Myers et al. (2007)
Gulf of Maine	Cod Sea urchins; lobsters Kelp	Trophic cascade	Density	Steneck (1997)
Southern California	Sheephead Black sea urchins	Urchin diel behavior	Trait	Nelson and Vance (1979)
Southern California	Sheephead Red sea urchins	Sea urchin population	Density and trait	Cowen (1983)
Southern California	Spiny lobsters Purple sea urchins	Urchin disease outbreaks Urchin wasting disease	Density	Lafferty (2004)
New Zealand	Lobsters and fishes Sea urchins Kelp	Trophic cascade	Density	Babcock et al. (1999)
Southeast Alaska	Sunflower stars Red and purple urchins Kelp	Trophic cascade	Trait	Duggins (1983)
Southern California	Planktivorous fishes Mesograzers Kelp	Trophic cascade	Density	Davenport and Anderson (2007)
South Africa	Lobsters Whelks Mussels Kelp	Predator–prey role reversal	Density	Barkai and McQuaid (1988)

The most detailed and numerous accounts are from the kelp forest ecosystems of southern California and northern Mexico, which support or once supported a diverse array of large-bodied consumers, including sea otters, lobsters, and various fish species. All of these consumers and many of their prey have been depleted or eliminated by human exploitation (Dayton et al. 1998; Jackson et al. 2001). A variety of experimental, comparative, and historical evidence indicates that these systems are or once were strongly influenced by top-down control.

Nelson and Vance (1979) provided some of the earliest evidence for top-down forcing effects by sheephead, *Semicocciphus pulcher*, a benthic feeding labrid fish in the warm temperate eastern North Pacific Ocean. These authors noted that sea urchins at Catalina Island retreated into cryptic habitats (substrate cracks and crevices) during the day when sheephead were active but moved onto the open reef to forage at night when sheephead were inactive. Sea urchins that were moved from cryptic to exposed habitats during the day had a high probability of being attacked and killed by sheephead. Cowen (1983) subsequently removed sheephead from an isolated reef at San Nicolas Island. Red urchin population density increased at 26 percent per year, in contrast with control sites that showed no change. More recently, Lafferty (2004) analyzed a 20-year data set that included information on predators, sea urchins, and sea urchin disease outbreaks from sixteen sites in the Channel Islands National Park that had been subjected to various fishing intensities. The protected sites contained higher predator (mostly lobster) densities, lower urchin densities, and reduced frequencies of disease outbreaks in the local sea urchin populations, presumably because disease transmission was impeded by the lower host (urchin) densities.

A trophic cascade involving predatory lobsters and fish, sea urchins, and kelp has been demonstrated in New Zealand by comparison of marine reserves with nearby unprotected areas (Babcock et al. 1999). The reserve sites contained larger and more abundant lobsters and fish, fewer urchins, and more kelp than the unprotected areas. However, the trophic cascade's influence on the distribution and abundance of kelp appears to be less in New Zealand than it is in the northeast Pacific. This may be because fishes and lobsters are less effective predators than sea otters, thus having led to a stronger coevolution of plant defenses, herbivore resistance to those defenses, and lower interaction strengths between New Zealand plants and their herbivores (Steinberg et al. 1995).

Predatory starfish also initiate trophic cascades. Kelp forests and sea urchin barrens co-occur as patchwork mosaics in parts of southeast Alaska where sea otters are absent. Duggins (1983) showed that this pattern is a trait-mediated effect (Werner and Peacor 2003) of predation by the starfish *Pycnopodia helianthoides*. *Pycnopodia* consumes sea urchins, but at low rates. Sea urchins nonetheless flee from nearby *Pycnopodia*, thus creating urchin-free patches into which kelps and other macroalgae can recruit. Algal detritus produced by the kelp patches probably provides adequate food for the remaining sea urchins (as shown by Harrold and Reed 1985 and Konar 2000), thus preventing them from attacking the living algae and helping to maintain the mosaic structure of the system.

The aforementioned examples all revolve around sea urchins and their well-known ability to denude reef ecosystems of various macrophytes. Temperate reefs support other potentially important herbivores, including fishes, gastropods, and various mesograzers (e.g., amphipods and mysids). Because mesograzers are consumed by a variety of microcarnivorous fishes, this predator–prey assemblage provides another potential top-down connection with autotrophs. Mesograzers are both at risk of predation and capable of damaging host plants (Hay et al. 1990). By experimentally excluding microcarnivorous fishes from small areas at Catalina Island off southern California, Davenport and Anderson (2007) demonstrated an increase in their mesograzer prey that in turn exerted a negative indirect effect on kelps. This trophic cascade may also be important in maintaining robust kelp forests. Massive mesograzer damage to giant kelp plants in southern California following an unusually large wave event that reduced the microcarnivorous fish populations supports this view (Tegner and Dayton 1991).

Our final vignette involves a predator–prey role reversal between lobsters and whelks at neighboring Marcus and Malgas islands (within 4 kilometers of each other) in South Africa (Barkai and McQuaid 1988). The shallow subtidal reefs at Malgas Island supported dense kelp forests and abundant lobsters, whereas at Marcus Island lobsters were absent, kelps were rare, extensive mussel beds covered the substrate, and whelks were comparatively abundant. The dearth of lobsters from Marcus Island was thought to have resulted from a localized anoxic event, although lobsters have been depleted elsewhere by fishing. In the absence of lobster predation, the mussel beds expanded (thus displacing kelps) and whelk numbers increased. The extraordinary feature of this example is the course of events that followed. After nine month-long caging studies conducted to ascertain that lobsters could indeed survive at Marcus Island, 1,000 lobsters were reintroduced. All of the lobsters were attacked and quickly killed (many within 15 minutes) by the now superabundant whelks, which attached themselves to and began consuming the lobsters whenever they settled to the seafloor. These various observations suggest that an alternate stable community developed with the loss of top-down forcing from lobster predation through prey population explosions and a predator–prey role reversal.

CONCLUSIONS

The depletion of apex predators is now almost ubiquitous. Nowhere is this more apparent than in the sea, where both the direct and unintended effects of whaling, sealing, and fishing have selectively stripped apex predators away from ocean food webs. And nowhere are the consequences of these losses more evident than in the higher-latitude coastal oceans, where the absence of species such as sea otters, sharks, cod, and lobsters have led to sweeping ecological changes.

Trophic cascades are the process by which apex predators commonly extend their influence to species other than their prey. Yet even trophic cascades are a gross oversimplification of the ways in which the influence of apex predators can penetrate food web structure and regulate food web dynamics. The kelp forest food web is so intimately interconnected that the loss of species such as cod and sea otters arguably extends in one way or another to all species. Similarly, the loss of great sharks may have provoked a range of nonintuitive ecosystem effects, including reduced water clarity and lowered production of benthic macrophytes, increased nutrient loading, the loss of biogenic reefs and nursery habitats for other fishes and invertebrates, and the rise or fall of various microorganisms. These highly serpentine interaction web effects of predators are not widely appreciated, in part because most ecologists haven't looked for them and in part because they are difficult to demonstrate.

One of the more intriguing indirect effects of predators is their potential link with disease. Predator reductions commonly lead to elevated prey densities. Not only might elevated prey density promote disease transmission between individuals, but prey species that historically lived at low densities would be more likely to lack resistance to parasites and pathogens for which the likelihood of infection is density dependent. These density-dependent processes could easily explain such events as the recent emergence of urchin diseases in various places around the world and abalone wasting disease in southern California. Unusually high host population densities coupled with high vulnerabilities to disease could lead in turn to the lack of ecosystem stability and extreme cyclic variation in population and community structure that has been observed recently in several kelp forest systems (Steneck et al. 2002).

If predators have strong influences on other species through direct and indirect interactions, it is not unreasonable to expect that these influences would appear as species-level characteristics when played out over evolutionary time scales. Predator-induced variation in the coevolution of plant defense and herbivore resistance may help explain why the northern and southern hemisphere kelp forests behave so differently, implying similar as yet unstudied processes as key factors in understanding the responses of other systems to the loss of apex predators. Geerat Vermeij (1977, 1987) invoked the rise of predators as the primary drivers of wholesale faunal changes in what he called the Mesozoic Marine Revolution. Very little imagination is needed to envision the deconstruction of these powerful and diverse forces with the selective loss of marine predators 150 million years later.

The implications of top-down forcing and trophic cascades for the management and conservation of temperate latitude coastal marine ecosystems are profound. Although most of the key predators in these ecosystems have been depleted, very few are globally extinct. Restoration is thus achievable through proper conservation and management. Many large apex marine predators are highly mobile, thus implying that the spatial scales of management must be large. For instance, even if the underlying reasons that killer whales began attacking sea otter populations in southwest Alaska were understood, the restoration of sea otters and coastal ecosystems in this region probably will necessitate actions that are broadly directed at the Bering Sea and North Pacific Ocean. Finally, the cascading effects of predators influence human welfare in numerous ways. These effects include impacts on regional economies, the maintenance or loss of entire industries and associated lifestyles, recreational opportunities, and even human health. Predators in the coastal oceans truly matter, but we are just now beginning to understand why and how much.

CHAPTER 4

Trophic Cascades in Lakes: Lessons and Prospects

Stephen R. Carpenter, Jonathan J. Cole, James F. Kitchell, and Michael L. Pace

Around 1980, three lines of thinking that were influential in aquatic sciences came together in the concept of trophic cascades. First, by the mid–1970s phosphorus was established as the currency of lake ecosystem functioning, and control of phosphorus inputs was shown to be effective in mitigating cultural eutrophication (Schindler 2006). Although phosphorus inputs explained patterns of phytoplankton biomass and production, a substantial amount of variance remained unexplained (Carpenter et al. 1991). As early as 1965, Brooks and Dodson (1965) suggested that size-selective predation could affect lake productivity. Second, biomanipulation of food webs was proposed to increase grazing rates and thereby control nuisance algae in lakes (Shapiro et al. 1975). The mechanisms through which biomanipulation improved water quality proved to be far more complex than grazing alone. Nonetheless, biomanipulation became an important management tool and a fertile research area (Scheffer 1998; Jeppesen et al. 1998). Third, synthesis of food web concepts integrated strong interactions (trophic linkages capable of completely reorganizing a food web) with flow of organic carbon and nutrients through networks (Paine 1980). In addition to introducing the term *trophic cascade*, Paine's paper suggested that massive changes in community structure could have implications for ecosystem processes such as biogeochemical flows.

After the introduction of the trophic cascade concept to limnology (Carpenter et al. 1985), a number of whole-lake experiments, cross-lake comparisons, and long-term studies showed that manipulation of top predators caused big changes in lake communities and ecosystem processes. In its simplest form, the trophic cascade addresses changes in a four-level food chain of piscivorous fishes, planktivorous fishes, zooplankton, and phytoplankton: A large increase (decrease) in piscivores causes a decrease (increase) in planktivores, an increase (decrease) in zooplankton, and a decrease (increase) in phytoplankton. This simple skeleton rapidly proved insufficient. Cascades depended on complex processes such as ontogenetic changes in diet and habitat use of fishes (Werner and Gilliam 1984), behavioral shifts related to foraging opportunity and predation risk (Werner et al. 1983), size-selective predation (Brooks and Dodson 1965), body size shifts among zooplankton (Pace 1984), nutrient recycling by zooplankton (Bergquist and Carpenter 1986), and stoichiometry of zooplankton (Elser et al. 1988). Nonetheless, whole-lake data from diverse studies established that big shifts of top predator biomass caused substantial changes in prey communities and ecosystem processes in lakes (Carpenter et al. 1987, 1991, 2001; Carpenter and Kitchell 1993; Jeppesen et al. 1998; Hansson et al. 1998; Schindler et al. 1997). Some important ecosystem changes included shifts in phytoplankton biomass, benthic plant biomass, nitrogen/phosphorus ratio of nutrient flow to phytoplankton, primary production, bacterial production, total ecosystem respiration, and direction and magnitude of net carbon dioxide exchange between the lake and the atmosphere.

Research in terrestrial, marine, and flowing-water ecosystems has also demonstrated effects of cascades on plant biomass and production (Pace et al. 1999). Lakes, streams, and marine ecosystems provide many striking examples of trophic cascades (as represented in this volume in Chapters 2, 3, 5, and 6). Although our chapter focuses on freshwater lakes, all the research themes we highlight are well developed in stream ecology: nonconsumptive effects of predators (Peckarsky et al. 2008), food web interactions with abiotic resources such as light (Wootton and Power 1993) and nutrients (Biggs et al. 2000), subsidies of organic matter and prey across ecosystem boundaries (Wallace et al. 1997; Nakano et al. 1999; Polis et al. 2004), and the importance of spatially extensive long-term study for understanding ecosystem dynamics (Power et al. 2005, 2008). An additional chapter on trophic cascades could easily be written based upon the literature of stream ecology.

In this short chapter, we present four important findings that emerged from studies of trophic cascades in lakes but were not expected when this research was initiated almost 30 years ago.

- Animal behavior—habitat choices and migrations of fishes and zooplankton—amplifies the rate and impact of trophic cascades.

- Nutrient enrichment intensifies effects of trophic cascades on primary producers, contradicting early expectations.
- Flow of terrestrial organic matter into lakes subsidizes consumers and thereby stabilizes alternative food webs established by trophic cascades.
- Whole-lake studies—comparisons, experiments, and long-term observations—spanning many years were crucial for measuring and elucidating effects of trophic cascades and often contradicted inferences from small enclosures studied for short periods of time.

These four findings are serendipitous discoveries that emerged as unintended benefits of trophic cascade studies yet have implications for a range of ecological research topics and ecosystem management. In this chapter we review these four discoveries. We close with a brief synthesis of current and future frontiers for research on trophic cascades.

ANIMAL BEHAVIOR AND CASCADES

Beyond the direct impacts of predator–prey interactions, trophic cascades can be either weakened or intensified by behavioral responses of both prey and predators. Animals at intermediate trophic levels may avoid top predators, with consequences for lower trophic levels and ecosystem processes. Behaviors of top predators that affect choice of foraging sites or prey preferences may also alter trophic cascades.

Nonconsumptive effects of predators (e.g., intimidation) can have strong effects on ecological communities (Peckarsky et al. 2008; Chapters 9 and 14, this volume). Predator avoidance behaviors are well known and are the focus of recent reviews (Lima and Dill 1990; Lima 2002) including one specifically focused on trophic cascades (Schmitz et al. 2004). Analogous development in the field of animal ecology is based on the ratio of benefit (growth) to cost (predation risk) at the individual, population, guild, or community scale and has also evoked a series of recent reviews (Werner and Peacor 2003; Pressier et al. 2005; Luttbeg and Kerby 2005). Theoretical work develops similar themes and a rich complexity of possible responses based around food chain or food web interactions (Abrams 1995). In a pioneering test of these ideas at the whole-lake scale, limnologists used acoustic sampling to demonstrate that behavioral responses accelerate the trophic cascade from piscivores to zooplankton (Romare and Hansson 2003). Thus the presence of a refuge changes interaction strengths and trophic cascades (Figure 4.1).

The importance of predator avoidance also emerged in fishery science

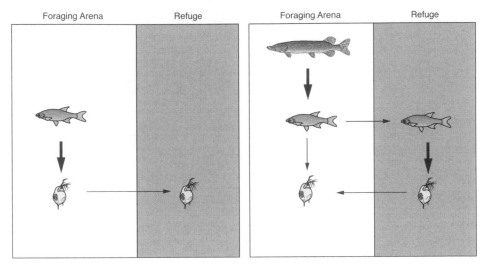

Figure 4.1. Romare and Hansson (2003) showed that in the presence of a refuge (shaded region) predator–prey dynamics are more rapid and complex than they are in the absence of a refuge.

through development of a theory for foraging arena effects (Walters and Juanes 1993) that modify the basic functional response of a predator–prey interaction. Nonlinear and threshold-like responses emerge in ways that influence fishery yields (Walters and Kitchell 2001; Walters and Martell 2004). Schmitz et al. (2004) summarize their review by calling for a new conceptual view of ecological systems as "landscapes of fear." Wirsing et al. (2008) refer to "seascapes of fear." The titles of Pressier et al. (2005) and Luttbeg and Kerby (2005) express much the same view.

Taken together, these reviews indicate that behavioral effects are more rapid and intense than predicted by simple predator–prey models. The consequences of behavior for trophic cascades are well expressed in aquatic ecosystems. Moreover, they have important implications for ecosystem consequences of trophic cascades and for applications of cascades in conservation, restoration, and biological control.

Schmitz et al. (2004) used some of our early work (Carpenter et al. 1987) as an example of indirect effects of behavioral responses that altered the trophic cascade in Peter Lake. We assembled a list of thirty-two specific predictions written in our grant proposals for whole-lake experiments on trophic cascades (Carpenter and Kitchell 1993). Of the thirty-two predictions, sixteen were confirmed, one was equivocal, and fifteen were not supported by the results. Some of the latter were caused by behavioral responses. When 90 percent of the largemouth bass were removed from Peter Lake and 49,601 zooplanktivorous minnows added shortly thereafter, the minnows behaved as expected and immediately began exploiting the large zooplankton as prey. That lasted about 2 weeks. Perception of predation risk caused by the remaining bass population rose, and by the end of the first month nearly all the minnows were densely aggregated in refugia (beaver channels), where they gradually starved, and many were eaten by piscivorous birds. That result was unexpected, and our monitoring program represented it only sparingly.

He and Kitchell (1990) conducted a whole-lake manipulation to measure the relative effects of behavioral responses and direct predation effects in a system that contained thirteen species of potential prey fishes but no piscivores. We expected that potential prey would aggregate in littoral refugia or leave the side where pike had been added (Figure 4.2). The response was stronger than expected. Emigration began immediately after a few pike were added and was led by the prey species whose size and morphology made them most vulnerable. Fish left the side with pike, and many left the lake through an outlet stream at the pike-free side. Pike did prey on some fishes, but over the course of the summer, emigration accounted for 50–90 percent of the total change in

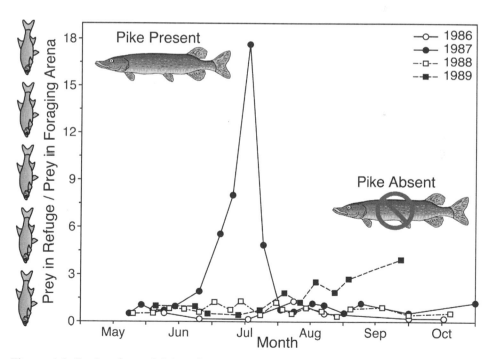

Figure 4.2. Ratio of prey fish in refuge to prey fish in foraging arena during a year when northern pike were absent and a year when northern pike were present (He et al. 1993).

biomass for individual species, exceeding the direct effects of predation (He et al. 1993).

NUTRIENT INPUTS AND CASCADES

Variation between lakes in many ecological properties including primary production is strongly related to loading of the limiting nutrients nitrogen and phosphorus. For this reason limnologists often study nutrient enrichment gradients, and a natural question concerns how trophic cascades vary in their potential to control phytoplankton biomass and productivity in oligotrophic (low productivity) relative to eutrophic (high productivity) lakes. High nutrient inputs to lakes lead to eutrophic conditions with noxious algae that are often too large for grazing, limiting the potential for herbivore control. Low nutrient inputs lead to oligotrophic conditions and dominance by smaller, more easily grazed algae. Therefore, it is plausible that nutrient enrichment could weaken trophic cascades in lakes.

Researchers proposed several alternative models to describe the variation of trophic cascades with enrichment. McQueen et al. (1986) argued from enclosure experiments and lake comparative data that trophic cascades were attenuated in eutrophic systems, limiting zooplankton control of phytoplankton, whereas in oligotrophic lakes trophic cascades facilitated zooplankton control of phytoplankton. This nutrient attenuation model in graphic form represents a wedge of potential lake conditions, with the greatest scope for trophic cascades at the low end of nutrient loading gradients (Figure 4.3a). Sarnelle (1992) used a simple predator–prey model and derived an alternative prediction that phytoplankton are strongly suppressed across nutrient enrichment gradients. His nutrient facilitation model was consistent with data on phytoplankton responses to large-bodied grazers, the cladoceran *Daphnia*, in enclosures and whole-lake manipulation studies. Graphically, this model also describes a wedge of lake conditions where the greatest scope for the effects of trophic cascades increases with nutrient enrichment (Figure 4.3b). Elser and Goldman (1991) studied the variation of phytoplankton–zooplankton interactions in lakes of contrasting nutrient levels using enclosures. Nutrients strongly limited zooplankton control of phytoplankton at low inputs, suggesting little scope for trophic cascades in oligotrophic lakes, whereas in more eutrophic systems phytoplankton size, palatability, and rapid growth rates limited grazer control. Lakes with intermediate nutrient loading (mesotrophic lakes) had the greatest scope for trophic cascades. This mesotrophic maximum model describes a set of lake conditions resem-

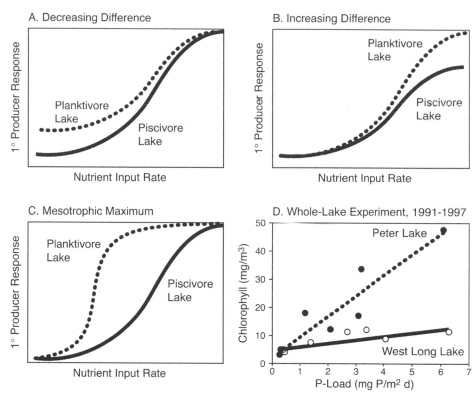

Figure 4.3. Hypotheses and experimental evidence about the interaction of nutrient input and trophic cascades. Each panel shows primary producer response and nutrient input rate for a lake with a planktivore-dominated food web (dashed line) and a lake with a piscivore-dominated food web (solid line). **(a)** Decreasing difference hypothesis: The difference between the lakes is greatest in low-nutrient conditions. **(b)** Increasing difference hypothesis: The difference between the lakes is greatest in high-nutrient conditions. **(c)** Mesotrophic maximum or banana hypothesis: The difference between the lakes is greatest at intermediate nutrient conditions. **(d)** Whole-lake experimental results from planktivore (Peter Lake) and piscivore (West Long Lake) dominated lakes.

bling a banana, where the potential for trophic cascade impacts is narrow at either end of the nutrient loading gradient and wider in the middle (Figure 4.3c).

These models were based on evidence from small–scale and short–term enclosure studies, lake observational studies, and a few whole-lake manipulations. However, the evidence from these studies was insufficient to provide a strong test of the interaction of trophic cascades and nutrients that would distinguish between the three models (Carpenter and Kitchell 1993). We conducted a series of food web and nutrient manipulations in lakes with contrasting fish communities that could then be enriched over a range of nutrient loads. Consistent with the views of some other researchers (e.g., Benndorf 1990), we reasoned that the "banana" model was most apt, especially because grazer control would be limited at high nutrient loads through destabilized predator–prey dynamics (Carpenter 1992). Over several years, we established Peter Lake as a planktivore-dominated system by removing piscivores and stocking minnows. We contrasted Peter Lake with the western basin of Long Lake, where the dominant fish was piscivorous largemouth bass. We enriched both Peter and West Long lakes with nitrogen and phosphorus for 5 years. Similar loadings were used for the two lakes within each year, but loads varied between years. These nutrient loads ranged from the low, natural rates of the study lakes to highly eutrophic systems encompassing much of the global variation in lakes (Carpenter et al. 2001). Corresponding to the expected differences resulting from size-selective planktivore predation, zooplankton in West Long were large-bodied grazers, primarily the cladoceran *Daphnia*, whereas the zooplankton of Peter Lake were a mixture of small-bodied grazers.

Phytoplankton primary production and biomass were strongly suppressed in West Long Lake at all nutrient loads, whereas phytoplankton biomass and productivity increased substantially across the same nutrient loading gradient in Peter Lake. The outcome clearly supported the nutrient facilitation model (Figure 4.3d). Manipulations in eutrophic Lake Mendota (Lathrop et al. 2002) and fish manipulation studies from a variety of eutrophic lakes (Hansson et al. 1998) support the result that trophic cascades can lead to strong limitation of phytoplankton through grazer control even in enriched systems. Through these studies, lake researchers have established that trophic cascades are evident across a range of lake conditions and largely independent of nutrient loading and primary production. However, it is possible that control of algae in highly enriched lakes can be destabilized by fishing or massive runoff events. The key food web features of lakes that promote cascades are stable and abundant populations of piscivores and large-bodied zooplankton grazers, especially large species of *Daphnia*.

LAKES IN THE LANDSCAPE

Cross-ecosystem subsidies are well documented in numerous landscapes (Polis et al. 2004). Lakes typically receive a significant amount of organic carbon from their surrounding watersheds. In fact, the loading of terrestrial organic carbon to these systems is typically as large as or larger than that from aquatic primary production (Caraco and Cole 2004). How and whether these terrestrial inputs influence trophic cascades depend on the magnitude of the input relative to local primary production, the mode of transport and physical form of the input, and the organisms in the receiving ecosystem.

There are three pathways by which terrestrial organic carbon can enter a lake ecosystem: as terrestrial dissolved organic carbon in inflowing ground or surface waters, as terrestrial particulate organic carbon in flowing surface waters or by aeolian deposition, and as terrestrial organisms that enter lakes, often accidentally from land. By far the largest input of terrestrial organic carbon to lakes is dissolved organic carbon, which is available only to microorganisms. Clearly, some bacterial respiration is fueled by terrestrial dissolved organic carbon. However, for terrestrial dissolved organic carbon to become a subsidy to the lake food web requires that bacteria assimilate it and pass this organic matter up the food web. Thus, bacterivorous consumers (flagellates, ciliates, some cladocerans) and higher consumers that feed on these ultimately use terrestrial dissolved organic carbon. Jansson et al. (2007) call this pathway heterotrophic energy metabolism and suggest that it is a subsidy especially to cladoceran zooplankton and possibly to the small fish that consume them, potentially setting up a subsidized cascade.

Terrestrial insects and other terrestrial prey items are a very minor component of organic carbon flow from land to lake ecosystems, generally less than 0.1 percent of the total supply of organic carbon to aquatic consumers from both terrestrial and aquatic sources. However, terrestrial prey are available to top and mid-level predators and can therefore have a large effect on trophic cascades. In small lakes, fish consume significant quantities of terrestrial prey (Hodgson and Kitchell 1987; Hodgson and Hansen 2005). Young-of-year fish are often planktivorous, and terrestrial prey are not significant for them. However, even for small (age 1+) fish, terrestrial prey averages about 20 percent of their total consumption. For adults, terrestrial prey averages nearly 40 percent (Figure 4.4a).

Particulate organic matter of terrestrial origin can enter lakes via stream flow and by aeolian deposition. This direct input of terrestrial particles is also a small component of the total organic carbon budget, on the order of a few percent. Particulate organic carbon can also be formed by flocculation of terrestrial dissolved organic carbon within the lake. Either way, these particles represent another pathway of a terrestrial subsidy to the lake ecosystem. In the water column terrestrial particulate organic carbon can be consumed by zooplankton. The terrestrial particulate organic carbon that reaches sediments can be consumed by a host of benthic invertebrates. Using Paul Lake as an example,

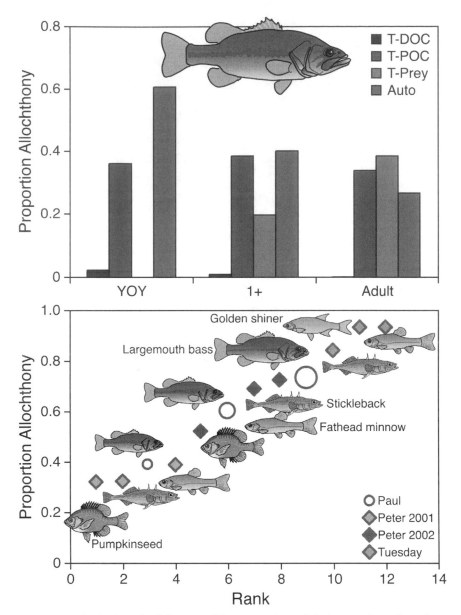

Figure 4.4. Fish obtain a high but variable proportion of their organic carbon from terrestrial sources (Cole et al. 2006). **(a)** Proportion of organic carbon derived from four sources: terrestrial dissolved organic carbon (T-DOC), terrestrial particulate organic carbon (T-POC), terrestrial prey (T-Prey), and autotrophic production in the lake (Auto) by three size classes of largemouth bass (young-of-year [YOY], 1+, and adult) in Paul Lake. **(b)** Proportion of terrestrially derived organic carbon in fish species from several whole-lake experiments.

Cole et al. (2006) found that terrestrial particulate organic carbon was a major diet item for both zooplankton (about 30 percent of consumption) and benthos (about 60 percent of consumption). The fish that feed on zooplankton and benthos are also subsidized, indirectly, by terrestrial particulate organic carbon. The terrestrial particulate organic carbon subsidy to fish averages about 30 percent consumption across age classes, but the pathway differs. Young-of-year fish consume terrestrial particulate organic carbon via zooplankton, whereas adult fish consume terrestrial particulate organic carbon via benthos and fish that consumed either zooplankton or benthos (Figure 4.4a).

In these small lakes, the terrestrial subsidy to fish, combining all pathways, is quite large, ranging from about 40 to 90 percent in lakes that were not eutrophied (Figure 4.4b). Even after artificial eutrophication of Peter Lake in 2002 (by the addition of nitrogen and phosphorus fertilizer over the course of 12 weeks) about 30 percent of fish biomass was made up of terrestrially derived organic carbon. In Paul Lake the average adult largemouth bass is made up of about 70 percent terrestrial organic carbon. Thus these piscivores can quickly reduce the population of planktivores to very low levels without suffering the consequence of losing their main source of organic carbon. In Peter Lake, before fertilization the populations of fat head minnows and pumpkinseed sunfish, which are about 40 percent terrestrial, can exert strong influence on the size of zooplankton and still be somewhat independent of the amount of zooplankton prey available to them.

Although these studies demonstrate the potential for predator biomass to be stabilized by cross-system subsidies of terrestrial organic carbon inputs, they do not directly address the effect of this stabilization on trophic cascades. Vander Zanden et al. (2005) used literature data and a model to show that benthic resources (subsidized by terrestrial organic matter) augmented piscivore biomass and intensified trophic cascades. A few studies have directly examined the intersection of cross-system subsidies and trophic cascades (Nakano et al. 1999; Polis et al. 2004). Knight et al. (2005) studied an intriguing case of a cascade that involved fish, dragonflies (a preferred prey of fish), bee flies (a favorite prey item for adult dragonflies), and the pollination of *Hypericum* by bee flies. In a series of nearly identical small ponds, Knight et al. (2005) selected four ponds with fish and four ponds without fish. Fishless ponds had a greater abundance of the aquatic larval stage of dragonflies, more adult dragonflies, and significantly fewer pollinator visits to pond-side *Hypericum* than did ponds with fish.

Trees that fall into lakes represent a cross-system subsidy of a different type with different implications for cascades. The logs create habitat for periphyton and invertebrates, which in turn are preyed on by fish. In addition they provide refuge for fishes of a range of sizes and foraging sites for sit-and-wait predators (see preceding section on "Animal Behavior and Cascades"). Sass et al. (2006) examined the food web of two basins of experimentally divided lakes, before and after the removal of fallen trees and logs from one basin. Before removal of the woody habitat, the food webs of both basins were essentially identical and fish consumed mostly aquatic prey. In the basin from which coarse wood was removed, the major predator, largemouth bass, consumed fewer fish and more terrestrial insects and grew more slowly than their counterparts in the reference basin.

In summary, terrestrial subsidies to aquatic food webs can help support high populations of top predators and thereby stabilize or even intensify trophic cascades. Fallen trees may also affect trophic cascades by providing food for benthic invertebrates that feed on periphyton and habitat and refuge for invertebrates and fishes of a range of body sizes. These findings show that trophic cascades cannot be fully understood by studying ecosystems in isolation from their surroundings. Landscape connections have important implications for trophic cascades.

WHOLE ECOSYSTEM APPROACHES

According to the title of a famous paper, lakes are microcosms of broader ecological interactions (Forbes 1887). Microcosms can be useful representations of complex ecological interactions, exposing key processes through simplification, in the same way that mathematical models clarify complex processes (Scheffer and Beets 1994). However, in ecology the problem of extrapolating results from one scale to another is pervasive (Levin 1992). The most straightforward solution is to conduct studies at the appropriate scale. In the case of trophic cascades in lakes, theory has been evaluated using whole ecosystem comparisons, experiments, and long-term studies. Our current understanding of trophic cascades is synthesized from all three kinds of studies of whole ecosystems. These whole ecosystem approaches have complementary strengths and weaknesses, so it seems reasonable to have more confidence in inferences that are consistent across all approaches (Carpenter 1998). Most importantly for this discussion, the whole-lake approaches sidestep many of the problems of extrapolation across scales that arise in microcosm experiments.

Extrapolations from microcosms to whole-lake results often fail. Pace (2001) evaluated microcosm and whole-lake responses of microbes and phytoplankton to nutrients and grazing and found that the microcosms were about equally likely to get it right or wrong. Carpenter and Kitchell (1988) found that microcosms and lakes showed opposite responses for 38 percent of phytoplankton species studied and consistent responses for only 34 percent of species. Schindler (1998) reviewed his experiences over about 30 years with microcosm and whole-lake experiments and found numerous cases in which microcosms would have led to incorrect scientific conclusions and faulty management decisions in the absence of evidence from whole ecosystem experiments. In view of the low reliability of microcosms, Carpenter (1996) suggested that overuse of microcosm experiments diverted human resources from useful research and wasted resources by training graduate students in irrelevant approaches. Limnologists are aware of a number of microcosm artifacts, such as reduced mixing, enhanced sedimentation, fish mortality, excessive predation, and overgrowth of periphyton (Pace 2001). These problems are sometimes downplayed in the literature because they complicate inference and perhaps because authors believe that everybody knows about them already. Such a tendency to minimize artifacts only perpetuates a questionable body of research.

It is more constructive to view microcosm experiments as a form of modeling, subject to the same kinds of assumptions and limitations as mathematical or simulation models in ecology (Scheffer and Beets 1994). Like mathematical models, microcosms can be an inexpensive way to build intuition about more complex systems. Microcosm experiments across a gradient of scales could even reveal insights about the effect of scale itself. As with mathematical models, we should be very cautious about extrapolating results of microcosm studies to whole ecosystems. Fortunately, it has been possible to study trophic cascades in lakes using whole-lake comparisons, experiments, and long-term data.

CONCLUSIONS

Nearly 30 years of research on trophic cascades in lakes has continued to expand our understanding of the phenomenon and our appreciation of its complexity. This short review has selected topics that have emerged in the course of ecosystem studies of trophic cascades in lakes and seem likely to continue to motivate research. These are topics in basic science, although they have implications for applied ecology. We close with some comments on additional research issues that are motivated by concerns about ecosystem management or conservation.

Trophic cascades are among the processes that cause large nonlinear changes in ecosystems, changes that may be difficult or even impossible to reverse (Chapter 17, this volume). Such changes have been identified as a major concern for ecosystem management and sustainability (Millennium Ecosystem Assessment 2005). Cascades can create large changes in ecosystems including living resources and ecosystem services, with implications for human well-being. Therefore it is important to understand the conditions that lead to trophic cascades—the subject of much of the research in this volume—and the indicators that a cascade may occur soon. Models demonstrate several kinds of statistical changes in time series that occur in advance of trophic cascades and thereby can serve as leading indicators (Carpenter et al. 2008). Research is under way to evaluate these indicators using retrospective time series analyses and ecosystem experiments. Interesting insights are likely to emerge from this line of research in the future.

The massive and sometimes surprising changes we have seen in whole-lake experiments and real-world lake management were usually not forecast from models and small container experiments. Managing food webs is extraordinarily difficult, even in small experimental lakes where we completely control exploitation and inputs. In the case of food webs, the notion of "managing" implies a degree of predictability and control that is not consistent with the empirical record. Nonetheless, food webs provide a unique set of possibilities for manipulating ecosystems, and managers are likely to try these tools even if the outcome is a gamble. Also, people will continue to manipulate food webs by fishing and by introducing species. It is clear that introductions or deletions of species can have effects on other species at distant positions in the food web and on ecosystem processes (primary production, carbon sequestration, carbon exchange with the atmosphere, nutrient cycling, nutrient limitation).

Examples of manipulations of food webs to achieve societal goals include management of water quality and contaminant biomagnification. Eutrophication can be mitigated by restoration of top carnivores (Lathrop et al. 2002; Hansson et al. 1998). However, in lakes subject to enormous and variable loads of excess nutrients the food web may become unstable, and the desired effects on water quality may break down. Food web manipulation can mitigate biomagnification of contaminants. Unfortunately, however, many real-world cases have tradeoffs. For example, in the Laurentian Great Lakes the food webs that best mitigate biomagnification are dominated by fast-growing steelhead trout, an invasive in those systems (Stow et al. 1995). Yet some other species invasions may exacerbate contaminant biomagnification by increasing the length of food chains (Vander Zanden and Rasmussen 1996).

In lakes, as in oceans, fishery management often focuses on one species at a time (Chapter 6, this volume). Such practices ignore trophic cascades (and other ecosystem phenomena) and are therefore subject to surprising break-downs when abrupt changes in food webs affect target species.

Over the past 50 years, ecosystems have changed more than at any previous time in the history of our species (Millennium Ecosystem Assessment 2005). The biota of both aquatic and terrestrial ecosystems has changed as a result of human activities including harvest and nonharvest mortality, land use change, habitat loss, and pollution. These changes often have disproportionately large effects on top predators because top predators often have large body sizes, long maturation times, large range needs, and complex life histories, which may necessitate multiple kinds of habitats or prey in the course of ontogeny. Thus it is reasonable to expect that trophic cascades played a role in the sweeping biotic changes of recent decades. Community and ecosystem structures that we see today may be legacies of lost cascades. Moreover, we expect that trophic cascades will be important in future changes as human action and natural processes continue to reorganize ecosystems. There is no foreseeable end to the need for research to understand and forecast the consequences of trophic cascades for ecosystems and their support of human well-being.

ACKNOWLEDGMENTS

Tim Essington and Jim Estes provided helpful comments on the draft manuscript. We are grateful for support from the National Science Foundation through an unbroken series of grants since 1982. Additional support came from the A. W. Mellon Foundation.

CHAPTER 5

Prey Release, Trophic Cascades, and Phase Shifts in Tropical Nearshore Ecosystems

Stuart A. Sandin, Sheila M. Walsh, and Jeremy B. C. Jackson

Tropical nearshore environments contain some of the most diverse marine ecosystems, including coral reefs, seagrass meadows, and soft bottom habitats. In general, high light levels, warm waters, and complex food webs characterize these habitats. Countless adaptations and symbioses among autotrophs and mixotrophs increase rates of primary productivity in the typically oligotrophic waters of the warm tropics (Odum and Odum 1955; Sorokin 1995). Similarly diverse strategies exist among heterotrophs, leading to high rates of herbivory and predation in the nearshore tropics (Sale 1991; Larkum et al. 2006).

A growing body of evidence highlights the strong trophic coupling characterizing nearshore ecosystems, especially in the historic past. Caribbean turtlegrass meadows were grazed by millions of green sea turtles, with trophic dynamics resembling the activity and turnover of the African savanna (Bjorndal and Jackson 2003; McClenachan et al. 2006). Analogously, intact coral reefs can support surprisingly productive fisheries because of the high trophic efficiency of these communities (Smith 1978; Russ 1991). Tropical nearshore ecosystems also support extraordinary biomass of predators. Unfished coral reefs, for example, can support fish assemblages with more than 50 percent of the total biomass in piscivorous, apex predators (Friedlander and DeMartini 2002; Brainard et al. 2005; Newman et al. 2006; Knowlton and Jackson 2008; Sandin et al. 2008). The inverted trophic pyramids on these reefs suggest high potential rates of predation and unique trophic dynamics.

Because of both the high productivity of these ecosystems and their location in the heavily populated coastal tropics, few tropical nearshore habitats remain unaffected by human disturbance. Essentially all seagrass meadows, coral reefs, and soft bottom habitats suffer the effects of fishing, pollution, and climate change. The conservation of these habitats depends on informed management policies and active political involvement. Fundamental to these efforts are sound ecological insights into the direct and indirect effects of the myriad anthropogenic disturbances. What are the consequences of removing dominant herbivores from seagrass habitats, and what are the consequences of extirpating predators from coral reefs?

In this chapter we review the evidence for trophic interactions in tropical nearshore habitats, building on a number of related essays on the topic (Roberts 1995; McClanahan 2005; Steneck and Sala 2005). We begin with a review of evidence for prey release, that is, the direct effects of altering densities of one functional group on the next-lower trophic level. We follow with a detailed description of four well-studied trophic cascades, in which removal of a predator results in increased abundance of its herbivorous prey and decreased abundance of the primary producer consumed by the herbivore. We then discuss some of the implications of these interactions for conservation in the tropics.

EVIDENCE OF PREY RELEASE AND TROPHIC CASCADES

Evidence of prey release and more complex three-level trophic cascades is mixed in tropical nearshore ecosystems. Herbivores generally exert strong control on algae, whereas the effects of predators on their animal prey are more inconsistent. Evidence for trophic cascades is limited, possibly because of high trophic flexibility and redundancy and the effects of historic human disturbance. Indeed, if we include effects of harvesting by humans, there is abundant evidence that trophic dynamics and anthropogenic effects are both strong factors structuring today's nearshore ecosystems.

Prey Release: Effects of Herbivores

The best evidence of prey release in tropical nearshore ecosystems is for herbivores and autotrophs. Experiments clearly demonstrate the very strong effects of herbivores on the abundance, biomass, and diversity of autotrophs. More than 40 years ago, Randall (1965) demonstrated that the ubiquitous 10-meter-wide bands of bare sand separating patch reefs from turtlegrass beds in the Caribbean were created by grazing by large herbivorous fishes that sheltered in the reef framework and foraged on the seagrass. Construction of small artificial reefs in the center of extensive seagrass meadows resulted in the formation of halos of bare sand soon after the arrival of herbivorous fishes to the reef (Randall 1965). Ogden et al. (1973) carried out similar observations and experiments to demonstrate comparable effects of grazing around Caribbean patch reefs by the formerly abundant sea urchin *Diadema antillarum.*

Experimental manipulations also have elucidated the primary role of herbivores in controlling autotrophs on coral reefs. In general, the exclusion of reef-based herbivores leads to an increase in cover and biomass of fast-growing, fleshy algae (Sammarco 1980, 1982; Carpenter 1986; Lewis 1986; Hughes et al. 2007). Factorial manipulations of herbivore abundance and nutrient concentration have revealed evidence of both top-down and bottom-up effects on reef algal biomass (Hatcher and Larkum 1983; Miller et al. 1999; Smith et al. 2001; Thacker et al. 2001; McClanahan et al. 2003). Across these studies, the effects of herbivore exclusion were similarly strong both with and without addition of nutrients, suggesting ubiquitous pressure imposed by herbivores in limiting algal growth. These studies span a spectrum of coral-dominated environments (Smith et al. 2001) and algal-dominated reef flats (Thacker et al. 2001), with strongly similar effects (Burkepile and Hay 2006).

Observational studies have also demonstrated strong effects of herbivorous fishes on benthic algae. Mumby and colleagues (2006) documented a clear relationship between herbivorous fish and the algae they consume by comparing rates of herbivory and associated algal cover on Bahamian reefs inside and outside marine reserves. In general, areas with higher herbivorous fish abundance realized higher rates of herbivory and lower fleshy algal biomass. Similar negative relationships have been documented between biomass of herbivorous fish and abundance of fleshy algae throughout the Caribbean (Williams and Polunin 2001; Newman et al. 2006). The strongest trophic consequences of the establishment of no-take marine reserves on Caribbean reefs are an increase in herbivorous fish biomass and a related decrease in fleshy algal biomass (Newman et al. 2006). However, all these Caribbean examples are based on data collected after the catastrophic die-off of the herbivorous sea urchin *Diadema antillarum* (Lessios et al. 1984) and thus describe trophic consequences in a simplified food web of reef-based herbivory.

Removal of herbivores also results in complementary shifts in composition and productivity of the producer assemblage. Crustose algae are largely lost on coral reefs in lieu of upright fleshy and filamentous turf algae (Miller et al. 1999; McClanahan et al. 2003). Similar compositional shifts in response to removal of herbivores, as well as addition of nutrients, have been observed across a variety of marine benthic ecosystems (Burkepile and Hay 2006). Cropping within species and shifts in species composition induced by herbivory cause large changes in biomass-specific rates of primary production (Carpenter 1986).

Prey Release: Effects of Predators

Predation in tropical marine environments is intense and is believed to be responsible for many of the somatic and behavioral adaptations common to resident animals (Sale 1991; Larkum et al. 2006). Cryptic coloration is common to many species as a way of limiting risk of predation (Hixon 1991; McFarland 1991). Mobile animals often survive disproportionately well in regions with high refuge availability (Ogden et al. 1973; Hixon and Beets 1993; Friedlander and Parrish 1998). Aggregation or schooling in fishes is one of the most ubiquitous strategies for minimizing per capita risk of predation, with individuals depending on the dilution of risk inherent to living in groups (Alexander 1974; Magurran 1990; Sandin and Pacala 2005). Despite such approaches for minimizing predation risk, most tropical marine species suffer high rates of predator-induced mortality before reaching sexual maturity, often exceeding 90 percent of the population (Shulman and Ogden 1987; Gosselin and Qian 1997; Hunt and Scheibling 1997). Therefore, we may expect to find strong evidence of prey release in response to reductions of predator density.

Predation affects all life stages of prey but seems to be most intensive in the early life history stages. For example, predatory fishes can account for more than 50 percent of the mortality of recruiting marine organisms. Through the experimental removal of predatory fishes, Carr and Hixon demonstrated dramatically altered intensity and patterns of mortality among juvenile reef fishes (Carr and Hixon 1995; Hixon and Carr 1997). Similarly, reductions in fish populations targeted by local fishing efforts in the Gulf of Thailand have led to increased survivorship of juvenile squid and recruitment of shrimp (Pauly 1982, 1985). Release from predation of juvenile squid in response to intensive harvest of their predators is the leading hypothesis to explain the transition from fish to cephalopod industries in tropical soft bottom ecosystems (Caddy and Rodhouse 1998).

Many marine species are believed to enjoy a size advantage in avoiding predation, with susceptibility to mortality dropping significantly through ontogeny (Hixon 1991; Gosselin and Qian 1997; Vermeij and Sandin 2008). The simple fact that larger organisms fit in the mouths of fewer predators is of primary importance. Gape limitation is a strong constraint for foraging among marine predators, leading to strong size-selective predation. Nevertheless, mature prey individuals suffer quantitatively important levels of predation and commonly increase in abundance when predator density is reduced. Across the

Great Barrier Reef, spatial restrictions on fishing through the establishment of no-take marine reserves have restored populations of formerly dominant predatory fishes, with more than fourfold increases in biomass of the coral trout *Plectropomus leopardus* (Graham et al. 2003). Densities of common prey fish vary twofold across sites, with significant reductions in abundance in protected areas of six of the eight species commonly preyed on by coral trout.

Less evidence of prey release has been found in multispecies assemblages of predators and prey. For example, variation of predator density across gradients of fishing activity in the Seychelles and Fiji resulted in no clear evidence of coupled variation in prey fish biomass (Jennings et al. 1995; Jennings and Polunin 1997). Similarly, dramatic variation in predator biomass across a gradient of fishing pressure in the northern Line Islands in the central Pacific was unrelated to variation in prey fish biomass (Sandin et al. 2008), and the density of gastropods and bivalves showed no significant relationship to the density of potential predators, including finfish and lobsters, in seagrass beds in Florida Bay (Nizinski 2007). It is possible that the large dietary breadth and potential flexibility typical of many tropical species (Randall 1967) may limit the realization of clear prey release among tropical predator–prey interactions.

A final example of prey release, which is almost certainly a trophic cascade, has been documented for populations of the reef-dwelling ovulid gastropod *Cyphoma gibbosum* in the Florida Keys. Across a mosaic of habitats, densities of *C. gibbosum* appear to decrease in areas protected from fishing (Chiappone et al. 2003). One of the main predators is believed to be a species of hogfish (*Lachnolaimus maximus*), a species known to prey on *C. gibbosum* (Randall and Warmke 1967) and to be targeted by recreational anglers (McBride and Richardson 2007). *C. gibbosum* is itself a predator, living on and feeding upon gorgonians, soft corals from the order Gorgonacea. Burkepile and Hay (2007) experimentally manipulated access of predatory fish to a number of gorgonian colonies in a marine protected area. Gorgonians in exclosures had densities of *C. gibbosum* almost twenty times higher than those exposed to fishes (including hogfish) (Burkepile and Hay 2007). Additionally, gorgonians in exclosures with higher densities of snails also showed evidence of greater tissue loss due to predation by *C. gibbosum*, consistent with models of trophic cascades. Thus, reduction of hogfishes by fishing appears to result in increases of *C. gibbosum*, which in turn results in reduced living tissue of gorgonians.

Trophic Cascades in the Tropics

A trophic cascade can be viewed as two adjacent and interconnected episodes of prey release (Chapter 2, this volume). In food web theory, this would be viewed as two strong interactions being linked to one another. For the sake of discussion, we limit our description of a trophic cascade to a three-level inter-action of nonhuman species, where trophic level $x+2$ has a negative, direct effect on trophic level $x+1$ and a positive, indirect effect on trophic level x. Therefore we are focusing on evidence of endogenous and tight trophic links between species in the ecosystem that dissipate slowly, thereby remaining strong across trophic levels (Figure 5.1).

Using an example from a Caribbean food web, Bascompte and colleagues (2005) identified that linkages coupling two strong interactions occur less often than expected by chance, thus buffering natural communities from trophic cascades. However, because humans disproportionately affect strongly interacting species (e.g., large predatory and herbivorous fishes), anthropogenic activities appear to selectively target these underrepresented combinations of strong interactions. Therefore, tropical food webs may be buffered from trophic cascades, but nonrandom human activities may weaken this resilience in heavily affected communities. Importantly, the study by Bascompte and colleagues was based on topological analyses of an idealized food web and did not consider dynamic interactions (e.g., prey switching, nonlinear functional responses) potentially associated with species reductions or removals.

Empirically, few clear examples of trophic cascades have been documented from shallow tropical seas compared with the temperate zone (Chapter 3, this volume). The examples that follow provide insights into factors that may favor trophic cascades and reasons why cascades are not more common. Importantly, we treat these examples as case studies and not as a comprehensive review of the tropical nearshore literature, which to be done justice would need a book of its own.

CASE STUDY #1: KENYAN CORAL REEFS (TRIGGERFISH → SEA URCHINS → REEF-BUILDERS)

Despite the high diversity typical of tropical nearshore ecosystems, there are a small number of keystone predators whose activities are particularly strong and noticeable. Echinoderms make up a disproportionate number of these keystone species, probably because of their typically narrow diets and their limited suite of predators (due to bodily specializations including long spines, calcareous exoskeletons, and toxic defenses) (Vermeij 1977; Steneck 1983; Pinnegar et al. 2000). An important secondary factor is the high potential growth rate typical of most echinoderms (Hunte and Younglao 1988). High intrinsic rates of growth are well known to exacerbate population fluctuations, leading to increasingly unstable dynamics (May 1974).

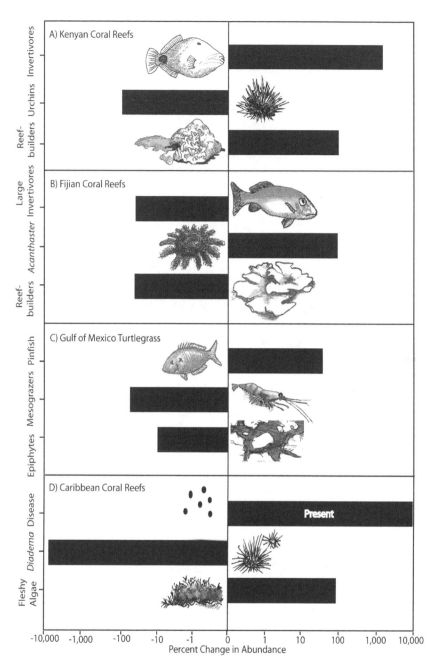

Figure 5.1. Summary of evidence showing trophic cascades in the nearshore tropics. Percentage change in abundance for each of three trophic levels for **(a)** protected Kenyan coral reefs: *Balistapus undulatus* and sea urchins (McClanahan 2000), reef-builders (McClanahan 1997); **(b)** fished Fijian coral reefs: predatory fish, *Acanthaster planci*, and reef-builders (Dulvy et al. 2004a); **(c)** experimental manipulations in Gulf of Mexico turtlegrass community: pinfish, mesograzers, and epibionts (Heck et al. 2000); and **(d)** Caribbean coral reefs post-1983: *Diadema antillarum* and fleshy algae (Hughes 1994).

Figure 5.2. Two states of Kenyan coral reefs. In the control state, triggerfish are sufficiently numerous to control sea urchin populations, and reef-building corals and coralline algae dominate the benthos. With overfishing of triggerfish populations, sea urchin populations expand, leading to the consumption and degradation of reef-building organisms and competitive release of turf algae.

On the coral reefs of Kenya, the important trophic role of sea urchins (predominantly *Echinometra mathaei*) has been documented in detail (Figure 5.2). McClanahan and colleagues have been recording changes in reef community structure inside and outside marine protected areas for more than two decades. In response to local fishery closures for more than 40 years, abundance and biomass of reef fishes have increased toward what may approximate their pristine biomass (McClanahan et al. 2007). Increases in the abundance of large triggerfish are particularly striking, including one of the dominant sea urchin predators, the red-lined triggerfish *Balistapus undulatus* (McClanahan and Shafir 1990; McClanahan 2000). The trophic consequences of triggerfish population increase include a strong increase in predation on sea urchins and a decrease in sea urchin density in protected areas (McClanahan 2000).

The role of sea urchins on coral reefs is varied, but in general sea urchins are effective herbivores that clear the benthos within a limited daily foraging area (Ogden et al. 1973; Carpenter 1984; Birkeland 1988). Using their scraping mouthparts, sea urchins remove all benthic organisms, leaving behind exposed calcium carbonate of the naked reef framework (Sammarco 1980). On Kenyan reefs, herbivory by sea urchins has been related to degradation of reef-building coral and coralline algal populations and a reduction in topographic complexity of the reef benthos (McClanahan and Shafir 1990). Additionally, on uncolonized substrates (e.g., settlement tiles), sea urchins have been shown to arrest successional development of the algal assemblage to the earliest stages of turf algal growth (McClanahan 1997). Therefore, dramatic differences in the benthic assemblages are apparent across boundaries of protected areas (McClanahan and Shafir 1990). Unprotected areas have low densities and biomass of fishes, including important predators on sea urchins, high densities of sea urchins that degrade coral and coralline algal development, and high cover of early successional, fast-growing turf algae. In contrast, protected areas have increased fish density, decreased sea urchin density, and a more heterogeneous benthos including coralline algae and some coral growth (Figure 5.2).

It is important to note that this three-group trophic pathway cannot wholly describe the community changes on these reefs. The reduction of fishing pressure in reserves leads to increases in most large groups of fishes, including herbivorous surgeonfish and parrotfish (McClanahan et al. 2007). In contrast to the effects of sea urchins, herbivory by fishes leads to greater successional development of the benthic community, favoring a mixed algal assemblage of turf and erect fleshy algae, as well as crustose coralline algal growth (McClanahan 1997). Complementing the trophic consequences of control of sea urchins in protected areas, herbivory is increasingly affected by fishes, favoring development of a more diverse benthos. Increased benthic heterogeneity, and especially the growth of crustose coralline algae, favors coral recruitment and growth (Morse et al. 1988; Harrington et al. 2004). Thus, the inherent complexity of tropical ecosystems belies and confounds the recognition of simple trophic pathways and cascades (McClanahan 2005).

The clear and dramatic effects of the removal or protection of predatory fish on sea urchins on Kenyan reefs is in striking contrast to the situation for the grazing sea urchin *Diadema antillarum* in the Caribbean. *Diadema* was apparently always superabundant, even when large predatory fishes were so common as to provide abundant food for the first European colonists (Jackson 1997). The similarly dramatic effects of *Diadema* in structuring the reef benthos are discussed in Case Study #4.

CASE STUDY #2: FIJIAN CORAL REEFS (PREDATORY FISHES → CROWN-OF-THORNS STARFISH → CORALS)

The corallivorous crown-of-thorns starfish (*Acanthaster planci*) is perhaps the most notorious echinoderm on coral reefs, famous for its episodic population explosions, formation of feeding herds of hundreds of thousands, and ability to denude all living corals across wide swaths of the Indo-Pacific (Chesher 1969; Birkeland and Lucas 1990; Saap 1999). One of the earliest hypotheses to explain the outbreaks concerned removal of predators, although very little was known of the life histories of the starfish and its biological interactions (Endean and Cameron 1990; Reichelt et al. 1990). Ormond et al. (1990) hypothesized that reduction of predator abundance should have a direct positive effect on the abundance of either juvenile or adult starfish, thereby increasing the probability of population outbreaks of *Acanthaster*. Because such outbreaks are linked to dramatic reductions of stony corals, the system potentially comprises a three-level trophic cascade.

Surveys of fish and starfish abundance across a gradient of fishing pressure among a series of Fijian islands support this hypothesis (Dulvy et al. 2004a, 2004b) (Figure 5.3). Abundance of *Acanthaster* increased and coral cover and crustose coralline algae decreased with increasing fishing pressure. Death of corals and coralline algae and overfishing of herbivores allowed fast-growing

Figure 5.3. Two states of Fijian coral reefs. In the baseline state, large invertivorous fishes are sufficiently numerous to control populations of crown-of-thorns starfish, and reef-building corals and coralline algae dominate the benthos. With the removal of large fishes, starfish populations can explode locally, leading to significant mortality of their favored prey, stony corals. After starfish outbreaks, the originally coral-dominated benthos is left dominated by fleshy algae.

fleshy algae to overgrow the newly empty space. Such a trophic cascade, combined with reduced herbivore populations, is contributing to the ongoing replacement of reef-building corals and coralline algae by opportunistic, fleshy algae on a global scale (McClanahan et al. 2002).

Closer inspection of these Fijian data reveals a more complex link between trophic dynamics and population growth. *Acanthaster* density did not increase monotonically across the gradient of fishing pressure; rather, the probability of a population outbreak increased (Dulvy et al. 2004a). Many of the islands in areas of light fishing activity (and high predator density) had no *Acanthaster*, whereas all islands with high fishing activity (and low predator density) had *Acanthaster*. These data, coupled with empirically approximated population growth rates, are consistent with a two-step model for *Acanthaster* outbreaks. First, predation pressure must be sufficiently low to allow starfish population growth to be greater than zero. If this criterion is not met, then the small population cannot increase. However, if predation is sufficiently low, then the starfish population may increase to a level determined by both food availability and predator density (Dulvy et al. 2004a).

The biggest question about this scenario concerns the ability of fish to limit *Acanthaster* population growth through predation. It is generally believed that the major predators of adult *Acanthaster* are large snails, in particular tritons, whereas predation by fishes appears to be limited to consumption of juveniles. Predation on juveniles can have dramatic structuring effects on adult populations (Caley et al. 1996; Hixon et al. 2002), but there are insufficient data for *Acanthaster* to evaluate the importance of this mechanism. However, the full potential of fishes to affect juvenile or adult *Acanthaster* remains unclear because of the extreme historical depredations on coral reef fishes (Jackson 1997; Pandolfi et al. 2003; Sandin et al. 2008). Although it appears that recovery of fish populations has led to a limited trophic impact in Fiji, even the most protected Fijian reefs have not attained anything like pristine abundance of fishes, especially of predators (Knowlton and Jackson 2008).

CASE STUDY #3: GULF OF MEXICO TURTLEGRASS (PREDATORS → MESOPREDATORS → EPIPHYTIC ALGAE)

Fishing and runoff of excess nutrients are the principal human impacts on seagrass habitats, where their effects may become so entwined as to be almost impossible to disentangle without careful experiments (Short and Wyllie-Echeverria 1996; Williams and Heck 2001). Runoff of excess nutrients increases water turbidity, which in turn decreases light availability, which disproportionately favors growth of macroalgal and epiphytic competitors of seagrasses (Hall et al. 1999; Williams and Heck 2001). The effects of fishing are more varied. Historical overfishing systematically removed large-bodied grazers such as turtles, manatees, and large fishes to the point of ecological extinction, with direct effects on seagrass production and trophic flow (Thayer et al. 1982; Jackson 1997; Duffy 2006; Heck and Valentine 2006, 2007; McClenachan et al. 2006). Thus, almost all studies of trophic dynamics of seagrass communities today are limited to smaller-bodied species. It is in this functionally altered ecosystem that trophic cascades have been suggested.

Today, fishing targets mostly small piscivorous fishes, allowing populations of smaller predators and omnivores to increase. Increases in these smaller species in turn reduce populations of mesograzers (small amphipods, isopods, and caridean shrimp) that feed predominantly on epiphytic algae living on the seagrasses, with more limited, direct consumption of the seagrasses (Figure 5.4). Synthesizing results from thirty-five studies in temperate and tropical seagrass habitats, Hughes et al. (2004) identified general patterns of nutrient and mesograzer effects on seagrass communities. Here we focus on the effects of fishing.

Figure 5.4. Two states of turtlegrass habitats in the Gulf of Mexico. In the presence of large piscivorous fishes, densities of smaller predators such as pinfish are maintained at low levels. Without small predators, mesograzers (e.g., caridean shrimp) reach high densities and efficiently graze epiphytic algae living on turtlegrass blades. Without large piscivores, pinfish densities are expected to increase, leading to reductions of mesograzers and overgrowth of epiphytic algae. Because epiphytes and turtlegrass are direct competitors for light and nutrients, the release of epiphytes will have a negative effect on turtlegrass growth or density.

In general, removal of mesograzers leads to an increase in epiphytic growth on seagrasses that appears to be responsible for a decrease in seagrass shoot density, although there are no consistent effects on seagrass biomass above or below the sediment surface (Hughes et al. 2004).

Despite the clear evidence of prey release consistent with trophic cascades in modern seagrass systems, there is less evidence to link the three trophic levels in the same place (Figure 5.4). Manipulation of nutrient availability and pinfish (*Lagodon rhomboids*) density in a factorial experiment in a turtlegrass ecosystem in Florida Bay demonstrated that pinfish can reduce densities of mesograzers, but epiphyte density did not respond in kind (Heck et al. 2000). Instead, the elevated density of pinfish was associated with reduced density of epiphytes and increased length of turtlegrass leaves, suggesting that pinfish are foraging across trophic levels, consuming both the invertebrate mesograzers and the algal epi-phytes (Heck et al. 2000). Thus, trophic complexity both within and between species appears to limit emergence of simple trophic cascades in these seagrass habitats (Polis and Strong 1996; Duffy 2002).

CASE STUDY #4: CARIBBEAN CORAL REEFS (PATHOGEN → SEA URCHINS → FLESHY ALGAE)

The sea urchin *Diadema antillarum* was the dominant invertebrate grazer in shallow-water Caribbean ecosystems until an undescribed pathogen killed more than 98 percent of the existing population in 1983 (Lessios et al. 1984) (Figure 5.5). Before the die-off, the importance of *Diadema* as a reef grazer was documented by numerous experiments in which exclusion of the sea urchin resulted in striking increases in macroalgal abundance (Sammarco 1980). Thus, as expected, the die-off of the sea urchin resulted in explosions in macroalgal abundance throughout the Caribbean. In Jamaica, algal overgrowth of reef sub-strates and living corals began within a few weeks of the die-off of *Diadema*, and fleshy algal cover increased from less than 10 percent to more than 90 percent within just 2 years (Hughes et al. 1987). Importantly, the fish assemblage of Ja-maica had suffered extreme degrees of exploitation long before 1983, leaving the *Diadema* as the primary reef herbivore (Hughes 1994). Thus, the redun-dancy of reef herbivores was greatly reduced before die-off of *Diadema*, simpli-fying the web of trophic pathways on Jamaica's reefs.

The effects of pathogens clearly can have cascading trophic consequences on ecosystems (Strong 1992; Pinnegar et al. 2000). However, an alterna-tive question is whether the emergence of pathogens can itself be a trophic

Figure 5.5. Two states of Caribbean coral reefs. Before 1983, many regions of the Caribbean had high population densities of the herbivorous sea urchin *Diadema antillarum* controlling cover of fleshy algae. Stony corals were favored competitors for space and dominated the benthos. In 1983 a basin-wide epizootic killed most *Diadema*, and in the years thereafter a number of hurricanes and other forcings killed many stony corals. Without control by *Diadema*, fleshy algae overgrew recently killed coral colonies, leaving many areas in an altered state of algal domination.

consequence. *Diadema* apparently were always extremely abundant in Caribbean shallow-water ecosystems (Jackson 1997). However, the removal of top predators, including those capable of consuming sea urchins, may have caused *Diadema* to become even more abundant and hence more susceptible to disease (Jackson et al. 2001). If that was the case, exploitation of predatory fishes may have been the linchpin that facilitated the rapid spread of the pathogen that eliminated the last remaining major herbivore in the system. A similar linkage between prey release and increased disease prevalence has been reported among sea urchins in kelp forests (Lafferty 2004) and may well be important in a number of other ecosystems.

PREY RELEASE, TROPHIC CASCADES, AND PHASE SHIFTS: ECOLOGY IN AN ALTERED SEA

The scars of human impacts in tropical nearshore habitats are so pervasive and profound that it is all but impossible to unravel cause and effect (Jackson et al. 2001; Knowlton and Jackson 2008). Large animals have been extirpated almost everywhere, dramatically altering the structure and function of food webs. Furthermore, pollution and climate change have changed the most basic of conditions in which the remaining organisms live. For example, water nutrient concentrations near human populations can be elevated by sewage and agricultural runoff, leading to blooms of fast-growing fleshy algae. Furthermore, rising temperatures and increased concentrations of carbon dioxide (with associated reductions in pH) have altered the fundamental physiology of reef organisms, particularly affecting symbioses between zooxanthellae and their coral hosts and calcification pathways for reef-building corals and coralline algae. Therefore, it is not surprising that evidence of prey release or trophic cascades is obscured by the past removals of species or by other stressors linked to human activities. Under such altered circumstances, predictions based on trophic theory provide only a rough guide to the possible consequences of manipulating the remaining predator populations.

Despite the limited evidence of well-defined trophic cascades, there are abundant data describing alterations in the composition of tropical shallow-water ecosystems (Hughes et al. 2003, 2004; Pandolfi et al. 2003; Dinsdale et al. 2008; Sandin et al. 2008), commonly called phase shifts or alternate stable states (Done 1992; Knowlton 2004; Chapter 17, this volume). Most of our insights into the role of trophic interactions in modern tropical ecosystems are based on extremely altered populations and communities or those that have been granted short-term reprieves in marine protected areas (Jackson 2001).

In contrast, surveys of coral reefs around remote islands and in otherwise well-protected areas are providing new insights into the structure and functioning of quasi-pristine habitats that are largely outside the reach of human activities (Knowlton and Jackson 2008) (Figure 5.6). Unsurprisingly, fish biomass is much greater and more dominated by large-bodied predators where fishing is reduced or absent (Jennings and Polunin 1997; Friedlander and DeMartini 2002; Brainard et al. 2005; McClanahan et al. 2007; Mumby et al. 2007; Sandin et al. 2008). More surprisingly, however, biomass of lower trophic level fishes appears to respond more to changes in fishing pressure than to changes in the abundance of their piscivorous predators (Friedlander and DeMartini 2002; McClanahan et al. 2007; Mumby et al. 2007). In contrast to biomass, the species composition, size structure, and trophic functioning of the prey assemblage is linked intricately to the biomass of predators (Dulvy et al. 2004b; DeMartini et al. 2005, 2008). In parallel, unfished reefs tend to have higher coral cover (Sandin et al. 2008), higher coral recruitment (Mumby et al. 2007; Sandin et al. 2008), and lower cover of competitive fleshy algae (McClanahan 1997; Mumby et al. 2006; Newman et al. 2006; Sandin et al. 2008). Trophic consequences associated with fishing probably precipitate these ecosystem-wide changes, although other factors including pollution and the synergistic consequences of climate change probably contribute further (Knowlton and Jackson 2008).

In addition, some of the most dramatic changes in the trophic dynamics of shallow-water tropical ecosystems caused by human impacts involve massive

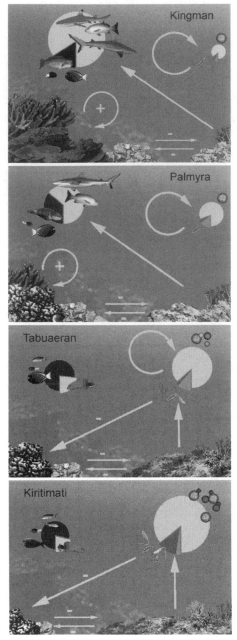

Figure 5.6. Schematic representations of the coral reef ecosystems of the four northern Line Islands, summarizing data from Dinsdale et al. (2008) and Sandin et al. (2008). Benthic cover (coral, crustose coralline algae, turf algae, and fleshy macroalgae, from left to right) is proportional to the length of the panels. Fish biomass, upper left pie diagram: Darker section denotes apex predators, and lighter section denotes other trophic groups. Microbial abundance, upper right pie diagram: Lighter section indicates the density of viruses, and darker section represents the density of Bacteria and Archaea split between autotrophs (lighter) and heterotrophs (darker). Solid lines denote trophic flow, and dashed lines denote nontrophic interactions (competition for space, coral recruitment, and coral disease), with the qualitative effect of interaction denoted by the sign (− for negative, + for positive).

changes in the ecology and relative abundance of microbes that are important agents of disease (Harvell et al. 2007). For example, outbreaks of coral disease in the Caribbean are concentrated in regions of moderate to high human impacts (Green and Bruckner 2000), suggesting a functional coupling between microbial dynamics and typical anthropogenic effects. There is also a strong correlation between fishing, microbial abundances, and prevalence of coral disease in the northern Line Islands (Dinsdale et al. 2008) (Figure 5.6).

The conventional explanation for increasing incidence of coral disease is that anthropogenic increases in sedimentation and nutrients are responsible (Harvell et al. 2007). However, recent evidence suggests that strong trophic interactions may link changes in the food web structure of the macrobiota to the dynamics of the microbiota. This trophic hypothesis predicts that increases in fleshy algal biomass associated with reef degradation lead to increased production and release of labile sugars that leak into the surrounding seawater from the algae, thereby fueling microbial growth that causes coral disease (Kline et al. 2006; Smith et al. 2006; Dinsdale et al. 2008). Laboratory experiments demonstrate that an elevated concentration of dissolved organic carbon increases coral mortality (Kuntz et al. 2005; Kline et al. 2006) and that proximity to fleshy algae leads to reductions of coral condition and increased mortality (Smith et al. 2006; Vermeij et al. 2009). Thus, release of reef algae from herbivory through reductions in fish biomass and grazing efficacy apparently leads to increases in microbial concentrations and prevalence of coral disease. As corals die from microbial activity, rapidly growing fleshy algae have more area available to expand, leading to increased algal growth and increased microbial activity, and so on in a self-reinforcing positive feedback loop (Figure 5.6). The feedback between the microbiota and macrobiota provides the functional reinforcement to create and maintain a macroalgae-dominated alternate stable state on coral reefs.

CONSERVATION IN AN ALTERED SEA

In diverse ecosystems that exhibit multiple stable states, past or current changes in the ecosystem may not be predictive of future changes in the ecosystem (Scheffer et al. 2001; Chapter 17, this volume). For example, the initial removal of top predators may have strong impacts on an ecosystem, but the magnitude of these impacts is likely to decrease as the functional role of top predators is taken over by a more diverse and weakly interacting assemblage of species. Consequently, the potentially large and irreversible impacts of the collapse of the last remaining predatory species past some threshold is unlikely to be predictable based on current marginal changes in the ecosystem.

To put this in context, a manager may observe that fishing of predators is imposing some ecological cost on an ecosystem. Based on linear models describing fishery dynamics, the manager may defer the imposition of regulations needed to slow exploitation because the costs of predator removal for other ecological processes are seemingly small. As the removal of the predators continues, the observable effects may attenuate, indicating that the ecological costs of fishing are even lower and no management need be done. What the manager cannot observe, however, is the increased risk that severe and potentially irreversible ecological costs may be incurred in the near future if the removal of the predators has passed some threshold. The epidemic mortality of *Diadema* and reef corals in the Caribbean with related population explosions of macroalgae and microbes is a cautionary example (Knowlton 1992, 2004; Hughes 1994).

Analyses of systems with unknown thresholds and common management goals, although complex, have resulted in simple policy prescriptions that are consistent with ecosystem-based management: Maintain resilience and apply the precautionary principle (Brozovic and Schlenker 2007). The application of this bet-hedging strategy (Costanza et al. 2000; Clark et al. 2001) motivates two areas of research. First, the most cost-effective strategies for maintaining the resilience of systems can be identified in part by studies that consider the marginal effects of different stressors rather than attempting to reject alternative stressors as the most important. For example, once some no-take marine reserves are established in a coral reef system, reducing nutrients by installing water treatment systems may be cheaper and more politically feasible than further reductions in fishing (Crépin 2007). It is therefore fundamentally important to identify accurately the context-specific benefits of candidate management actions. The second, less invasive approach is to undertake very large-scale studies of the dynamics of ecosystems over extreme spatial or temporal gradients of relevant stressors to more effectively identify and respond to human impacts (Knowlton and Jackson 2008). This approach is especially important because appropriate experiments are generally unethical or impractical on ecologically relevant scales.

The key point is that adaptive management is not particularly informative or useful in systems with thresholds because small changes in policy will not identify the location of a threshold, and larger changes using experiments will simply produce the exact state the manager is trying to avoid. Instead, a few well-chosen sites should be used to conduct careful studies of system dynamics. The last remaining quasi-pristine sites and the few well-functioning marine reserves can provide important data to describe pathways of degradation and recovery, respectively. However, inherent in systems with thresholds is a mismatch in the time scale of different functional groups (Pandolfi et al. 2005; Crépin 2007). To expedite data collection, space may be used as a proxy for time if the treatment variable is well identified. In addition, with proper identification, marginal effects of different stressors that vary together can also be determined. The major challenge is to develop careful and creative natural experiments that are inherently interdisciplinary.

CONCLUSIONS

There is good evidence for prey release but only limited evidence of true three-level trophic cascades in tropical nearshore habitats. The strongest examples involve herbivory, and with few exceptions herbivores have direct and quantifiable impacts on their algal or plant prey (Hughes et al. 2004; Burkepile and Hay 2006). Thus, alteration of densities of particular herbivores or guilds of herbivores will commonly have profound effects on the composition, biomass, and productivity of assemblages of primary producers (Carpenter 1986). There is much less consistency in the effects of predators on heterotrophic prey, probably because of the high degree of trophic overlap and complexity inherent to tropical nearshore habitats (Jackson et al. 2001). Thus, if trophic complexity is the best descriptor limiting the emergence of consistent evidence of prey release among higher trophic levels, then trophic complexity among herbivores must be less than for predators. Indeed, analysis of fifty-eight food webs demonstrated that omnivory across trophic levels is common among predators but that herbivores are typically limited to single trophic levels (Thompson et al. 2007). Moreover, omnivory among predators was most common in marine food webs relative to terrestrial and freshwater environments (Thompson et al. 2007).

Strong trophic responses, including prey release and trophic cascades, are hypothesized to be more common in linear food chains, whereas increased complexity of trophic interactions including omnivory and ontogenetic dietary switching are believed to dampen strong trophic effects and weaken trophic responses (Strong 1992; Polis and Strong 1996). For example, data from pelagic fisheries along a latitudinal gradient suggest that top-down control is more common in colder waters, whereas bottom-up control is more common in warmer waters (Frank et al. 2007; Chapter 6, this volume). Because food web complexity generally increases with water temperature in the sea, this pattern is consistent with the hypothesis that predator control (and thus prey release) becomes less common in increasingly complex food webs (Frank et al. 2007).

Alternatively, evidence of trophic cascades should be most apparent in cases composed of two strongly interacting pairs of species or functional guilds that are trophically linked (Paine 1980; Pace et al. 1999; Bascompte et al. 2005; Chapter 2, this volume). Because food web complexity limits the strength of trophic interactions, trophic cascades should be more common among subsets of species that are particularly specialized. The disproportionate number of observed cascades involving echinoderms and gastropods that are heavily defended against predators (Pinnegar et al. 2000) appears to support this hypothesis. The majority of case studies in the nearshore tropics demonstrating trophic cascades include echinoderms (Figure 5.1).

Trophic cascades may have been more prevalent in historically complex food webs dominated by large-bodied and strongly interacting herbivores such as sea turtles, manatees, and bumphead parrotfish and slow-growing predators such as sharks, monk seals, and crocodiles (Kenyon 1977; Jackson et al. 2001; Pandolfi et al. 2003; Dulvy and Polunin 2004; McClenachan et al. 2006; McClenachan and Cooper 2008). Modern food webs lack a number of the major linkages that historically dominated trophic interactions, with a consequent drop in trophic redundancy at each trophic level and increased vulnerability to alternative community states (Knowlton 2004). We know very little, and unless we make an immediate and concerted effort to study the last few "wild" places, we may never know.

ACKNOWLEDGMENTS

We are grateful to Mark Vermeij for drawing the figures and to Julia Baum, Nancy Knowlton, and Jennifer Smith for their many helpful suggestions. S. Sandin was supported by E. Scripps and S. Walsh by a National Science Foundation predoctoral fellowship.

CHAPTER 6

Trophic Cascades in Open Ocean Ecosystems

Tim Essington

The study of oceans and ocean ecosystems has long been in the domain of physical science. Indeed, it's difficult to experience an oceanographic research cruise without immediately appreciating the power of the physical processes that shape those systems. Biological components of these ecosystems have generally been viewed as actors whose only role is to respond to the dynamic, powerful changing environment around them. That is, biology is the passenger, geophysics is the driver. This view began to shift as information from benthic marine ecosystems (reef, estuarine, intertidal, and kelp forests) provided compelling evidence that animals in these food webs play key roles in shaping these ecosystems (Estes et al. 1998; Pinnegar et al. 2000; Myers et al. 2007, Baum and Worm 2009).

How strong is the evidence for consumer control in open ocean ecosystems? Can we generalize from studies in lakes, estuaries, and nearshore systems to predict the strength of consumer control and trophic cascades in large, dynamic open ocean systems? Thus far, the weight of evidence suggests not: On the whole, there is far less evidence of strong consumer control and trophic cascades than in other aquatic systems. Here I explore the evidence for weak trophic cascades in open ocean ecosystems and reflect on whether this represents practical limits in measuring and detecting species interactions or whether it represents properties of these ecosystems that weaken trophic cascades. Throughout, I approach this question from the background of a system ecologist whose primary

91

work is on upper trophic level organisms. As a result, this review is quite different from that which would be produced by one with expertise on trophic interactions between phytoplankton and their grazers. It is notable that many of the conclusions I reach here were recently echoed in an independent review of trophic cascades in marine environments (Baum and Worm 2009).

DYNAMICS OF OPEN OCEAN ECOSYSTEMS

Any consideration of open ocean ecosystems must begin with a synthesis of the types of geophysical forcing that drive these ecosystems and the time–space scales over which they operate. These properties are highly germane to predicting the context in which we might expect trophic cascades to develop. Moreover, recognizing the importance of decadal-scale variability and its role in shaping marine communities provides an important context with which to view the role of consumer control.

The phrase *open ocean ecosystem* is potentially misleading; it suggests a commonality to all ecosystems in the marine realm, much in the way *forest ecosystem* belies the diversity of forest ecosystems. Considering that most surface area on this planet consists of open ocean, it is not surprising that scientists have tried to make sense of this diversity by recognizing ecologically unique and significant units. One approach has been to classify ocean ecosystems on the basis of biogeography and prevailing oceanographic features. For instance, Longhurst (1998) identified four major pelagic oceanic biomes and within them fifty-one provinces. Spalding et al. (2007) used similar criteria to further specify coastal and shelf areas (a single biome under Longhurst's classification), yielding 12 realms, 62 provinces, and 232 marine ecoregions. An alternative way to consider open ecosystems is to classify them on the basis of their dominant physical drivers. Mann (2000) takes such an approach, producing a simpler set of ecosystem types: estuaries, rocky shores and beaches, and shelf. The latter is then divided into continental shelf, coastal upwelling, and coral reefs.

Not only can we classify ocean ecosystems on the basis of physical drivers, we can also classify physical drivers on the basis of the time–space scales at which they operate. Small time and space variation (meters and days) is notable because it is the scale at which primary producers sequester nutrients, collect the sun's energy, and build up sufficient production to support food webs. Mesoscale variation (tens of kilometers and months) is reflected in the form of intermittent oceanographic features such as eddies that can temporarily attenuate or promote production or El Niño events that shift the distribution of warm surface waters in the Pacific Ocean. Longer, decadal-scale variations (e.g., Pacific Decadal Oscillation, North Atlantic Oscillation) are even longer-term patterns of variability that have profound impacts on populations and communities (Anderson and Piatt 1999; Beaugrand et al. 2003; Chavez et al. 2003). Lastly, some cycles of ocean biophysical forcing occur on time scales exceeding 100 years. Notable examples include high-magnitude cycles of Atlantic bluefin tuna production at scales of 100 years or more and swings in North Pacific sockeye salmon production at cycles exceeding 250 years (Ravier and Fromentin 2001; Schindler et al. 2006).

Why consider marine ecosystem typology and the dynamics of physical forcing? First, it's clear that the physical characteristics of marine ecosystems and the organisms that reside in them are extraordinarily variable, making broad generalizations difficult. Furthermore, it reminds us that marine populations and communities can oscillate greatly in response to short and long time scale variability in these forcing variables. But perhaps most importantly, I suggest in this chapter that the relative importance of trophic control depends in part on the physical processes that characterize these ecosystems. This hypothesis may help identify systems where we might expect strong or weak trophic control.

EVIDENCE FOR CONSUMER CONTROL

Although there is little doubt that geophysical interactions are responsible for much of the observed variability in ocean ecosystems, several lines of evidence suggest that consumer control also plays important roles in the ecology of the biota. Verity and Smetacek (1996) argue that the evolution of zooplankton morphology and behavior has clearly been shaped by predation and that therefore predation must be a key organizing process. Banse (1994) argues that grazing is a dominant control on phytoplankton production. Several mesocosm studies suggest conditions in which grazing copepods can exert control on phytoplankton production (Stibor et al. 2004; Sommer and Sommer 2006). Although biological interactions certainly shape the evolution and population trajectories of marine organisms, their roles in limiting productivity and driving ecosystem processes are less obvious.

Predation is a dominant source of mortality among upper trophic level organisms (fish, large invertebrates, marine mammals, and seabirds), particularly for juvenile stages. There is a burgeoning literature comprising studies that apportion the total mortality of populations among the various predators that feed on them (Pauly et al. 2000; Tsou and Collie 2001). These studies reveal that a handful of species may be responsible for much of the mortality that juvenile stages experience (Collie and DeLong 1999; Tsou and Collie 2001) and thereby identify the interactions that are most likely to be consumer controlled. The next, more challenging step is confirming that these relationships produce strong consumer control through direct observations of predator and prey abundances. For the most part, this enterprise in open ocean ecosystems has been based largely on time series analysis. For instance, Shiomoto et al. (1997) provided evidence for alternating, inverse patterns of abundance of pink salmon, zooplankton, and surface chlorophyll as a result of the alternating strong and weak years of pink salmon reproduction (itself a consequence of the 2-year life history of this species). Ward and Myers (2005) looked for longer-term trends in apex predator and mesopredator fish species' abundances by combining a historical scientific study conducted in the mid-twentieth century with contemporary catches of longline fisheries in the same region, and their findings suggest an increased catch rate of several mesopredator species coinciding with the depletion of apex predators. Both studies revealed patterns in communities and food webs that supported the hypothesis of consumer control in open ocean food webs.

The reliance on time series analysis is due to the paucity of adequate controls for comparative analyses and the practical constraints on experimentation at the proper spatial and temporal scales. Time series analysis faces important limitations as well; chief among them is the potential confounding effect of environmental drivers that may produce a "consumer control mirage." Such effects are driven by associations of multiple species with hidden, unmeasured environmental drivers leading to spurious correlations that may be misinterpreted as indications of interactions. For instance, the data presented by Ward and Myers (2005) are consistent with the hypothesis that there has been predatory release of mesopredators, but they cannot determine whether differences between mid-century survey and current fishery catches are related to changing climate or differences in sampling methods between the two data sets. That is, given the commonness of long-term patterns of variation in biophysical coupling in ocean ecosystems, it will always be challenging to discern strongly interacting species from time series data. A powerful solution to this problem is to use multiple time series of predator and prey abundances and combine these into meta-analyses of correlation coefficients between predator and prey abundances (Worm and Myers 2003). In this way, one can systematically test several alternative hypotheses and potentially avoid spurious correlations.

EVIDENCE FOR TROPHIC CASCADES

Several recent meta-analyses have indicated that consumer control is less likely to cascade down through multiple trophic levels in open ocean systems. Micheli (1999) used a synthesis of mesocosm experiments and time series analysis from pelagic marine ecosystems to demonstrate that zooplanktivores do suppress zooplankton but that phytoplankton were regulated primarily by bottom-up processes. A series of related papers comparing the strength of trophic cascades among ecosystems found that herbivores tend to exhibit modest responses to predator removal in marine pelagic ecosystems, but this response tends to have little impact on phytoplankton abundance (Shurin et al. 2002). Borer et al. (2005) subsequently assessed whether variance in trophic cascades could be attributed to characteristics of the fauna that make up food webs, but again marine pelagic ecosystems stood apart from comparable aquatic ecosystems. Notably, marine benthic systems (data gathered primarily from kelp forests, coral reefs, and intertidal rocky shores) showed the strongest tendencies toward trophic cascades.

The absence of cascading consumer control from these analyses introduces the question, "What elements of these marine ecosystems might preclude trophic cascades?" Several authors have speculated whether specific ecosystems have unique characteristics that foster trophic cascades (Strong 1992; Polis 1999). Borer et al. (2005) succinctly summarize these attributes: Trophic cascades should be most common in systems with linear food chains with little omnivory or functional redundancy, small-bodied primary producers that have little investment in structural tissues or secondary compounds, and highly efficient predators, measured through either searching efficiency (ability to find prey) or numerical response (the ability to turn consumed prey into new born predators). One element not included, one that is particularly germane to ocean ecosystems, is that there must be spatial and temporal overlap between areas of production and areas of consumption for consumer control to be important and to cascade down through the food web.

This checklist of qualities promoting trophic cascades provides a framework for considering which, if any, of these elements can best explain the generally weak trophic cascades in marine ecosystems. Three possibilities seem most relevant: the complex life histories of apex predators (primarily teleost fishes) that can dampen density dependence, the high degree of functional redundancy in marine food webs, and the strong role of advection in marine planktonic systems. I describe each of these in turn and consider the explanatory power of each.

Numerical Response of Marine Apex Predators

By far the most numerous and biomass-dominant apex predators in marine ecosystems are teleost fishes. These include the large-bodied tunas, marlins, gadids (cod, haddock, hake), and several other taxonomic groups that are the targets of some of the largest fisheries in the world. A striking feature of all these stocks is the apparent lack of relationship between the production of new offspring ("recruits" in fishery science terminology) and the abundance, biomass, or egg production by adult stages. For most fishes, only at very low population sizes does the production of eggs appear to be a limiting factor affecting recruitment (Myers et al. 1999). In general, this is thought to represent the presence of critical periods during early life history stages (egg, larval, or juvenile) that largely control reproductive success in a given time period. The leading hypothesis to explain these critical periods is that very early life stages depend on abundant and suitably sized food at key time periods (often immediately after feeding commences). Several lines of evidence have been used to support this notion. The first is the recognition that spatially or temporally averaged food conditions are usually insufficient to prevent starvation of many larval fishes. Thus, larval fish survivorship depends in large part on chance encounters with patches of abundant food. A related idea is that survivorship depends on a match between the timing of first feeding and secondary production (Cushing 1996). Beaugrand et al. (2003) demonstrated that this process was responsible for the surge and subsequent collapse of gadoid populations in the North Sea over the past three decades. Because fishes have little ability to control for these conditions, most marine fishes have evolved "bet-hedging" life history strategies that involve large body sizes at maturity and multiple reproductive events, each yielding many very small offspring (Winemiller and Rose 1992, 1993).

The consequence of these environmental controls and the life history traits they produce is that the abundance of the adult life history stages that occupy the top of the food web is partially decoupled from the availability of their prey. That is, trophic interactions occurring low in the food web play a significant role in dictating the abundance of large, high trophic level species. As a result, several years of good reproduction can produce an enormous standing stock of predators, despite the fact that these predators might deplete their own preferred prey. Alternatively, poor conditions for juveniles can limit adult stocks below the inherent capacity of the environment.

What are the implications of this type of numerical response for trophic cascades? Marshall (2007) developed theoretical models to explore conditions that might promote strong top-down control in marine ecosystems. She found that the decoupling of births and the availability of adult-stage prey may promote strong top-down control. In effect, recruitment acts as a sort of energy subsidy to the adult population (by providing energy to the adult stock that is independent of the availability of adult prey). In general, these types of energetic subsidies strengthen top-down control (Vander Zanden et al. 2005) and stabilize these interactions (Huxel and McCann 1998). One suggestion that follows from this consideration is that the unique life history attributes of marine apex predators may promote, rather than retard, strong consumer control on their prey.

Functional Redundancy

Whether because of the diversity of habitats represented in marine ecosystems, the presumed openness of marine populations, a long geologic history for evolutionary diversification to take place, or the spatial and temporal heterogeneity induced by climate and oceanographic properties, food webs in open ocean ecosystems are astonishingly diverse. As a result, some have argued that there are no keystone species because of the high degree of functional redundancy inherent in these food webs (although Verity and Smetacek [1996] provide notable exceptions to this perspective). What is the evidence for this diversity retarding trophic cascades?

Essington (2006) simulated how the open ocean ecosystem of the central Pacific may have responded to a half century of commercial whaling and industrial fishing. Over this time period, the abundances of most major apex predators (toothed whales, tunas, marlins, and sharks; Figure 6.1) were all reduced to less than 50 percent of their initial abundances. This modeling analysis suggested that the total biomass of the apex predator guild remained unchanged over this time period, largely because of a predicted compensatory increase in the abundance of large squid, which also prey voraciously on smaller fishes and squids. No long-term data on squid abundance are available in this region to validate this model prediction. However, the prediction implies that the food web shifted from a community of long-lived, slow-turnover species to one dominated by short-lived, fast-turnover species. Because squid populations are notoriously variable and driven by environmental conditions, this prediction also implies that the shift in species composition of the apex predator guild should also be accompanied by decreased stability of the guild over time and space.

How does this functional redundancy affect the potential for trophic cascades? Here I performed a similar model simulation analysis of the eastern Pacific ocean food web, developed by Watters et al. (2003) and Hinke et al. (2004). Commercial fishing targeting tunas and billfishes has shifted in both intensity

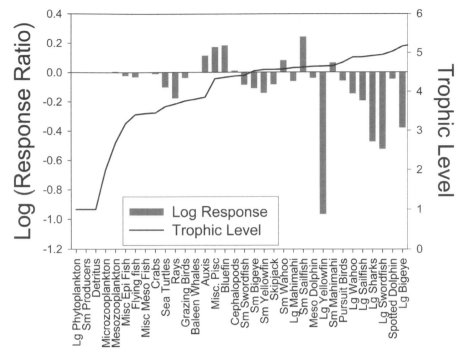

Figure 6.1. Simulated response of food web components to 50 years of commercial fishing development in the eastern tropical Pacific. Targeted species (mostly on the right-hand side) are less abundant now than in the mid-20th century. Some species of mesopredators (auxis, mahimahi) are predicted to increase in abundance. Despite these alterations to the top of the food web, low trophic level components showed little response.

and techniques over the past 50 years. This has resulted in predictable reductions of most of the upper trophic level predators in this system (Figure 6.1). The food web model suggests that the depletion of apex predators may have cascaded down to produce increased abundance of mesopredators, such as mahimahi (*Coryphaena hippurus*). This model prediction is supported an independent analysis by Ward and Myers (2005) based on data collected in the central Pacific Ocean. However, functional groups lower on the food web (trophic level 3 or less) were predicted to show only minor shifts in response to this large-scale manipulation at the top of the food web. No modeling analysis can capture all the complexities and interactions of real food webs. But we have full knowledge of the underlying model structures that are responsible for the predicted outcome, so that we can fully understand the (assumed) process that gave rise to the (predicted) phenomenon. In this model, mesopredators prey on a diverse guild of epipelagic and mesopelagic fishes, who in turn feed on primary consumers (zooplankton). Predation on the secondary consumers is divided between many predators so that the effect of any single predator on these guilds is small. As a result, the epipelagic and mesopelagic trophic guilds are not predicted to undergo drastic changes despite increases in some mesopredator populations.

Importantly, this model aggregates many species together into these functional groups, so the model cannot possibly capture trophic cascades instigated by selective predation on keystone species within these groups. Indeed, it is precisely this type of size selectivity that is thought to produce the striking trophic cascades in pelagic systems of lakes (Carpenter et al. 2001). Unfortunately, our knowledge of the dynamics of these guilds and the unique ecological roles of species within them is limited. Consequently, our understanding of functional linkages between these trophic levels is woefully insufficient to permit development and parameterization of food web models that would include the processes that produce strong trophic cascades observed in lentic ecosystems.

Advection

Our classic conceptual and mathematical models of predator–prey interactions and trophic cascades are based on underlying simplifying assumptions of spatial homogeneity or "well-mixed" systems (Hairston et al. 1960; Oksanen et al. 1981). The role of spatial heterogeneity in consumer control and trophic cascades has been considered in some detail (Holt 1984; Van de Koppel et al. 2005), but few have considered the nature of food chain dynamics in advective ecosystems, that is, those in which plankton exhibit a directed movement with oceanic currents. Oceanographers have long considered advection's role in the formation and dissipation of algal blooms in nearshore upwelling ecosystems, where the prevailing currents carry plankton offshore. These upwelling systems are ecosystems on a "conveyor belt," where production in one area is shuttled offshore. Initially, this conveyor belt has plenty of nutrients delivered from the deep ocean depths and few phytoplankton or grazers. As this parcel of water moves down the conveyor belt, the phytoplankton sequester nutrients and exhibit growth that outpaces the grazer's ability to feed on them. As this mass of water continues to move offshore, eventually the grazer's numerical response catches up to the phytoplankton, allowing them to reach sufficient densities to have a significant grazing impact (at about this same time phytoplankton begin to run out of limiting nutrients). The consequence of this conveyor belt effect is that the peak in primary production occurs upstream from the peak in grazing pressure. Thus, areas of grazing and production are spatially decoupled and thereby limit grazers' ability to directly reduce phytoplankton production (Figure 6.2).

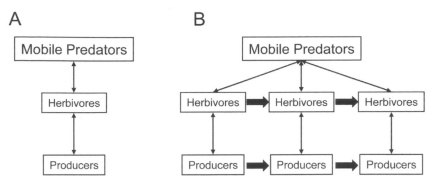

Figure 6.2. Conceptual differences between models of trophic cascades in **(a)** well-mixed food webs and **(b)** advective food webs. In **(a)**, herbivores can exert consumer control on prey because their spatial distribution overlaps that of producers. In **(b)**, herbivores can exert limited consumer control on their prey because their abundance is maximized downstream of producers.

Advection is not only important for trophic interactions among the plankton: Many upper trophic level predators rely on secondary production that has been advected from elsewhere. In some cases this involves actively moving to local convergence zones where plankton are concentrated. For example, skipjack tuna in the western Pacific Ocean warm pool (a highly unproductive biological desert) are somehow able to maintain extraordinarily high production rates despite the low productivity of lower trophic levels. One leading hypothesis for this apparent contradiction is that skipjack rely on allocthonous-derived forage advected to a convergence zone that defines the eastern edge of this region (Lehodey et al. 1997). Indeed, the skipjack population and the fisheries that target them move over a thousand kilometers in as little as 6 months during El Niño events when the convergence zone is displaced eastward, highlighting the importance of this region to the skipjack tuna population.

WHERE SHOULD TROPHIC CASCADES OCCUR?

These considerations suggest that trophic cascades are most likely to be observed in systems with low functional redundancy or weak advection. In this chapter I consider primarily the latter. Obviously, benthic systems are not driven by advection because production is attached to the seafloor (e.g., reefs, kelp forests). Perhaps it is not surprising that these ecosystems demonstrate some of the most pronounced trophic cascades of any ecosystem on Earth (Shurin et al. 2002). Semienclosed seas, with physical boundaries to water movement, might also have reduced impacts of advection. The assumptions of nonspatial models may also be more applicable in nonupwelling, continental shelf ecosystems that are dominated by tidal mixing. Here I explore the evidence for trophic cascades in these ecosystems.

Evidence in Semienclosed Seas

Semienclosed seas are commonly subjected to a host of human impacts: Fishing, invasive species, and eutrophication are all important ecological concerns. As a result, identifying trophic cascades that extend from upper trophic level predators (large fish or mammals) to primary producers is often confounded by the temporally coincident effects of overfishing and eutrophication. Nonetheless, there is a substantial body of evidence from the Baltic and Black seas that consumer control can be quite strong in these ecosystems.

The Baltic Sea is a large (415,000 square kilometers), deep fjord-like sea whose shores are shared by ten independent states (Jansson and Velner 1995). The Baltic has been a center of human activity for centuries, but the most recent environmental problems have been overfishing, eutrophication, hypoxia, and contaminants (Osterblom et al. 2007). Given the myriad human impacts on this system and the fact that these impacts are directed at both the top (fishing) and the bottom (eutrophication) of the food web, it is not surprising that it has been difficult to ascertain the relative importance of each. Despite these difficulties, the Baltic Sea has also been closely monitored for years, providing some of the best long-term time series data on fish and fisheries, zooplankton, and primary producers. Here I describe the evidence pointing to trophic control within the pelagic fish community and the evidence suggesting that these impacts are felt in the zooplankton community as well.

The Baltic Sea supports three main commercial fisheries: cod (*Gadus morhua*), the apex predator in the pelagic system, and herring (*Clupea harengus*) and sprat (*Sprattus sprattus*), both primarily zooplanktivores. On a per-pound basis, cod is the most valuable fish to the fishery, herring is of intermediate value, and sprat is a low-valued fish used primarily in the production of fertilizer. Cod populations are overfished and are still subject to very high fishing mortality. As a consequence, cod biomass has been reduced to a small fraction of its historical maximum. While cod populations dwindled, fisheries began directing more effort towards the clupeid species, preferentially targeting the more valuable herring over sprat. As a result, herring populations have declined over the past decades, but sprat populations have surged. Detailed analysis of the trophic control of sprat and herring by cod confirmed that the recent abundance of sprat is due to the relinquishing of predation pressure by cod

(Essington and Hansson 2004). Möllman and Köster (2002) explored in detail the shifts in abundance of three dominant copepod species and concluded that the reduction of the previously dominant species, *Pseudocaloides elongatus*, was probably caused by the increased predation pressure from sprat and that the simultaneous increase in *Temora longicornis* abundance would have been even more pronounced had the sprat not contributed to keeping this population in check.

Like the Baltic, the Black Sea is subject to a suite of human impacts: over-fishing of piscivorous and zooplanktivorous fishes, introduction of gelatinous zooplankton species, and eutrophication. Two recent syntheses of time series of fishery catches, fish abundance indices, zooplankton, and ocean color suggest that this system has been profoundly altered by anthropogenic forces and may have experienced a regime shift (Daskalov et al. 2007; Oguz and Gilbert 2007). These alterations began with the initial depletion of large piscivorous fishes, followed by depletion of zooplanktivorous fishes that coincided with outbursts of gelatinous zooplankton. Daskalov et al. (2007) conclude that the early dynamics reflected a trophic cascade, where the effects of overfishing piscivores cascaded down to zooplankton and phytoplankton, but the recent dynamics reflect a different ecosystem where gelatinous zooplankton play a dominant functional role. Oguz and Gilbert (2007) examined these and other times series from the interior Black Seas but expanded their analysis to consider bottom-up forcing such as nitrogen loading and climate shifts. These authors reach largely the same conclusions: The serial depletion of fishes and the rise of gelatinous zooplankton have fundamentally altered the nature of consumer and resource control in this ecosystem. Importantly, Daskalov et al. (2007) suggest that the Black Sea is in a new stable state that might prevent recovery to historic conditions (see Chapter 17, this volume).

Evidence in Continental Shelf Ecosystems

Continental shelf ecosystems support most fishery landings worldwide. Some of these ecosystems are dominated by coastal upwelling, where prevailing wind conditions create a net offshore movement of water, thereby drawing up deep, nutrient-rich water. Other shelf ecosystems are productive because the continental shelf that extends from shore out to the continental slope interacts with tidal flow of water above it to permit mixing of the water column. The former are highly advective ecosystems, where we would not necessarily expect to see consumer control cascading through multiple trophic levels down to primary producers. Ware and Thomson (2005) proposed this explanation to suggest why zooplankton and fishery catches are positively correlated with surface chlorophyll along the coastal northeast Pacific. Here I discuss evidence for consumer control and trophic cascades on continental shelf ecosystems.

Worm and Myers (2003) provided the clearest line of evidence that trophic control, not environmental bottom-up forcing, was responsible for the surge in shrimp (*Pandalus borealis*) abundance throughout the North Atlantic Ocean after the collapse of several cod stocks. This work was unique in that they identified a conceptual way of dealing with the "consumer control mirage" problem: Rather than looking at a single site, where the depletion of cod and the change in environmental conditions may be related, they instead looked at multiple sites where cod depletion occurred at different times. This analysis produced the statistical power to discriminate the effects of cod abundance from environmental controls (in this case, temperature) and permitted them to conclude that the highly productive shrimp stocks in the North Atlantic are the consequence of overexploitation of their predators.

In a series of subsequent papers, it became clear that consumer control is not uncommon in the North Atlantic shelf ecosystems. Frank et al. (2005) collated time series data from the eastern Scotian Shelf ecosystem from multiple trophic levels and argued that the dynamics of each are explained as a trophic cascade. The cascade began with the intense exploitation of cod and haddock, which was shortly followed by a rise in abundance of many of cod's prey, such as snow crabs, shrimp, and pelagic fishes. Although data records of zooplankton and primary producers (as measured through ocean color) were not as extensive, their changes in time were consistent with consumer control.

In a subsequent analysis, Frank et al. (2006) explored the spatial and temporal consistency of consumer and resource control. Here, similar time series data were used from sites that spanned a latitudinal gradient, bounded on the south by Georges Banks and the north by the Labrador Shelf. Once again, the contrast in food web structure was created by fisheries targeting cod and haddock. They found that at broad spatial scales, fish production was generally positively correlated with primary production. However, there was a latitudinal gradient in the nature of consumer control. In southern regions, there was no evidence of consumer control; predator and prey populations were positively correlated through time, indicating strong bottom-up limitation of consumers. In contrast, northern areas showed negative correlations between piscivore and planktivore biomass, suggesting a dominance of top-down regulation. They speculate that these differences are attributable to differences in species richness in the fish community; southern sites had more diverse communities of apex predators and as a result were able to maintain function despite the depletion of a few species. However, because the spatial gradient in species richness was also

confounded with a spatial gradient in mean annual temperature, it was not possible to definitively point to species richness as the sole causative agent.

Frank et al. (2006) also noted that the correlation between abundance measures of large, benthic fish and smaller pelagic fishes was nonstationary and that
food webs shifted from being predominantly resource controlled to consumer
controlled during the period of intense fishing. This suggests the possibility that
these ecosystems, if left unperturbed, would be limited by nutrients but that
fishing on top trophic level species fundamentally altered this relationship.
Hunt et al. (2002) also suggest that the nature of trophic control varies over
time, in this case corresponding to changes in climate regimes that either promote or retard the early life history survivorship of large predatory fish in the
Bering Sea. These and similar findings, suggest that the nature of trophic control can be highly context dependent.

CONCLUSIONS

It is remarkable that less than 10 years ago it was not uncommon to find assertions in the literature that marine fish do not regulate the abundance of their
prey. These statements now sound as dated as Thomas Huxley's famous procla-
mation that the "great sea fisheries" were inexhaustible. In fact, the closer we
look the more we find evidence of consumer control of fish prey. Identifying
cases in which these effects cascade further through food webs has proven to be
more elusive.

One must ask whether trophic cascades in open ocean ecosystems are rare
or just hard to detect. There are at least three reasons why it has proven difficult
to identify trophic cascades in open ocean ecosystems. The first is the obvious
problem that appropriately scaled experiments are impossible to conduct in
ocean ecosystems. As a result, our lines of evidence come from exploring the
unintended indirect effects that follow from human alteration of marine food
webs. Human impacts on marine ecosystems are diverse and often include simultaneous perturbations to the top (fishing) and bottom (eutrophication) of
marine ecosystems. Even the effects of fishing on food webs are difficult to assess because in most regions of the world fisheries are now exploiting multiple
levels of marine food webs simultaneously (Essington et al. 2006). The second is
that climate and large-scale oceanographic processes also exert a strong control
on marine populations and food webs so that distinguishing between alternative hypotheses from time series data is challenging. Moreover, we are now
learning that consumer control is context specific in that certain ocean regimes

foster consumer control, whereas others diminish its importance. The third problem is a cultural one: There is not a rich tradition of integrated research in open ocean ecosystems of the kind that has proven successful in lakes and other aquatic realms. That is, the culture of scientific enterprise in open oceans tends to foster work within disciplinary boundaries (e.g., biological oceanography, chemical oceanography, fish ecology).

The future for this line of research in open ocean ecosystems looks bright. With each year that passes, our time series become longer and our ability to choose between alternative hypotheses improves. Ocean observing systems are growing in spatial extent and in the level of biological data they can collect. No-take fisheries areas are being created around the world, and these will provide much-needed contrasts for comparative studies. We continue to develop novel analytical approaches to synthesizing complex data. Perhaps most importantly, the notion of consumer control in open ecosystems, once dismissed as an ecological oddity, is now gaining growing acknowledgment (Baum and Worm 2009). One challenge in moving forward is to remind ourselves that as scientists, we need to test several competing hypotheses, acknowledge cases in which we can't choose between two or more alternatives, and identify key lines of evidence that would enable us to choose between them. We must continue to be aware that the dynamics of biophysical forcing can abruptly and fundamentally alter structuring processes, so that our understanding of these ecosystems may be highly context specific. Lastly, we must endeavor to inform policymakers of the potential indirect consequences of large predator removals and the ecological risks associated with dramatic alterations of marine food webs.

PART II

Terrestrial Ecosystems

Distinct differences occur in ecosystem structure, composition, and process as one moves from the aquatic to the terrestrial realm. Perhaps the most striking of these is the appearance of a spectacularly varied assemblage of vascular plants, which are essentially absent from the world's oceans and of comparatively modest taxonomic and structural diversity in freshwater systems. A correlated life history trait is the markedly different floral generation times, which range from days to years in the aquatic realm but from years to centuries or even millennia for woody vegetation on land. Consequently, ecosystem response times to shifts in trophic interactions between plants and their herbivores, and by extension between plants and apex predators via trophic cascades, are typically much longer on land than they are in water. Other differences are apparent between aquatic and terrestrial ecosystems. For instance, the main physical disturbances to terrestrial systems are fire, drought, and flood, processes that either do not exist or are of only localized importance in aquatic systems. The density of water is roughly 1,000 times greater than that of air, and therefore aquatic environments have much larger capacities for convective movement of nonmobile entities. The list could go on and on.

Despite these many differences, there is a similarity of method and result between the aquatic examples in Part I and the terrestrial examples in Part II. In both, predator effects are explored via perturbations, by contrasting systems with and without predators through spatial comparisons, time series, and experimental manipulations. And in both, top-down forcing and trophic cascades emerge as recurrent phenomena. Chapter 7 (Marquis) provides an overview of the diverse ways in which herbivores influence terrestrial plant species and assemblages on ecological and evolutionary time scales. This is important because so much attention in terrestrial ecology has focused on plant–herbivore interactions, plant–herbivore interactions are an essential part of most trophic cascades, and the large apex predators that ultimately mediate these trophic cascades have been lost from most terrestrial ecosystems. Chapter 8 (Terborgh and Feeley) considers the tropical realm, where species diversity is high and complex interactions, most weak but some strong, come together as an intricate web of life that might seem resilient to the loss of a few species. Yet despite the complexity of these systems and the overall lack of single keystone predators (as we have seen in lakes and coastal oceans), powerful trophic cascades become evident when the assemblage of large predators is lost, with diverse bounceback effects that influence many other species and ecosystem processes. In Chapter 9 Ripple and colleagues extend this view to temperate and boreal systems. These authors establish that ungulate populations irrupt to levels where they damage the vegetation wherever large predators have been lost, and here again the trophic cascade has important effects on many other species and ecosystem processes. Chapter 10 (Oksanen and colleagues) further extends these same top-down processes to arctic and subarctic systems. Like Terborgh and Feeley, these authors focus on the findings from their own work in Finland and northern Sweden. Chapter 11 (Schoener and Spiller) consider trophic cascades on islands, in part as a synopsis of their own elegant experimental studies and in part through a review of the relevant literature from various islands throughout the world's oceans. These island studies are reminiscent of Paine's account (Chapter 2) of the rocky intertidal (because of their experimental nature) and Carpenter and colleagues' (Chapter 4) account of freshwater lakes (because they were done at the level of entire ecosystems). Finally, Chapter 12 (Wardle) explores connections between aboveground and belowground components of terrestrial ecosystems. Top-down forcing processes are shown to occur belowground, and the resulting trophic cascades interact between belowground and aboveground components of the ecosystem.

CHAPTER 7

The Role of Herbivores in Terrestrial Trophic Cascades

Robert J. Marquis

Herbivores play determining roles in trophic cascades by affecting both the strength of cascades (i.e., the degree of plant response) and the scale, from single plants to entire communities (Polis et al. 2000). By definition, the strength of a cascade can be no greater than the effect of herbivores on plants. These effects often are dramatic. At the extreme, individual plants may be deformed or killed (Whitham and Mopper 1985), understories of forests decimated (Russell et al. 2001), wide swaths of trees killed (Bjørnstad et al. 2002), and grasses and grass relatives grazed to ground level (Kotanen and Jeffries 1997).

According to Hairston, Smith, and Slobodkin (HSS, 1960) experimental exclusion of herbivores should have little or no effect on plants when the carnivore level of a tritrophic system is intact. In contrast, exclusion of carnivores should be followed by major increases in herbivory. Hairston (1991) stressed that this scenario applies to herbivores alone (not omnivores), in situations in which the carnivore assemblage is in intact (no effect of hunting and not in successional vegetation), and only to native plant species.

Herbivore impacts are extremely variable from habitat to habitat and biome to biome (Crawley 1997). The roles of these mitigating factors (Hairston 1991; Sih 1991) and of the third trophic level itself have gone largely unexplored by those reviewing the impacts of terrestrial herbivores on plants (e.g., Hendrix 1988; Weis and Berenbaum 1989; Lindroth 1989; Louda et al. 1990; Marquis 1992; Strauss and Zangerl 2002,; see also Schmitz 2008b). Exceptions are plants

Table 7.1. Characteristics of the major classes of herbivores relevant to their impacts on plants and susceptibility to trophic control (after Crawley 1989).

Characteristic	Invertebrates	Vertebrates	Fungus–Rearing Insects
Body size	Small	Large	Small but with large colonies
Metabolic rate	Low	High	Low
Population density	Large	Small	Low to high
Food specificity	High	Low	Low
Bite size	Small	Large	Small
Mobility	Low to high	Low to high	Medium
Starvation tolerance	Low to high	Low to high	High
Susceptibility to predation	Generally high	Low especially at high body mass	High

that have evolved rewards to attract ants as protectors (Rico-Gray and Oliveira 2007).

The overall goal of this chapter is to review what we know about the role of herbivores in terrestrial trophic cascades. My first objective is to examine the range of known effects of herbivores. I then consider complicating factors that may influence the great variability in impact seen from system to system, as noted by Crawley (1997). These factors include those that might make trophic control "leaky" and conditions that allow herbivores to have significant impacts on plants despite intact predator guilds. The guild of herbivores is a mix of very different kinds of organisms (Table 7.1). I point out the potential influence that herbivore taxon has on the aforementioned issues, recognizing that there is also great variability within major herbivore groups. Finally, I suggest what the world might be like with no third trophic level. Through active mismanagement or benign neglect we apparently are heading rapidly toward such a world. My last objective is to speculate about the consequences of such mismanagement.

IMPACTS OF HERBIVORES

Herbivores affect their communities in two major ways, through trophic and nontrophic pathways (Figure 7.1; see also Schmitz 2008b). First, in their more traditionally considered trophic role, herbivores are predators of plants, influencing individual plants directly by consuming plant tissue. The overwhelming

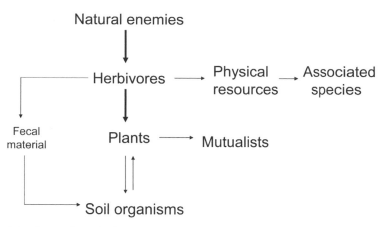

Figure 7.1. Pathways by which the third trophic level (natural enemies) may influence associated organisms in their environment by affecting herbivores. The thick arrows constitute the trophic cascade. The pathway through physical resources to associated species constitutes ecosystem engineering. The pathway through fecal material is further elaborated in Bardgett and Wardle (2003), and that through mutualists in Marquis (2005).

majority of studies have quantified herbivore impact by excluding herbivores rather than by manipulating the third trophic level. Second, herbivores influence other species through nontrophic pathways. Here I consider their impact as ecosystem engineers, changing their physical environment (including the food plant), with the result that resource availability for other organisms is modified. Another pathway not considered here is through deposition of fecal material, which then influences nutrient cycling and breakdown of decayed plant tissue, with attendant effects on plants and soil organisms (Bardgett and Wardle 2003). A third pathway is through effects on associated mutualistic webs (Marquis 2005).

Trophic Impacts of Herbivores as Predators of Plants

In this section I present examples of the diversity of herbivore impacts, from individual plants to plant communities, involving both vertebrates and invertebrate herbivores.

Population Dynamics and Species Range

Invertebrate herbivores, both folivores (Rausher and Feeny 1980; Doak 1992; Ehrlen 1995) and seed predators (Louda and Potvin 1995; Ehrlen 1996; Kelly and Dyer 2002; Chapter 8, this volume), affect plant population dynamics. A classic example of the impact on local population dynamics and distribution is that of the coastal sage shrub *Haplopappus squarrosus* of California (Louda 1982). Higher flower production by coastal plants would predict greater abundance on the coast, but abundance increases from the coast to the interior. This pattern results from geographically variable predispersal seed predation, which is greatest along the coast. Vertebrates also can affect plant population dynamics, but studies are fewer. White-tailed deer (Russell et al. 2001) and mule deer (Binkley et al. 2006) can have devastating effects on vegetation in the absence of large carnivores (Chapter 9, this volume). Rooney and Gross (2003) and Knight (2004) provide some of the few detailed studies of the impact of vertebrates on plant population dynamics.

PLANT SUCCESSION

Herbivores affect plant succession, either speeding it up or slowing it down depending on their preference for earlier or later successional species (Brown and Gange 1992; Davidson 1993). Vertebrates influence tree seedling establishment in herbaceous vegetation (de Steven 1991), whereas insect herbivores influence the establishment of early successional species in primary succession (Fagan and Bishop 2000) and in secondary succession (Brown and Gange 1992). A comparative study of old field succession in the United Kingdom and the United States highlights the impacts of insect herbivores. Insect herbivores slowed succession in the English old field (Hendrix et al. 1988b), whereas they had no effect in the Iowa old field. The latter was composed mostly of European invasives that had arrived without their herbivores (Hendrix et al. 1998a).

Plant species of early succession are of higher quality, resulting in higher rates of insect herbivory (Coley 1983; Hunter and Forkner 1999). Deer, elk (Collins and Urness 1983), and moose (Leptich and Gilbert 1989) often gravitate to early successional vegetation, but preferences vary depending on season, gender, and habitat availability. Hairston (1991) predicted less trophic control in early succession, but a recent test of this hypothesis (Richards and Coley 2008) demonstrated greater predation on caterpillars in light gaps than in adjoining understory.

PLANT SPECIES RICHNESS AND VEGETATION COMPOSITION

Herbivores can increase or decrease plant species richness, depending on herbivore abundance, tradeoffs between plant species in competitive ability and herbivore resistance, and herbivore preference patterns (Lubchenco 1978). In tropical systems, both increases and decreases in plant species richness have been noted when herbivores are excluded (increases: Ickes et al. 2001 [pigs, *Sus scrofa*], Sherman 2002 [land crabs]; decrease: Dirzo and Miranda 1991 [mammals]). These contrasting results may arise because plant species richness is highest at intermediate herbivory levels (Lubchenco 1978), decreasing when moderate herbivore abundance is reduced (Dirzo and Miranda 1991) and increasing when intense local herbivore pressure is eliminated (Ickes et al. 2001; Sherman 2002). Alternatively, impacts may vary with soil nutrients (Proulx and Mazumder 1998) or precipitation (Chase et al. 2000).

Exclusion of both small (*Microtus*: Howe et al. 2006) and large vertebrate herbivores (Serengeti megafauna: McNaughton 1979a) can completely change plant species community composition in 5–7 years, from communities dominated by edible species to those dominated by inedible species, without discernible effects on species number. Herbivores released from top-down control on small islands in Lago Guri, Venezuela transformed closed-canopy forests to treeless scrubs of shrubs and lianas, drastically reducing plant diversity (Terborgh et al. 2006; Chapter 8, this volume). Experiments, both controlled and natural, suggest that the important herbivores in this system are leaf-cutting ants (Terborgh et al. 2006; Lopez and Terborgh 2007).

VEGETATION BOUNDARIES

Shifts in vegetation type are most often attributed to plant responses to changes in abiotic factors. However, herbivores can override the abiotic determinants of vegetation, influencing the local distribution of multiple species simultaneously (Chapter 16, this volume). In today's Africa, experimental exclusion of mega-herbivores and correlation of changes in their abundance with changes in vegetation demonstrate their ability to convert woodland into grassland (Dublin 1995; Sinclair 1995; Augustine and McNaughton 1998; Rutina et al. 2005). The herbivores do not have to be large to have such effects, however. Exclusion of kangaroo rats (*Dipidomys* spp.) transformed desert shrub vegetation into arid grassland after 13 years, mostly as a result of increased abundance of tall grasses through changes in seed predation pressure (Brown and Heske 1990). Fine et al. (2004) found that selective herbivory was responsible for restricting tree species to either clay or white sand soils in Amazonian Peru. Reciprocal transplant experiments revealed that the growth performance of clay soil specialists equaled or exceeded that of white sand species under all conditions, except when they were exposed to herbivores on white sand soils. Clay species are unable to invade neighboring white sand soils because they are vulnerable to insect herbivore attack on those soils. At an even broader landscape scale, the absence of large-hooved herbivores from the steppe regions west of the Rocky Mountains in North America in presettlement times is correlated with the dominance of grazing-intolerant caespitose grasses. East of the Rockies, where bison were abundant through the last 10,000 years, the grasses are rhizomatous and fast growing (Mack and Thompson 1982).

PLANT EVOLUTION

Evolutionary pressures imposed by both vertebrate and invertebrate herbivores can be significant. In the Serengeti where megaherbivores are abundant, plants of the grass *Themada triandra* are short and fast growing, whereas in regions where grazing is low the plants are tall and slow growing. These presumably genotypic differences are maintained in the greenhouse (Hartvigsen and McNaughton 1995). The wild parsnip (*Pastinaca sativa*) and its main herbivore, the parsnip webworm (*Depressaria pastinacella*), were introduced into the United States from Europe. The plant escaped into the wild in about 1630, whereas the insect arrived some 250 years later (Zangerl and Berenbaum 2005). Assays of seeds from herbarium specimens revealed that levels of four of five furanocoumarin types increased significantly not long after the first evidence of occurrence of the webworm in the United States. Thus, herbivore-driven evolution of defenses can occur in a short time.

RELATIVE IMPACTS OF VERTEBRATES AND INVERTEBRATES

When both vertebrate and invertebrate herbivores are manipulated factorially, sometimes the vertebrates have greater effects (Hulme 1994, 1996; Palmisano and Fox 1997; Sessions and Kelly 2001; Gomez and Gonzalez-Megias 2002), and sometimes the invertebrates do (Tscharntke 1997; Prittinen et al. 2003; Goheen et al. 2004; Lopez and Terborgh 2007). Even in the Serengeti, where megaherbivores would seemingly overwhelm any impact by insect herbivores, amounts of tissue eaten by the two classes are essentially equal (Sinclair 1975). We live in a world in which vertebrate carnivore populations are most often greatly reduced, and vertebrate herbivore populations are themselves either greatly reduced (due to habitat destruction and hunting) or extremely high (due to loss of top-down control). Conclusions from meta-analyses about the relative roles of vertebrate and invertebrate herbivory (e.g., Bigger and Marvier 1998; Broitman et al. 2005) must be tempered by the fact that studies rarely take into consideration the status of the vertebrate fauna at a particular study site. Researchers have very different views (e.g., contrast Oksanen and Oksanen 2000 with Schmitz et al. 2000) about the contribution of vertebrates and invertebrate herbivores to trophic cascades. These differences probably arise because the research is conducted in places that differ greatly in how much the vertebrate fauna is intact. To test the HSS hypothesis in a terrestrial system most effectively, one would want to conduct experiments in systems in which the vertebrate carnivore fauna (and, by extension, the vertebrate herbivore fauna) is intact (Krebs et al. 1995; Chapters 8, 10, and 11, this volume).

Nontrophic Effects: Impacts as Ecosystem Engineers

HSS viewed herbivores solely as consumers of plant tissues, but more recently ecologists have come to appreciate that certain herbivores and other organisms can modify their environment as ecosystem engineers (Jones et al. 1997; Buchman et al. 2007). Jones et al. (1997) predicted that engineering impacts would be greatest for species that produced persistent modifications of the environment and those with larger body sizes, because body size would positively correlate with impacts.

The classic example of an herbivore that has a large impact as an engineer is the beaver *Castor canadensis* (Wright et al. 2003). As herbivores, beavers fell trees and saplings to feed on leaves, buds, and twigs, and they do so selectively with respect to plant species. However, as engineers they modify the environment upstream of the dams they construct, turning a stream environment into a pond, marsh, or wetland environment. Associated plants and animals change as a result. Regional species diversity increases because a new habitat type has been added to the landscape. No doubt, the effect of *C. canadensis* as an engineer on its associated community far outweighs its impact as an herbivore. Thus, when wolves, wolverines, bear, lynx, and mountain lions prey on beavers, they may exert indirect effects on many other organisms.

Activities of other vertebrate herbivores have potentially major engineering effects, including digging by porcupines (Shachak et al. 1991; Wilby et al. 2001), pikas (Lai and Smith 2003), pocket gophers (Reichman and Seabloom 2002), and prairie dogs (Bangert and Slobodchikoff 2004; VanNimwegen et al. 2008), trampling by wildebeests (McNaughton et al. 1997), bison (Polley and Collins 1984), seed eating by rodents (Brown and Heske 1990), and rooting and nest building by wild pigs (Ickes et al. 2001).

The quintessential destroyer of vegetation, the elephant *Loxodonta africana*, has major direct impacts as a plant predator, debarking and delimbing trees and reducing seedling survivorship (Laws 1970; Dublin 1995; Chapter 16, this volume). Because of their size and the resultant impact on vegetation, elephants, as ecosystem engineers, also have multiple indirect effects on associated organisms (Pringle et al. 2007b; Pringle 2008). For example, by breaking branches of trees and often splitting them down the center, elephants create crevices that can be colonized by other organisms. The arboreal gecko *Lygodactylus keniensis* preferentially colonizes elephant-damaged trees over nondamaged trees. When crevices in broken branches and stems are experimentally filled, lizards quickly abandon those trees. In the absence of elephant-damaged trees, gecko abundance is much reduced.

In contrast to the role of terrestrial vertebrate herbivores, the impact of invertebrate herbivores as ecosystem engineers has received little attention. By changing soil properties associated with nests, seed-harvesting ants influence local plant community composition on mounds (Wilby et al. 2001). Leaf-cutting ants also affect local soil properties by concentrating organic matter, resulting in higher fine root biomass from surrounding trees (Moutinho et al. 2003). Their nests are also home to lizard and beetle inquilines (Riley et al. 1986), the latter associated with detritus cavities (Moser 1963). Terrestrial insect herbivores often engineer their host plants, thereby affecting the abundance of other arthropods (Marquis and Lill 2006). Constructs made by insect herbivores include silk tents, leaf tents, leaf ties, leaf mines, leaf rolls, galls, and bored holes, and these can affect the diversity and number of associated arthropod species (Lill and Marquis 2003, 2007).

VARIABILITY IN SUSCEPTIBILITY TO THIRD TROPHIC LEVEL CONTROL

The impact of the third tropic level on herbivore numbers, and in turn on plants, will depend on the susceptibility of herbivores to their predators and parasites. Body size, chemical and morphological defense traits (especially in insects), background matching, immune responses, and herbivore behavior all will influence susceptibility to third trophic level attack.

The Importance of Body Size

Extant terrestrial herbivores vary greatly in size, from aphids to elephants, a range in mass of ten orders of magnitude. Body size is important because herbivore selectivity is inversely related, and herbivore appetite positively related, to herbivore size. Importantly, from the perspective of trophic cascades, the number of potential predators an herbivore must evade is a strongly declining function of body size (Sinclair 2003).

A reasonable deduction would be that herbivore impact increases with body size (Crawley 1989; Danell and Bergström 2002). Invertebrate herbivores show little overlap in body size with vertebrates (Table 7.1). Larger herbivores take bigger bites, are more mobile, and tolerate starvation better. However, confounded with body size is degree of specialization. For example, most insect species are specialized, at most able to feed on plants of closely related species, and often restricted to a single species, and then to a single plant individual during the larval stage (many Lepidoptera) (Robinson et al. 2002). This is not to say that broad generalists do not exist in the insect world. Grasshoppers and locusts are known for their broad diets. Even among Neotropical Lepidoptera, for which specialization is the rule, some very broad generalists can be found (Dyer et al. 2007). With a few exceptions such as pandas (bamboo feeders), koalas (eucalypt feeders), and some *Neotoma* woodrats (Boyle and Dearing 2003), most vertebrate herbivores have a broad diet, feeding on plant species from a range of unrelated plant families. Moreover, vertebrate herbivores usually have greater mobility than their invertebrate counterparts. These contrasting characteristics suggest that invertebrates may have greater impacts on individual plants and single plant species than vertebrates, whereas vertebrates might have greater impacts at the level of the plant community.

Large-bodied vertebrate herbivores vary greatly in their relative abundance among current terrestrial ecosystems, with more large species in savannas and wooded savannas, in tropical rather than in temperate climates, and in ecotones rather than in either closed or open habitats (Fritz and Loison 2006). This current pattern is only a shadow of the Pleistocene 20,000 years ago, when megaherbivores were found throughout the globe. During the Pleistocene, at least eighteen extinct genera of mammalian herbivores roamed North America, all of which were 150 kilograms or larger as adults (Martin 1984). Herbivores over 150 kilograms are essentially immune as adults to death by predation (Chapter 15, this volume). Megaherbivores not only were found in North America but were distributed globally; their large body size would have provided immunity from predation as adults. Extinction of these herbivores in eastern Russia is hypothesized to have caused a shift from dry grassland to mossy tundra (Zimov et al. 1995). It is very likely that the Pleistocene mammalian herbivore fauna, much of which was immune to predation as adults, had major impacts on vegetation of that time (Fritz and Loison 2006). Similar effects occurred in Australia, only earlier, about 45,000 years ago (Miller et al. 2005).

Today, the largest species are intermediate grazers and browsers rather than specialists on either grasses or woody plants (Fritz and Loison 2006), although there are exceptions. Rhinos are either browsers or grazers, and hippos are grazers. If indeed impact scales with body size, then the largest of the large megaherbivores would potentially have the greatest effect because they could switch between woody plants and grasses. Left unchecked, megaherbivores can have devastating effects on vegetation, converting woody vegetation to grassland in a few seasons when hunting reduces the predators of smaller species (Ripple and Beschta 2006a, 2006b; Chapter 9, this volume) or when population size increases in the larger, predator-free species (Chapter 16, this volume).

This not to say that insect herbivores have no or little impact because of their small body size. Individual insects can kill seeds and seedlings, especially of smaller plants through stem boring. Woody plants can be killed when multiple insects attack single trees, through multiple defoliations or destruction of the cambial layer. Multiple generations of high populations can destroy large swaths of forests (Powers et al. 1999; Logan and Powell 2001).

Fungus-growing attine ants in the New World (Howard et al. 1988) and termites that harvest green leaf material in Africa are herbivores that are functionally intermediate between invertebrates and vertebrates (Aanen et al. 2002) (Table 7.1). Although they are small in individual body size, colony size can be huge. Furthermore, these species are generalized in their feeding habits, potentially increasing their impacts at the community level.

The Trophic Crunch: Between the Devil and the Deep Blue Sea

One of the most indelible images in the plant–herbivore literature is that of herbivores trapped between the Devil and the deep blue sea (Lawton and McNeil 1979), caught between the poor food quality of the first trophic level and the teeth, mandibles, sucking stylets, or ovipositors of the third trophic level. Herbivores can be poisoned outright or slowly starved to death and eaten whole or piece by piece from the outside, slowly ingested from the inside by parasites and parasitoids (parasitic insects), or quickly digested by enzymes injected for that very purpose. When the HSS article first appeared (1960), it had only been 6 months since Fraenkel's (1959) seminal article outlining the arguments for secondary compounds in plants as adaptations to reduce herbivore attack and feeding intensity. This may explain in part the emphasis of HSS on third trophic level control.

We now know that numerous plant traits can have profound impacts on plant susceptibility to both invertebrate and vertebrate herbivores. These traits include secondary compounds (Rosenthal and Berenbaum 1992); low nutrient concentrations, especially nitrogen (Feeny 1970); and physical factors such as thorns, hairs (Lill et al. 2006), leaf toughness (Coley 1983), architecture (Marquis et al. 2002), and phenology (Aide 1988; Kelly and Sork 2002). Many of these traits, if not all, can be changed (induced) by previous damage, often to the detriment of current and subsequent herbivores (Karban and Baldwin 1997) and associated species (Palmer et al. 2008). Some of these traits may be

selective in their impacts: Thorns, spines, and stinging hairs would be most effective against vertebrate herbivores (Pollard 1992), antifungal agents against
fungus-rearing ants and termites (Howard et al. 1988), and leaf pubescence
against insect herbivores (Lill et al. 2006). In contrast to many physical defenses, a given secondary compound can be effective against both vertebrate
and invertebrate herbivores (Rosenthal and Berenbaum 1992). See Rosenthal
and Berenbaum (1992), Seigler (1998), Danell and Bergström (2002), and
Strauss and Zangerl (2002) for initial entry into the vast literature on terrestrial herbivore–plant secondary compound interactions.

Not all secondary compounds are detrimental: Compounds can be sequestered by larval stages and then passed on to adults, providing protection
against both invertebrate (Rowell-Rahier et al. 1995) and vertebrate predators
(Brower 1988). Sequestration of plant compounds by vertebrate herbivores
does not occur. However, both vertebrates (Pinto da Silveira 1969; Jisaka et al.
1993) and insects (Singer et al. 2004) practice "pharmacophagy," wherein the
selective digestion of certain plants increases defenses against natural enemies.

If noxious plant traits are the Devil, then carnivores, as the natural enemies
of herbivores, represent the deep blue sea. A few herbivorous vertebrates may be
able to fight back (e.g., porcupines), but perhaps the best defenses against natural enemies in vertebrates are group herding behavior and large body size.

Insect herbivores also vary in their susceptibility to predation and parasitism. Birds will not bother with very small caterpillars (Strong et al. 2000), and
most bird species avoid hairy and spiny species (Marquis and Whelan 1994), although there are hairy caterpillar specialists such as yellow- and black-billed
cuckoos (Barber et al. 2008). Insect herbivores also are differentially susceptible
to attack by parasitic insects (parasitoids) (Le Corff et al. 2000). Concealed feeders (leaf miners and shelter builders) often have higher parasitism levels than
free feeders, but shelter builders enjoy some protection against birds (personal
observation) and arthropod predators (Jones et al. 2002).

Role of Herbivore Behavior

Herbivore behavior is important in two ways for understanding the nature of
trophic cascades. Mobility probably affects impacts on individual host plants
and communities, and response to and avoidance of natural enemies will modify the impact of the third trophic level (Chapters 9, 14, and 18, this volume).

Genotypes of some insects, particularly leaf miners, scale insects, and galling
insects, are associated with the same individual food plant because there is little
or no dispersal at any life stage (Mopper and Strauss 1998). But most insects
have a mobile dispersing stage. Locusts and grasshoppers (and at least some

Lepidoptera) disperse over large areas in response to food shortage. In favorable years, locusts can saturate local predators, escaping top-down regulation to explode into huge nomadic swarms (Dempster 2008). Elephants in Africa migrate seasonally with changes in food quality, as affected by rainfall (McNaughton 1985), and several ungulates (bison, elk, caribou, pronghorn) migrated between summer and winter ranges in North America before fencing of the land. Ten species each of terrestrial and arboreal vertebrates migrate in response to multi-annual fruiting cycles (masting) of dipterocarps in Southeast Asia (Curran and Leighton 2000). Fragmented Bornean forests are not extensive enough to satiate migratory bearded pigs that feed on mast, so intensive seed predation has severely curtailed dipterocarp establishment (Curran et al. 1999).

Safety in numbers appears to apply to both vertebrate and invertebrate herbivores. Herding behavior increases the ability of vertebrate herbivores to stave off attack by predators (Caro 2005) and is more common in grazers than browsers (Fritz and Loison 2006). Gregariousness in insect herbivores, sometimes associated with parental care (Cocroft 2002), decreases the likelihood of predation (Hunter 2000). This behavior apparently has consequences for population dynamics, as gregariousness at the larval stage is the only trait that distinguishes tree-feeding species of North American Lepidoptera that go through periodic outbreaks from those that do not (Hunter 1995).

Predators can indirectly affect plants both by reducing the numbers of their herbivore prey, as described earlier, and by changing prey behavior. The behavior of both invertebrates (Schmitz 2008a; Chapter 18, this volume) and vertebrates (Chapters 9 and 14, this volume) can be sufficiently affected by the mere presence of their natural enemies to cause cascading effects on vegetation composition. Absence of the predator would increase herbivore impact on the plant even without an attendant increase in the number of herbivores. This makes understanding the reactions of herbivores to predators critically important because behavioral responses can occur so quickly (Werner and Peacor 2003).

What Makes Trophic Control "Leaky"?

Given the preceding discussion, a number of characteristics may decrease the control of an intact third trophic level. In vertebrates, large body size and gregariousness decrease predator success. In invertebrates, gregariousness, sequestration of secondary compounds or the de novo production of defensive compounds, morphological traits such as hairs and urticating spines, and construction of shelters increase escape from predators. Herbivores may escape trophic control early in succession if they arrive before their predators. Given the high quality of early successional vegetation, however, we would expect that once natural enemies do colonize, their experimental exclusion would show a greater impact on herbivore numbers and the vegetation than later in succession. Climate should also contribute to leaky trophic control. Outside the tropics, ant species richness and abundance are reduced, resulting in lower ant predation on insect herbivores (Novotny et al. 2006). Unpredictability in climate (Stireman et al. 2005) coupled with unpredictability in the timing of leaf production in seasonal environments (Forkner et al. 2008) leads to lower trophic control by parasitoids.

CONCLUSIONS

HSS envisioned a green world, one in which trophic control on terrestrial herbivore communities is strong so that herbivores are only rarely able to inflict noticeable damage on their food plants. Experimental exclusions of predators at the community level are few for terrestrial systems (Krebs et al. 1995; Chapters 8 and 11, this volume), but initial evidence suggests that loss of predators, for whatever reason (overhunting, habitat fragmentation, introduction of competing species, or climate change), would have significant effects on lower trophic levels.

What would a world look like without the third trophic level? Lamentably, imagining such a loss is not just a thought experiment. Our world is changing so quickly that current terrestrial mammal assemblages may seem as different from those in the near future (Ceballos and Ehrlich 2002; Cardillo et al. 2005) as the Pleistocene fauna 12,000–20,000 years ago would appear to us now. In addition to the widespread extirpation of large carnivores (Terborgh 1992; Fritz and Loison 2006), we are seeing declining populations of many insectivorous bird species (Terborgh 1989) and the alteration of native parasitoid (Henneman and Memmott 2001) and arthropod predator assemblages (Landis et al. 2008).

In the absence of vertebrate carnivores, arthropod predators, and parasitoid insects, I would predict that there would be major changes in plant species distribution, community composition, and vegetation structure. We would see a shift to fewer plant species, with those remaining heavily defended by very toxic compounds, tough leaves, digestibility-reducing compounds, heavy pubescence, and spines and thorns. This would be an herbivore's greatest nightmare. Plant species richness would most assuredly decrease. Tradeoffs between plant resistance (few plant species would be resistant against all herbivore species) and plant competitive ability would help determine the actual number of plant species occurring at any one location (Tilman 1997). Heterogeneity in soil nutrients (Tilman 1982), facilitation between plant species (Hacker and Bertness 1999), fire frequency, and topography would be mitigating factors.

Mobile generalist herbivores would contribute the most to de-diversification and outright shifts in vegetation structure. In a world without carnivorous mammals and birds, the impact of small-bodied vertebrate species free from predator control would be added to that of megaherbivores. Mammalian herbivores (grazers, browsers, and seed predators), browsing lizards, and giant tortoises (grazers and browsers) would be affected. The unfettered impact of leaf-cutting ants would be unleashed in the Neotropics, as well as that of green leaf harvesting termites in the Old World, although much less is known of their potential impact. Highly mobile, generalist insect species would also select for a few well-defended plant species. Increased attack by specialist insect species on seeds and seedlings would tend to maintain diversity of the plant species they attack if those insects search in a density- or distance-dependent manner (Janzen 1970; Connell 1971). Thus, increased herbivore pressure by specialist insect species would only add to the competitive advantage of plant species not subject to attack by these kinds of insects.

The greatest changes would occur in aseasonal tropical forests where high parasitism, high ant predation, and high bird predation, in combination with high levels of defense and low host plant density, keep herbivorous insect populations in check. Small-bodied vertebrate herbivores are kept in check by multiple vertebrate carnivores where their populations are still intact. Without predator control, populations of these vertebrate herbivore species would increase with little controlling effect of abiotic factors. Leaf-cutter ants would decimate understories of Neotropical forests (Terborgh et al. 2006). Herbivorous insects would abound throughout the canopy with the loss of ant and avian predators. Plant species that provide food and shelter to ants would be particularly vulnerable. Other species that depend on these plants would disappear as well, putting the web of interacting species (Gilbert 1980) at risk of collapse. These forests probably would come to be dominated by tough-leaved monocots (Dominy et al. 2008; Grubb et al. 2008) interspersed with a few very toxic dicotyledonous plant species.

We would expect the least amount of change in ecosystems where current trophic control is already leaky. Ecosystems with large vertebrate herbivores would change little, remaining grasslands. The most variable climates would change the least with respect to the impact of parasitoids, as parasitism levels are lowest where annual variation in rainfall is the highest (Stireman et al. 2005). Temperate and boreal forests would become more species poor than they already are. Vegetation would become more savanna-like as tree seedling establishment declined and understories were decimated. Even in the far northern reaches of Eurasia and North America, however, vertebrate predators (wolves, foxes, and owls) are sufficiently abundant that their loss would have cascading effects.

An uncertain factor in all of this would be the role of disease, but my intuition suggests that disease would only slow the process of vegetation change, not prevent it. Another process slowing vegetation change would be evolution. Intense herbivore pressure would select for well-defended genotypes, maintaining species locally. However, this response is likely to be sufficiently fast only in short-lived herbaceous species. We have a glimpse of a world without carnivores in the case of the gypsy moth (*Lymantria dispar*) in the absence of its natural predators in United States, where it has defoliated entire forests. Insufficiently regulated white-tailed deer are altering forests throughout eastern North America (Côté et al. 2004; Waller and Rooney 2008; Chapter 9, this volume). At the producer level, exotic plant species that arrive without their co-evolved herbivores can be analogous to superdefended species in a world of only two trophic levels.

Terrestrial herbivores have multiple and extensive effects on the communities in which they live. They do so through trophic effects, as predators of plants. Their nontrophic effects (e.g., as ecosystem engineers) are only just becoming appreciated, and are not yet incorporated into tritrophic theory. Most importantly, only a few experimental and comparative studies have been conducted at the community level to determine the impact of third trophic level on herbivores and, in turn, the impact of those herbivores on their food plants. As a result, we are far from understanding the conditions under which trophic control in terrestrial ecosystems is strong and when it is not so strong. Fleshing out the picture will allow us to understand the nature of population controls in natural systems and how to restore natural communities when they have been disrupted.

ACKNOWLEDGMENTS

I thank Jim Estes and John Terborgh for the opportunity to participate in this venture. I gratefully acknowledge the stimulating discussions with all, especially John Terborgh, Lauri Oksanen, Bob Holt, and one anonymous reviewer. I thank Nick Barber for his comments on an earlier version and Bob Paine for his encouragement.

CHAPTER 8

Propagation of Trophic Cascades via Multiple Pathways in Tropical Forests

John Terborgh and Kenneth Feeley

Textbooks have long pointed to an abundance of mutualisms as a distinguishing feature of tropical forests. A high frequency of mutualisms in the tropics is often attributed to a benign climate and long periods of environmental stability that favor the evolution of elaborate adaptations. In this chapter we develop the thesis that mutualisms, as viewed in the light of trophic cascades, not only set low-latitude forests apart from high-latitude forests but play a central role in organizing responses to perturbations of the trophic system.

Among the most ubiquitous mutualisms of tropical forests are those involved in plant–animal interactions. These include pollination and primary and secondary seed dispersal. Many authors have commented that these interactions are critical to the functioning and integrity of tropical forest ecosystems (Howe and Miriti 2000; Wang and Smith 2002), but the extent to which this is true is still underappreciated. Quite simply, a tropical forest cannot maintain itself without these mutualisms because animals mediate vital processes in plant reproduction (Terborgh et al. 2008).

In contrast, the trees of high-latitude forests are largely free of dependence on animals for reproductive services. Although derived from insect-pollinated early angiosperms, genera such as *Alnus, Betula, Ostrya, Fraxinus, Populus, Salix,* and most conifers are both pollinated and dispersed by the abiotic agency of wind. In tropical forests, few species are dispersed by wind and even fewer are wind pollinated. Among more than a thousand species of trees at our field site

in the Peruvian Amazon, there is only one, a species of *Salix*, that is both pollinated and dispersed by wind. Instead, more than 90 percent of tropical
trees, perhaps more than 99 percent, need animal services to complete their
reproduction.

Other classes of plant–animal interactions are of a negative/positive character, harming the plant but benefiting the animal. Seed predation and herbivory
fall into this category. Although not mutualisms in the usual sense of positive/
positive interactions, these negative/positive interactions also involve links between trophic levels. Seed predation and herbivory are ubiquitous in terrestrial
ecosystems from the Arctic to the tropics. What distinguishes tropical ecosystems from high-latitude ones is the prevalence of positive/positive interactions
linking trophic levels. Both positive/positive and negative/positive interactions
serve as pathways for transmitting trophic perturbations through terrestrial
ecosystems.

We shall now demonstrate this through the elaboration of a case study, one
in which a trophic perturbation results in a great excess of animals having negative interactions with plants and deficiencies of animals having positive interactions with plants. We shall see that an imbalance in the animal community
leads to gross distortions in the pattern of plant recruitment with accompanying changes in species composition and loss of diversity.

CONSUMER RELEASE: THE LAGO GURI ISLANDS

Our case study is that of a mega-experiment, fortuitously initiated in 1986
when the completion of a giant hydroelectric dam flooded the Caroní Valley of
Venezuela to a depth of 170 meters (Morales and Gorzula 1986). Rising water
isolated hundreds of hilltops and turned them into islands. Islands less than
about 20 hectares in area were too small to support predators of vertebrates yet
large enough to support a variety of vertebrate consumers. Many such islands
were far out in the impoundment, as much as 7 kilometers from the mainland.
We noticed that some vertebrates, such as red howler monkeys (*Alouatta senicu-
lus*), were absent from all small, near islands but were present on most far islands
of the same size, suggesting that they would venture to swim only so far to escape. Similarly, we saw the occasional mammal-eating raptor on near islands
but never on a far island (more than 2 kilometers from a large landmass).

The combined effects of area reduction and isolation resulted in major
losses of species diversity and accompanying distortions in the functional organization of the residual fauna. Already in 1990, only 4 years after flooding, the

faunas of islands smaller than 20 hectares had lost more than 75 percent of the vertebrates found in similar dry forest habitat on the nearby mainland. Entirely absent were all predators that normally prey on vertebrates, including jaguar (*Panthera onca*), puma (*Puma concolor*), lesser mammalian predators, and a number of large raptors and snakes. Most species able to persist on small islands increased dramatically into a condition we called hyperabundance. From a functional perspective, the persistent fauna was strongly imbalanced. Some pollinators, most seed dispersers, and large seed predators were missing from most islands. Most persistent vertebrates belonged to one of three functional groups: small generalist rodents, predators of arthropods (birds, lizards, amphibians, spiders), and generalist herbivores (Terborgh et al. 1997a, 1997b; Shahabuddin and Terborgh 1999; Lambert et al. 2003; Rao 2000; Aponte et al. 2003). Prominent among the latter were red howler monkeys, common iguanas (*Iguana iguana*), and leaf-cutter ants (*Atta* spp., *Acromyrmex* sp.).

These many changes in the fauna of the dry forest ecosystem included disruptions and distortions of both mutualisms and top-down interactions between plants and animals, implicating multiple pathways for the propagation of trophic perturbations through the system. For example, small islands lacked social bees, perhaps because the numbers of trees present (300 to 400 in several instances) were not sufficient to support a colony. We did not investigate the effectiveness of persistent insects in pollinating trees, but the absence of an important class of pollinators (social bees) is unlikely to have been without effect (Aizen and Feinsinger 1994; Kearns and Inouye 1997). Small and medium islands also lacked phylostomatid bats and nearly all frugivorous and omnivorous birds, making these islands severely deficient in seed dispersers (Terborgh et al. 1997a, 1997b). Red howler monkeys, the only generalist seed dispersers present on most islands, defecated in communal latrines, concentrating rather than scattering seeds (Stokstad 2004; Feeley 2005). Hyperabundant rodents may have removed most of these seeds, and leaf-cutter ants may have defoliated any seedlings derived from surviving seeds because there was no obvious recruitment of saplings around latrine sites (personal observation). Although altered pollination and seed dispersal provided pathways for a trophic perturbation to cascade to the producer lever, the dominant effect was exerted by hyperabundant herbivores.

HERBIVORY AS A DRIVER OF ECOSYSTEM CHANGE

We investigated herbivory via an experiment that used three classes of islands or landmasses, designated as "small," "medium," and "large." Small islands ($N = 6$)

ranged in size from 0.25 to 2.0 hectares and uniformly lacked armadillos, a predator of leaf-cutter ants. Medium islands were somewhat larger (3–12 hectares, $N = 4$) and uniformly supported armadillos. Large landmasses served as controls ($N = 4$). These consisted of designated sites on two large islands (88 and 190 hectares) and two locations on the nearby mainland. Islands or sites on large landmasses served as replicates. The treatments consisted of distinct levels of herbivore density, for which island size served as a convenient surrogate. Leaf-cutter ants were most abundant on small islands (mean of 4.5 mature colonies per hectare), much less abundant on medium islands (0.20 mature colonies per hectare), and least abundant on large landmasses (less than 0.05 mature colonies per hectare). Howler monkey densities varied inversely with island size in a parallel fashion (Terborgh et al. 2001).

To assess the impacts of herbivory associated with the three size classes of islands, we monitored plant demography and used exclosure treatments (Terborgh et al. 2006; Lopez and Terborgh 2007). Tree saplings of two size classes ("small," at least 1 meter tall and less than 1 centimeter in diameter at breast height, and "large," from 1 and less than 10 centimeters in diameter at breast height) were numbered, mapped, measured, and identified in 225-square-meter (15 × 15 meters) subplots in larger encompassing plots within which all trees at least 10 centimeters in diameter at breast height were similarly enumerated. After 5 years, all stem size classes were recensused (Table 8.1).

Mortality of both sapling size classes was greater on islands supporting hyperabundant herbivores than on large landmasses, but contrasts with controls mostly fell within a factor of 2. Much larger contrasts were observed with re-

Table 8.1. Demography of small and large saplings on small and medium landmasses at Lago Guri, Venezuela, 1997 to 2002. Values given are relative to those observed on the large landmasses that served as controls (modified from Terborgh et al. 2006, Table 2, p. 257).

	Relative Number of Stems/225 m², 1997	Relative Proportion Died, 1997–2002	Relative Proportion Recruited, 1997–2002	Relative Number of Stems/225 m², 2002
Small Sapling Class				
Small islands	0.36	1.53	0.19	0.25
Medium islands	0.79	1.31	0.33	0.64
Large Sapling Class				
Small islands	1.24	2.07	0.32	1.04
Medium islands	1.57	1.60	0.39	1.47

cruitment, which was reduced by more than a factor of 2 in both sapling size classes, especially that of small saplings on small islands. The recruitment rate of these was only 0.19 of the control value. Saplings growing on islands in the presence of hyperabundant herbivores thus experienced both elevated mortality and depressed recruitment, especially on small islands where the densities of leaf-cutter ants were greatest. Mortality exceeded recruitment in the sapling cohorts of more than 90 percent of the woody species present on small islands. In the remaining few species, mortality and recruitment were roughly in balance. No species appeared to be at an advantage under the circumstances. The demographic collapse of sapling cohorts was attributed to the actions of leaf-cutter ants (Figure 8.1). Howler monkeys and iguanas (also present on all islands but not studied because of their cryptic habits) were not suspected because they forage mainly, if not exclusively, in the canopy. An alternative hypothesis, that sapling survival and recruitment on islands were suppressed as a consequence of

Figure 8.1. (Top) Normal dry forest understory on a large (control) landmass. **(Bottom)** Understory of an herbivore-impacted small island.

exposure to desiccating trade winds, was examined in several ways and discarded (Terborgh et al. 2006).

We complemented demographic monitoring with exclosure treatments in which seedlings of six species of common trees were set out under fine wire mesh that was impervious to leaf-cutter ants and most other defoliating arthropods. Seedlings survived significantly better over the 90-day period of the experiment under fine mesh than under 1-centimeter mesh that freely admitted arthropods or under cages with openings that allowed rodents as well as arthropods to enter (Lopez and Terborgh 2007). Again, the implication was that leaf-cutter ants were the principal agent responsible for the low survival of unprotected seedlings (Terborgh et al. 2006; Lopez and Terborgh 2007).

TROPHIC BOUNCEBACK

The trophic cascade on predator-free Lago Guri islands did not stop at the producer level. An unexpectedly strong signal propagated beyond the producer level into the soil and rebounded back up to the predator level. The details are contained in two published papers (Feeley and Terborgh 2005, 2006). We shall provide only a synopsis here.

Small Lago Guri islands harbored up to three generalist herbivores: red howler monkey (some islands only), common iguana (all islands), and leaf-cutter ants (all islands). Iguanas dwell in the canopy, where they are highly cryptic, so we were obliged to ignore them. As ectotherms weighing about 1 kilogram, they should process much less plant matter per individual than a howler monkey weighing 4–5 kilograms. In any case, the roles of the two species in nutrient cycling should be positive and additive. In contrast, the role of the third generalist herbivore, leaf-cutter ants, appears to be negative, for the following reason. Howlers and iguanas digest plant material in situ and deposit urine and feces directly onto the forest floor, whereas leaf-cutter ants harvest plant matter and carry it into underground chambers that can be as much as 5 meters below the surface, below the zone of fine roots (Haines 1983). Thus, the effect of hyperabundant howlers and iguanas should be to accelerate nutrient cycling, whereas the expected effect of leaf-cutter ants will be to sequester nutrients beyond the reach of plants. To investigate this further, we compared processes driven by bottom-up forces on nineteen small islands (0.2–2.4 hectares), ten without and nine with howler monkeys (range, 0.4–8.6 individuals per hectare; all islands supported hyperabundant leaf-cutter ants).

Carbon/nitrogen ratios varied positively ($r^2 = .38$) with howler monkey density (Figure 8.2; Feeley and Terborgh 2005). Tree growth showed an even stronger correlation ($r^2 = .77$) with carbon/nitrogen ratios from the corresponding islands (Figure 8.3). However, the strongest relationship ($r^2 = .82$) was that between tree growth and howler monkey density, with tree growth being on average six times greater on islands supporting howlers. In contrast, growth was very slow on islands lacking howlers, approaching zero in the most extreme cases. These results suggest that howlers, complemented by iguanas, transmitted a bottom-up stimulus to tree growth, presumably via rapid return of foliar nutrients to the soil surface as urine and feces.

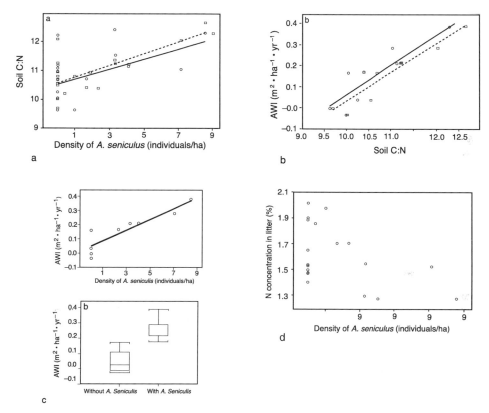

Figure 8.2. Ecosystem-level effects associated with hyperdense populations of the red howler monkey on Lago Guri islands. **(a)** Soil carbon/nitrogen (C:N) ratio as a function of the density of red howler monkeys. **(b)** Annual woody increment (AWI) of trees at least 10 centimeters in diameter at breast height on Lago Guri islands with and without red howler monkeys in relation to soil C:N ratio. **(c)** AWI as a function of the density of red howler monkeys on Lago Guri islands. **(d)** Nitrogen concentration in the leaf litter of islands as a function of the density of red howler monkeys.

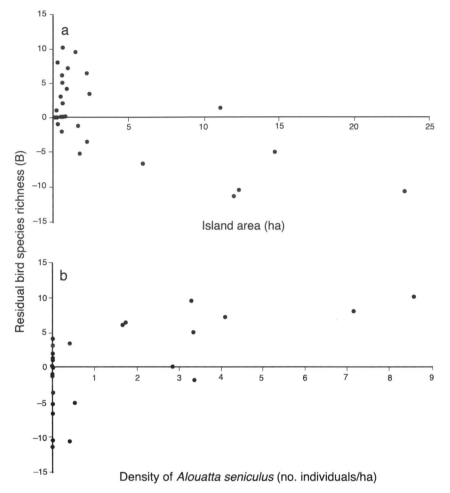

Figure 8.3. The residual bird species richness on islands **(a)** decreases with island area but **(b)** increases in relation to the density of red howler monkeys.

The bottom-up signal generated by howler monkey and iguana folivory propagated beyond producers to two additional levels, primary consumers and predators, as inferred from counts of birds resident on the same islands (Feeley and Terborgh 2006). We analyzed a quantity we called the residual bird species richness, defined as the difference between the observed and expected number of bird species for a given island. The expected species richness was determined by sampling pairs at random from the pooled list of all resident pairs from all twenty-nine islands censused (the pair was the sampling unit because breeding birds were assumed to be paired). For each island, the number of pairs sampled at random was proportional to the area of the island. The number of species in the random sample was counted, and the procedure was iterated 10,000 times. We used the median species richness predicted for each island.

Using linear regression, we determined that residual bird species richness was negatively associated with island area ($r^2 = -.58, p < .005$) and positively related to howler monkey density ($r^2 = .59, p < .001$) (Figure 8.4). The two trends are consistent because howler monkeys occurred at hyperabundant densities only on small islands. Small islands supporting high howler densities harbored surprising numbers of bird species, for example, nine species on a 0.3-hectare island for which the predicted number was two. Most of these birds were obligate insectivores; the rest were omnivores or nectarivores (Feeley and Terborgh 2006). On some islands supporting high howler densities, passerine birds maintained territories in the range of 0.3 to 0.5 hectares, whereas on the nearby mainland the same species typically held territories five to ten times as large (Terborgh et al. 1997b). We attempted to assess prey availability on some of the islands but found the process of sampling small arthropods in the canopy to be too complex, variable, and physically risky to accomplish successfully.

Thus, the pathway of the bottom-up forcing that resulted in such high bird densities could only be inferred. Intimate familiarity with the islands provided some important details. The semideciduous forest of the region leafs out in May, about the time the annual rains begin. Insectivorous birds initiate breeding somewhat earlier. The timing allows adults feeding young to benefit from the peak in arthropod abundance that follows bud break. At herbivore densities prevalent on the mainland, few or no crowns are defoliated and leafing out is

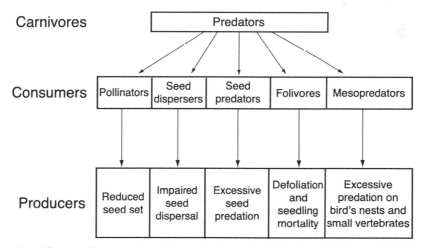

Figure 8.4. The trophic cascade in Neotropical dry forest as suggested by results from Lago Guri, Venezuela.

completed by the end of May. In contrast, on herbivore-impacted small islands, crowns are commonly and repeatedly defoliated, so that recovering crowns continue to initiate new leaves at dates well beyond that at which leafing activity slows on the mainland. Repeated leafing out, presumably abetted by rapid nutrient turnover, could expand the availability of folivorous arthropods, thereby making the affected islands attractive to insectivorous birds and accounting for the observed fivefold to tenfold reduction in territory size.

Obviously, any such scenario warrants further confirmation. Regardless, the remarkable numbers of birds (up to a dozen species) we found breeding on half-hectare islands with howlers and the few (or even none) on similar-sized islands without howlers provide strong prima facie evidence of bottom-up forcing. The link from howlers plus iguanas to tree growth is better substantiated and supports the notion of a rebound of the initial top-down perturbation via nutrient inputs and soil processes (Chapter 12, this volume).

The bottom-up bounceback observed on Guri islands supporting howler monkey and iguana populations is surely a transient phenomenon. Repeated defoliation of canopy crowns by vertebrate herbivores entails a shortcutting of the normal nutrient recycling mechanism by which photosynthate is reprocessed through decomposers. Accessible nutrients in dung and urine are readily leached out of the system, and recycled nutrients are subject to being sequestered via leaf-cutter ant burial. Accelerated recycling therefore cannot persist for long (de Mazancourt and Loreau 2000).

MULTIPLE LINKS AND DIFFUSE NETWORKS

The community-wide responses documented at each level examined (sapling demography, tree growth, avian hyperabundance) demonstrate that the trophic cascade on Lago Guri islands doesn't trickle; rather, it rushes (Halaj and Wise 2001). The species in each trophic level (consumers, producers, etc.) do not respond uniformly, but they do respond mostly in parallel. The dry forest ecosystem thus appears to react to strong trophic signals as a whole rather than as a set of partially independent to independent modules or subsystems (Paine 1980), thereby lending support to theoretical treatments that consider entire trophic levels as coherent units (Hairston et al. 1960; Oksanen et al. 1981). In our work, and in that of Oksanen et al. (1981), we restricted our observations to vertebrates (plus leaf-cutter ants). Most folivorous arthropods exhibit varying levels of host plant specificity, a fact that undoubtedly narrows the community-level impact of trophic cascades mediated by arthropod herbivores (Schmitz et al. 2000).

In the tropical forest there appear to be few strong vertical links between carnivore and consumer populations, such as those that are so prevalent at high latitudes, including lynx–snowshoe hare, snowy owl–lemming, marten–red squirrel, fisher–porcupine, wolf–moose, and wolf–caribou. The closest approximations in the New World tropics would be jaguar with capybara (*Hydrochaeris hydrochaeris*), tapir (*Tapirus terrestris*), and peccaries (*Tayassu* spp.) (Emmons 1987). Whether puma would or could turn to these as major prey species in the absence of jaguar is unknown.

Instead, high diversity ensures redundancy of trophic pathways. Vertebrate predators are legion in the Venezuelan dry forest, including at least five species of felids, one canid, two mustelids, more than a dozen medium to large raptors, several owls, boa constrictor (*Constrictor constrictor*), anaconda (*Eunectes murinus*), tropical rattlesnake (*Crotalus durissus*), lesser serpents, and a large carnivorous lizard (*Tupinambis teguixin*). Forest predators tend to be searchers rather than pursuers, so prey selectivity is low (MacArthur and Pianka 1966; Emmons 1987; Terborgh 1992). This ensures wide overlap in the prey base of species in the carnivore trophic level. The food web is stabilized by multiple and redundant links between levels, so, unlike Paine's (1966) *Pisaster* system or lakes in which largemouth bass is the only piscivore (Carpenter and Kitchell 1993), the dry forest fauna appears to consist mainly, if not entirely, of weak interactors.

Just as high-diversity tropical systems display redundancy in links between the carnivore and consumer trophic levels, so do they display redundant linkages between the consumer and producer levels. As mentioned earlier, some of these links are mutualistic (pollination, seed dispersal), whereas others are of a negative/positive type (seed predation, herbivory). Redundancy is the rule among both mutualistic and predatory links, resulting in broadly overlapping connecting webs for the transmission of trophic cascades. Just as exclusive predator–prey interactions are rare in the tropics, so are exclusive plant–animal mutualisms, best known in the case of orchids and orchid bees (Walter 1983). More commonly, pollination of large plants such as trees is carried out by myriad species of bees, lepidoptera, and other insects, less often by bats, hawkmoths, or hummingbirds (Janzen 1983). Seed dispersal is similarly a diffuse function, being carried out by birds, bats, and both arboreal and terrestrial mammals. Again, exclusivity in disperser–plant relationships is the exception rather than the rule (Gautier-Hion et al. 1985).

Similar conclusions pertain to the negative/positive interactions of seed predation and vertebrate herbivory. Seed predators in the Neotropics range from large mammals (peccaries, agoutis) through a host of smaller mammals (squirrels, rats, mice) to the adults and larvae of invertebrates and fungi (Asquith et al. 1997; Pringle et al. 2007a). Moreover, there is evidence that smaller seed predators can compensate demographically for the absence of larger guild members (Asquith et al. 1997; Wright et al. 2000; Wright 2003). Similarly, herbivory is carried out by a wide range of vertebrates and invertebrates, with a general tendency for larger herbivores to be less selective than small ones (Bell 1971; Owen-Smith 1988; Novotny et al. 2002). Herbivory is especially complex because seedlings and the adult foliage of trees are likely to come under attack from different suites of herbivores as trees grow from the forest floor to the canopy (Smythe 1982).

In Paine's (1966) landmark experiment, exclusion of a keystone top predator resulted in reduced diversity (and somewhat increased biomass) in a space-limited intertidal community. However, in the open nearshore benthic community, release from top-down forcing by sea otters resulted in a massive increase in consumer (sea urchin) densities (Estes and Palmisano 1974; Estes 2005). Sea urchin release resulted in the decimation of macrophyte producers, as postulated by the ecosystem exploitation hypothesis (Oksanen and Oksanen 2000). Whether a drastic reduction of edible plant biomass, similar to that observed in kelp forests, will take place on Lago Guri islands is unknown at present, but collapse of the entire forest of some islands, apparently in response to herbivore pressure, may provide an early glimpse of the future (Terborgh et al. 2006). The conclusion that emerges from these observations is that altered top-down forcing (either decreased or increased) can reduce diversity at the consumer or producer levels, respectively. Species are adapted to live within a circumscribed range of bottom-up and top-down forces. Novel forcing conditions that fall outside the historical range carry the risk of eliminating species from a system.

ALTERNATIVE STATES?

Does the trophic cascade, as manifested by the dry forest ecosystem, exhibit alternative states as aquatic systems do (Paine 1966; Estes and Palmisano 1974; Carpenter and Kitchell 1993)? We shall venture a qualified "yes" to the question. The Guri islands yielded compelling evidence of a drastic change in plant composition in the presence of food-limited herbivores. Tree saplings surviving on *Atta* colony sites about 15 years after inundation comprised a small subset of those present at control sites and exhibited dramatically altered relative abundances (Terborgh 2009). One can only speculate that under the continued press of hyperabundant herbivores, the vegetation would eventually yield to something approaching a steady state composition. If the preliminary indications manifested in our results are interpreted at face value, this steady-state vegetation would consist of a low-diversity assemblage of the most herbivore-resistant plant species in the local flora (Lubchenco 1978). One could consider such a fundamentally altered vegetation to constitute an alternative state (sensu Scheffer et al. 2001).

Such a state would resemble the alternative states exhibited by aquatic systems in that it would be constituted out of the regional species pool via large shifts in the relative abundances of species. Maintained on a large enough scale for long enough, it would probably lead to a raft of local extinctions of both plants and animals. It would differ from the aforementioned aquatic cases in that the perturbation needed to force the state shift involved the exclusion of an entire trophic level (predators of vertebrates and leaf-cutter ants) rather than the manipulation of a single keystone species. Removal of any single predator species from the Guri system, even jaguar, would be unlikely to produce such a dramatic state shift, for reasons discussed earlier. Indeed, small perturbations of the type induced by adding or deleting single weakly interactive species would be expected to generate discernible responses but generally responses limited to one or a few species (Paine 1980; Asquith et al. 1999; Cordeiro and Howe 2003; Silman et al. 2003; Wyatt and Silman 2004). Thus, in principle one could perform myriad manipulations with single species and generate myriad minor changes in the state of an ecosystem, but it would be a stretch to claim that each of these constituted a state change. There is thus likely to be a broad gray zone in which graded perturbations of trophic levels lead to graded responses at other levels. Major state changes, at least in diverse systems incorporating high levels of functional redundancy, may be expected to occur when an entire trophic level, or major parts of it, is severely perturbed.

A particularly familiar case involving perturbation of part of a trophic level is that of eastern continental North America in which wolf (*Canis lupus*) and puma have been widely extirpated, but not lesser carnivores such as black bear (*Ursus americana*), bobcat (*Lynx rufus*), and coyote (*Canus latrans*). The result has been a dramatic overabundance of large herbivores (white-tailed deer [*Odocoileus virginiana*], beaver [*Castor canadensis*]) but not smaller ones (cottontail [*Sylvilagus virginiana*], squirrels [*Sciurius* spp.], smaller rodents) (McShea et al. 1997; Chapter 9, this volume). Similarly, the disappearance of jaguar, puma, and harpy eagle (*Harpia harpyja*) from Barro Colorado Island, Panama has released the medium- to large-sized prey of these species into hyperabundance without noticeably affecting the abundances of smaller consumers, such as small primates, squirrels, and the spiny rat (*Proechimys semispinosus*) (Terborgh 1988, 1992; Glanz 1990). Given the selective persecution of large carnivores over much of the inhabited portion of the planet, grossly perturbed mammal communities have become the norm rather than the exception.

ANTHROPOGENIC ALTERATION OF TROPHIC RELATIONSHIPS

Human interventions into the natural environment unavoidably affect trophic interactions. This is true whether the intervention is physical (logging, mining, contamination, habitat destruction, fragmentation) or biological (hunting, persecution of predators, introduction of exotic species and livestock). Should this be a matter of concern to conservationists? Obviously, yes. Trophic cascades are powerful agents of biological change and appear to be ubiquitous across all the world's major ecosystems. Moreover, large shifts in forcing, whether bottom-up or top-down in polarity, predictably lead to changes in the species composition of the affected communities and often to losses in diversity.

Ecosystems have evolved over millennia to operate within certain ranges of biological and physical conditions. When these conditions are fundamentally altered, change is likely to ensue. To a degree, management can substitute for nature (e.g., expanding seasons and take limits for white-tailed deer), but management can only target one or a few species at most, whereas natural communities are composed of hundreds of species. Nature can manage itself better than humans; it always has. But if nature is to manage itself, it must be provided with the necessary conditions, one of which, as Aldo Leopold so presciently observed, is to retain all the parts. This is a simple message, perhaps too simple for the tastes of some managers, because the essential advice is to keep hands off and let nature have its way.

CONCLUSIONS

Tropical forests are distinct from temperate and, especially, boreal forests in the prevalence of mutualisms, particularly those linking the consumer and producer trophic levels via such processes as pollination and seed dispersal. More than 90 percent of tree species in tropical forests depend on animals for these crucial reproductive services. Mutualisms involved in pollination and seed dispersal are rarely exclusive. More commonly they form diffuse webs involving numerous species at both producer and consumer levels. Diffuse webs create weak species-level links, such that the addition or removal of any given species is likely to result in only minor adjustments elsewhere in the system. A number of authors have concluded that diversity and its concomitant, a prevalence of weak or redundant links between levels, implies that trophic cascades are weak in tropical forests and in high-diversity ecosystems in general (e.g., Strong 1992).

We take strong exception to this conclusion. The trophic cascade can be just as powerful in a tropical forest as in the intertidal zone of Tatoosh Island, Washington; Yellowstone National Park; or the nearshore zone of California and Alaska. The difference is that highly diverse tropical systems lack keystone species analogous to the starfish *Pisaster*, gray wolf, or sea otter. Predators are no less important in the tropical forest; the difference is that there are many of them. Experimentally, this presents a daunting challenge. Fortuitously, the challenge was overcome with the flooding of the Caroní Valley of Venezuela to create 4,300-square-kilometer Lago Guri and hundreds of small landbridge islands. The invariant rule that predators need to forage in much larger areas than their prey created the experiment by ensuring that predators of vertebrates could not survive on Lago Guri islands smaller than about 20 hectares.

We took advantage of this fortuitous experiment to study bottom-up systems created by predator-free islands. Lack of top-down control led to hyperabundance of persistent vertebrate species and some invertebrates. Among the hyperabundant groups were predators of invertebrates (birds, lizards, anurans, spiders), rodents of four genera, and generalist herbivores: howler monkeys, common iguanas, and leaf-cutter ants. Sapling mortality was elevated and sapling recruitment was depressed in the presence of hyperabundant herbivores to the extent that mortality exceeded recruitment in nearly every plant species, signaling a massive shift in the vegetation. Thus, despite the high diversity of the tropical dry forest ecosystem and the occurrence of numerous parallel linkages between the consumer and producer levels, the top-down trophic cascade was powerful and ecosystem-wide in its effects.

Impacts on the producer level conformed to those of other well-documented trophic cascades in that a major shift in community composition was observed, not so much from extinctions and immigrations of new species as from major alterations in the relative abundances of species already present. Our results document transitory states in a process that we expect to continue for decades more, one that will eventually lead to an alternative state bearing little resemblance to the top-down system that prevailed before flooding. Over the interim, the dynamics of change are slow, playing out on the scale of tree lifetimes. Extinctions and immigrations may further modify community composition at a later date, as species vulnerable to herbivory are driven to extinction and replaced by herbivore-resistant or tolerant ones (see Chapter 10, this volume).

The ubiquity of top-down trophic cascades and their potential for generating alternative ecosystem states poses a huge threat to biodiversity because humans have altered the trophic functioning of ecosystems all over the world through persecution of top predators, habitat fragmentation, introduction of alien species, and other means. Although not yet widely recognized for the threat they pose to biodiversity, trophic cascades loom at least as large as climate change in their capacity to destabilize the mechanisms that sustain biodiversity. Of the two threats, climate change and the disruption of trophic relationships, the latter is the more urgent because trophic relationships have been altered for decades or even centuries over much of the earth, whereas climate change has only recently begun to distort ecological processes.

ACKNOWLEDGMENTS

The Guri project could never have succeeded without the participation of numerous people who identified plants, birds, mammals, reptiles, amphibians, butterflies, ants, and more. We warmly thank Electrificación del Caroní (EDELCA), the Venezuelan hydroelectric company that manages the Guri impoundment, for authorizing the project. Within EDELCA, we are especially grateful to Luis Balbas for indispensable support and sage advice. Financial support from the MacArthur Foundation and National Science Foundation (DEB97-09281, DEB01-08107) is gratefully acknowledged.

CHAPTER 9

Large Predators, Deer, and Trophic Cascades in Boreal and Temperate Ecosystems

William J. Ripple, Thomas P. Rooney, and Robert L. Beschta

Historically, humans have modified many boreal and temperate ecosystems by decimating native animal populations and often substituting domesticated stock, thereby influencing food webs and simplifying interactions between species. Large predators, especially, have been subject to worldwide persecution. The profound ecological implications of losing top predators are only now beginning to be understood. Across a variety of environments, predator extirpation can lead via trophic cascades to habitat degradation at multiple trophic levels, species loss, and even ecosystem collapse (Terborgh et al. 1999; Ray et al. 2005).

In the 1800s and early 1900s, gray wolves (*Canis lupus*) and other large predators were the target of widespread eradication efforts throughout much of the United States (Boitani 2003). Extirpation of these predators and subsequent deer irruptions (abrupt population rise) generally occurred earlier in the eastern United States than in the West. The last wolves and cougars (*Puma concolor*) were killed by the 1880s in New York State, and deer subsequently irrupted in the 1890s. By the early 1900s, deer irruptions also occurred in Michigan, Pennsylvania, and Wisconsin (Leopold et al. 1947). In western states, the fate of the wolf was sealed in 1915 when the U.S. Congress authorized eliminating any remaining wolves and other large predators. As part of this program, the U.S. Biological Survey systematically killed wolves, extirpating them from the western United States by the 1930s. Deer irruptions soon followed, with most population increases in the West taking place between 1935 and 1945 (Ripple and Beschta 2005). Herein we refer to cervids collectively as "deer" in the European manner, unless referring to individual cervid species, which we identify by species name.

Concerns over large predator loss became an important scientific issue in the 1920s and 1930s when leading biologists opposed federal predator extermination programs (Dunlap 1988). Charles Elton created the concept of the food pyramid in his seminal 1927 book *Animal Ecology*, and was the first to hypothesize that wolf extirpation would cause widespread increases of ungulate populations to unsustainable levels. Inspired by Elton (1927), Aldo Leopold (1943) was among the first to investigate how the removal of wolves and cougars across large portions of the United States was a precursor to deer irruptions.

Leopold's views on the importance of large predators were influenced by Charles Elton and by trips in 1935 to Germany and Mexico. In the predator-free forests of Germany, he observed extensive plant damage resulting from overabundant deer (Leopold 1936). In contrast, in the Sierra Madre mountains of northern Mexico he found intact ecosystems, representing reference conditions, where he observed healthy relationships between predators, prey, and their plant communities (Leopold 1937). In 1947 (p. 176), Leopold et al. wrote,

> Irruptions are unknown in Mexico, and we know of only two in Canada. Both Canada and Mexico retain wolves or cougars, except in certain settled areas. Since irruptions coincide both in time and space with greatly reduced predation by wolves or cougars, and since they are not known to have occurred in the presence of these predators, there is a strong presumption that over-control of these predators is a predisposing cause.

Leopold argued that predator loss set the stage for deer irruptions, followed by degradation of habitat and eventual reduction in carrying capacity. He supported these views by summarizing reports of irruptions from more than a hundred deer ranges in various portions of the United States where large carnivore populations had been removed or suppressed. From these case studies, Leopold et al. (1947) concluded that irruptions led to overbrowsing and subsequent ecosystem damage, such as reduced biodiversity, habitat loss, and accelerated soil erosion.

To evaluate the Leopold hypothesis, we surveyed a range of studies on trophic cascades related to deer browsing, covering both boreal and temperate ecosystems. In this chapter, we focus on trophic cascades involving relationships between large mammalian carnivores, deer, plants, and ecosystems. First, we describe several predator–prey systems of the boreal forests of North America and Eurasia. Next, we discuss trophic relationships in the temperate zone where predators have been extirpated, with special emphasis on introduced deer, and finally we recount some case studies from western North America. We close with a discussion of how trophic cascades might facilitate ecological restoration and present conclusions emerging from the synthesis of our field studies and the research of others.

BOREAL FORESTS

The boreal region is one of the world's best for examining trophic dynamics, because large carnivores still abound in much of this vast northern realm. Circumpolar in extent, the boreal region encompasses large portions of northern North America, Russia, and Scandinavia. One of the classic studies of top-down versus bottom-up determinants of population dynamics was conducted in the boreal forest. Krebs et al. (1995) used a large-scale replicated experiment to test the response of an herbivore (the snowshoe hare, *Lepus americanus*) to manipulations of food availability and mammalian predators. The principal mammalian predators (lynx [*Lynx rufus*], coyote [*Canus latrans*]) were excluded by a single strand of electrified wire that allowed hares to pass freely. Hare densities doubled in the predator reduction plots but increased elevenfold when food addition was coupled with predator reduction, indicating a strong interaction between the two variables.

Consistent with Hairston, Smith, and Slobodkin's (HSS 1960) green world hypothesis, Crête (1999) found that deer populations in northern North America remained nonirruptive in the presence of wolves. However, south of the wolf range, deer biomass increased with increasing primary productivity, a relationship predicted by the exploitation ecosystems hypothesis (see Chapter 10, this volume). Crête and Dangle (1999) found that deer biomass in portions of North America that deer share with wolves (British Columbia to southern Quebec) was approximately a fifth of that in areas where wolves are absent (maritime provinces), rare, or recolonizing (northwest United States). Evidence is now mounting that an overabundance of deer is imposing unprecedented browsing pressure on plant communities across much of wolf-free North America (McShea et al. 1997; Crête 1999). Crête and Manseau (1996) found moose (*Alces alces*) densities in eastern North America to be seven times as high in a region without wolves (1.9 moose per square kilometer) than in one with wolves (0.27 moose per square kilometer), even though primary productivity was higher in the area with wolves. High moose densities in the absence of predation can have a significant impact on forests and biodiversity. In Sweden, after predator extirpation in the twentieth century, high moose densities have also been a significant source of damage to boreal forests (Hörnberg 2001). For example, Angelstam et al. (2000) compared moose densities in the boreal forests without wolves (Sweden) and in those with wolves (Russian Karelia). Both areas have similar climate, vegetation, and timber harvesting history, but moose densities were five times as high in Sweden (1.3 moose per square kilometer) as in Russia (0.25 moose per square kilometer).

In Isle Royale National Park in Lake Superior, wolves have not prevented periodic moose irruptions and resulting damage to balsam firs (McLaren and Peterson 1994). Moose colonized the island around 1900, with wolves colonizing it 50 years later. The balsam fir component of forests on the island has been reduced from 40 percent in 1848 to less than 5 percent today. The absence of bears on the island could be important, as the coexistence of wolves and bears may prevent the irruptions of ungulate populations (Gasaway et al. 1992; Messier 1994).

Overall, studies of boreal ecosystems have shown that deer irruptions are extremely rare where predator populations remain intact (Gasaway et al. 1983; Flueck 2000; Peterson et al. 2003). Deer irruptions in boreal forests typically occur only after wolf extirpation (Corbett 1995; Peterson et al. 2003).

TEMPERATE ECOSYSTEMS

The temperate region is found in the northern and southern hemisphere. The temperate region's climates are milder, human populations are higher, and predator ranges are more fragmented than in the boreal zone.

Deer Introductions

Researchers investigating why invasive species are so successful often highlight the absence of natural enemies. Within its native range, any species of deer contends with a suite of factors, such as predators, hunters, pathogens, and primary productivity, that can limit population densities. When released in a new region without its coevolved competitors and disturbance regimes, a species can reach unprecedented densities. The "enemy release hypothesis" (Keane and Crawley

2002) appears to accurately characterize such invasions and represent a specific case of trophic cascades.

New Zealand provides a renowned example. Although native deer (and, by extension, native deer predators) were originally absent, since 1800 eight species of deer have been introduced. After subsequent irruptions, New Zealand began aggressively culling them in the 1930s (Caughley 1983). Hunting of wild deer there (and elsewhere) probably prevented outright starvation and subsequent population crashes. Over time, even reduced numbers of deer continued to alter species composition in forests, creating plant communities dominated by both browse-tolerant and unpalatable species (Veblen and Stewart 1982; Stewart and Burrows 1989; Husheer et al. 2003). Even where deer densities have been reduced, regeneration of palatable trees remains limited. In some cases, unpalatable ferns and other plants create an environment inhospitable to seedling establishment and growth of trees (Bellingham and Allan 2003; Coomes et al. 2003). Deer-induced changes have altered New Zealand's forests to such an extent that it may no longer be possible to restore plant community composition to a pre-deer state (Coomes et al. 2003). Similar impacts have been observed on Patagonia's Isla Victoria, where introduced red deer (*Cervus elaphus*) have significantly depressed maqui (*Aristotelia chilensis*) while favoring recruitment of the thorny shrub Darwin barberry (*Berberis darwinii*) and sedges (*Uncinia* spp.) (Veblen et al. 1989).

Black-tailed deer (*Odocoileus hemionus*) have been introduced to several islands in British Columbia's Haida Gwaii (Queen Charlotte) archipelago. Because other islands still lack deer, this archipelago provides an ideal natural laboratory to evaluate deer impacts within these temperate coastal rainforests. Stockton et al. (2005) reported that predator-free islands with deer for more than 50 years have 85 percent less shrub and herbaceous cover and 20–50 percent fewer plant species than deer-free islands. This dramatic transformation has cascaded to birds and invertebrates. Bird densities were greatly diminished on islands with deer, and ground-dwelling invertebrates were reduced by 90 percent (Allombert et al. 2005a, 2005b). Deer influence has been so strong in the archipelago that it upended a central tenet of island biogeography. The smallest islands furthest from the mainland have the greatest species richness, because they are the least likely to have introduced deer (Gaston et al. 2006).

On Quebec's Anticosti Island, introduced white-tailed deer (*Odocoileus virginianus*) have greatly depressed regeneration of balsam fir (*Abies balsamea*) (Potvin et al. 2003) and eliminated nearly every deciduous shrub or seedling in the browse zone (Tremblay et al. 2005). Through excessive browsing of berry-producing shrubs, deer may have triggered the local extinction of the black bear (*Ursus americanus*) (Côté 2005). Arboreal lichens and broken balsam fir twigs that make up winter litterfall increasingly serve as an alternative food source for deer (Tremblay et al. 2005).

Hyperabundant Native Deer

Consistent with HSS, the decimation or extirpation of natural enemies enables deer to thrive, which in turn reshapes plant communities and generates far-reaching effects that ripple through the ecosystem. Even where deer are native, predator reduction or elimination can generate many of the same effects observed with introduced deer.

Sika deer (*Cervus nippon*) is indigenous to eastern Asia, including the Japan archipelago. After the extinction of Japanese gray wolves more than a century ago, sika deer populations greatly increased and now are altering forest composition. On Kinkazan Island for example, densities of more than sixty sika deer per square kilometer have adversely affected Japanese beech (*Fagus crenata*) such that seedlings can now only grow where protected from herbivory (Takatsuki and Gorai 1994). On Honshu Island's Kii Peninsula, overabundant sika deer are favoring the spread of sasa grass (*Sasa nipponica*) and inhibiting tree regeneration (Ando et al. 2006; Itô and Hino 2007). If this process continues, sasa-dominated grasslands could gradually replace forests.

In Britain, the wolf was extirpated south of Scotland in the 1500s, effectively releasing native red and roe deer (*Capreolus capreolus*) from predation. Simultaneously, gamekeepers guarded deer against poachers because game animals were owned by the Crown and commoners had no hunting rights (Munsche 1981). In their study of the New Forest in southern England, Peterken and Tubbs (1965) found a perplexing age structure: Canopy trees all dated to the 1750s, 1850s, and 1930s. It turns out that these periods corresponded to three periods of release from browsing pressure. In the 1750s, oak was planted to provide timbers for the British Navy, and plantations were fenced to keep deer out. An 1851 act of Parliament mandated removal of all deer from royal forests. Lastly, the Great Depression of the 1930s brought increased hunting pressure on deer (Putman 1996). Deer populations again increased in the latter half of the twentieth century, augmented by the introduction and spread of Chinese muntjac deer (*Muntiacus reevesi*) and sika deer (Fuller and Gill 2001). Today, the combined effects of native and exotic deer in British woodlands are depressing numbers of woodland herbs (Kirby 2001) and altering forest structure, whereas deer exclusion results in an increase in density and cover of understory herbs, shrubs, and saplings and leads to a threefold increase in migratory bird densities and a fourfold increase of birds that nest or feed in the shrub layer (Gill and Fuller 2007).

In Scotland, more than 99 percent of the Caledonian pinewoods have been lost (Baines et al. 1994), in part because Scots pine (*Pinus sylvestris*) is failing to regenerate (Palmer et al. 2007). Although pine seedlings are often abundant, they fail to recruit into larger size classes because of excessive browsing by red deer. Watson (1983) notes that recruitment failure dates to the late 1700s, when wolves were extirpated (Nilsen et al. 2007) and the local earl enacted draconian policies to deter deer poachers. Using exclosure experiments in Scotland's native pinewoods, Baines et al. (1994) determined that deer browsing resulted in a fourfold reduction in caterpillar larvae and raised questions about potential cascading effects to insect-feeding birds.

In the eastern United States, the wolf was relentlessly persecuted and effectively extirpated by 1900. These efforts coincided with the felling of the last old-growth forests and hunting of white-tailed deer to very low levels (Whitney 1994). By the mid-twentieth century, forests had regrown and deer populations had recovered to levels that supported recreational hunting in most eastern states (Warren 1991). However, since then deer populations have continued to increase (Ripple and Beschta 2005; Figure 9.1), suggesting that contemporary hunting pressure (combined with other sources of mortality, such as vehicular

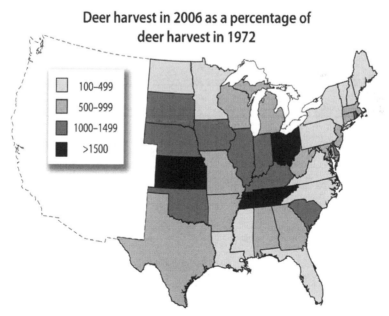

Figure 9.1. Eastern United States showing 2006 deer harvest, by state, relative to harvest levels in 1972. Data from Shead (2005) and unpublished agency deer harvest records.

collisions) has been insufficient to keep irrupting deer populations in check or to preserve biodiversity.

Changes in vegetation over the past century on the Allegheny Plateau in northwestern Pennsylvania provide an important case study of hyperabundant native deer. Old-growth eastern hemlock (*Tsuga canadensis*) and northern hardwood forest once covered this area. Because of its rot-resistant wood, high whole-tree transpiration rates, and influence on soil acidity and moisture, hemlock is a keystone species that contributes to regional biodiversity (Ellison et al. 2005). After wolf and cougar extirpation in the late 1800s, heavily hunted deer populations reached a record low around 1900 (Redding 1995). Deer protection allowed deer densities to exceed sixteen animals per square kilometer by the 1930s. Between 1935 and 1942, deer eliminated most regenerating eastern hemlock and deciduous shrubs from remaining old-growth stands, opening up the forest understory (Hough 1965). Whitney (1984) found that between 1929 and 1978 the recruitment of eastern hemlock saplings more than 30 centimeters tall declined to essentially zero, a decline that coincided with increasing deer densities. The ecological impacts of increased numbers of deer extended beyond eastern hemlock to biodiversity in general. For example, in 1995 Rooney and Dress (1997) replicated a 1929 floristic survey (Lutz 1930) and found a 75 percent decline of shrub and herbaceous species richness in an old-growth stand located in a national scenic area.

Perhaps the most damning findings come from a Pennsylvania study that used large-scale (13- or 26-hectare) deer enclosures. Density control treatments maintained deer at four levels (four, eight, fifteen, and twenty-five deer per square kilometer) for 10 years in replicated plots. Like Krebs et al. (1995) this study is one of the few well-replicated, large-scale, and long-term experiments on terrestrial trophic cascades. Over time, deer reduced sapling density and growth rates while altering species composition in favor of browse-tolerant or -resistant species (Horsley et al. 2003). Furthermore, shrub–nesting birds present at the end of the experiment exhibited the lowest richness and abundance in plots with more than eight deer per square kilometer (de Calesta 1994). Increasing evidence suggests that deer are pushing Allegheny Plateau forests into an alternative stable state, one characterized by a dense cover of regionally common grasses and ferns that inhibits germination and establishment of tree seedlings (Royo and Carson 2006). In the long term, these forests could become similar to bracken grasslands, largely devoid of trees (Stromayer and Warren 1997) and the terrestrial equivalent of sea urchin barrens.

The large predator and deer history of northern Wisconsin parallels that of northwestern Pennsylvania. Predators were extirpated, deer densities declined from intensive hunting, deer received protection, and deer populations irrupted. In 1945, researchers surveyed more than 3,000 plots widely distributed throughout Wisconsin's northern forests to examine potential deer damage to regenerating trees (De Boer 1947; Swift 1948). They found eastern hemlock recruitment only on tribal lands, the only part of the region where deer were subject to greater year-round hunting pressure. Similar trends occurred with northern white cedar (*Thuja occidentalis*), as 70 percent of observed stems exhibited browse damage.

Both exclosure and comparative studies relying on tribal lands as reference areas indicated that the absence of eastern hemlock recruitment in Wisconsin was a consequence of excessive deer browsing (Anderson and Loucks 1979; Anderson and Katz 1993). Similarly, a geographically extensive multivariate analysis revealed deer browsing to be the primary cause of seedling mortality in the 30- to 100-centimeter height class (Rooney et al. 2000). Even if browsing pressure were eliminated, it would take more than 70 years to eliminate the eastern hemlock recruitment gap (Anderson and Katz 1993). Large, long-lived species are vulnerable to deer browsing only for the brief seedling stage of the life cycle, and because they can produce seed for hundreds of years, populations can persist for centuries with no recruitment. Other palatable species with shorter life cycles do not have this advantage.

Researchers have also documented 50-year changes in the forest herb and shrub layers in northern Wisconsin, largely attributable to excessive deer browsing. Sixty-two presettlement-like reference stands showed an 18 percent decline in native species richness between 1949 and 1999, with declines four times higher in state parks and other properties without deer hunting (Rooney et al. 2004). Many of the declining species were known to be susceptible to deer browsing, whereas many of the increasing species were either tolerant or resistant to browsing (Wiegmann and Waller 2006).

Throughout the 1970s and 1980s, wildlife biologists believed that recreational hunting could control deer populations in most places (Warren 1991). They observed excessive deer impacts in some regions, such as northern Wisconsin and Michigan (Figure 9.2) and parts of Pennsylvania, and in state and national parks where deer hunting was prohibited, but at that time they viewed these effects as anomalous. Little research on deer impacts was conducted outside these areas (Russell et al. 2001), so broader generalizations were elusive. However, evidence of widespread deer damage continues to accumulate (Côté et al. 2004). Wildlife managers are increasingly recognizing that control of deer numbers through hunting is limited by declines in hunter numbers, increases in areas off limits to hunting, and a reluctance of hunters to engage in antlerless harvests or other population control strategies (Brown et al. 2000). Indeed, white-tailed deer harvest rates have increased dramatically in the eastern

Figure 9.2. A 16-year-old white-tailed deer exclosure in Vilas County, Wisconsin reveals a recovery of deciduous shrub and tree species inside the fence. Note also the high proportion of grasses and sedges that make up the ground-layer community outside the fence.

United States over the past 30 years (Figure 9.1), even with a decline in the number of deer hunters over the same interval (U.S. Fish and Wildlife Service 2006). In addition, hunting by humans is typically not functionally equivalent to predation by large carnivores due to factors involving risk and the ecology of fear (Berger 2005; Chapter 14, this volume). Human hunters exhibit a Type 1 functional response to deer numbers, whereas large carnivores exhibit Type 2 or Type 3 functional responses (Van Deelen and Etter 2003).

Wolf eradication in the western United States took place a half-century or more later than in the East, not being completed until the 1930s. Although the problem of deer overabundance has often been considered an eastern phenomenon, recent studies have begun to reveal the extent to which predators indirectly structure plant communities and ecosystem processes in the western United States. In the sections that follow, we summarize results of two case studies that took advantage of unplanned landscape-scale natural experiments spanning more than a century. The focus of these studies was to evaluate the status of ecosystems where large mammalian predators had been extirpated or their populations greatly diminished. The first is a wolf–elk (*Cervus elaphus*) system in Yellowstone National Park; the second, a cougar–mule deer (*Odocoileus hemionus*) system in Zion National Park. In both instances, comparisons occurred across temporal (predator removal) and spatial (browsing refugia) treatments. Additionally, these investigations extended beyond the classic three-level trophic cascade of predator–consumer–producer by evaluating the indirect influence of predators on the physical integrity of riparian streambanks and channel morphology. Cascades involving the overbrowsing of streamside vegetation can degrade riparian and aquatic ecosystems and can eventually cause the loss of beaver.

Yellowstone's Wolves and Elk

Long-term trophic linkages have been identified in the northern range of Yellowstone National Park (Wyoming, USA) between wolves, elk, and deciduous tree species—aspen (*Populus tremuloides*) and cottonwood (*Populus* spp.)—via the use of tree ring analyses to establish the age structure of remnant stands. Tree recruitment occurred regularly in the northern elk winter range when wolves were present but became extremely rare after wolf elimination. However, growth of young aspen within exclosures and young cottonwoods within browsing refugia (areas with little browsing) indicated that climate, fire regimes, or other factors probably had little influence on long-term patterns of recruitment (Ripple and Larsen 2000; Beschta 2005).

In an attempt to replicate findings in the northern range studies, Ripple and Beschta (2004b) evaluated the status of wolves, elk, and woody browse species in the upper Gallatin elk winter range in the northwestern corner of Yellowstone National Park. As in the northern range, wolves were present here until about the mid-1920s and generally absent until reintroduction in the mid-1990s. Analysis of historical reports and a chronosequence of photographs along the Gallatin River indicated that willows (*Salix* spp.) eventually began to die after the removal of wolves, apparently because of unimpeded browsing by elk.

More recently, Halofsky and Ripple (2008b) investigated potential trophic cascades by evaluating aspen recruitment during two time periods in the upper Gallatin winter range: an intact tritrophic cascade of wolf–elk–aspen (pre-1930s and post-1995) and a bitrophic cascade of elk–aspen (1930s to 1995). Abundant aspen recruitment occurred throughout the early period with wolves present, followed by a sharp decline in aspen recruitment subsequent to wolf extirpation (Figure 9.3b). In contrast, aspen recruitment occurred continuously within a fenced exclosure after its construction in 1945 (Figure 9.3a), confirming

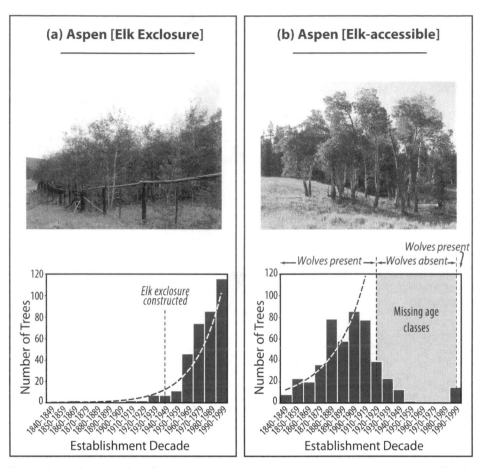

Figure 9.3. Photographs of aspen stands and corresponding age structure: **(a)** inside the Crown Butte elk exclosure and **(b)** outside the exclosure in an area accessible to elk browsing. Both locations are in the upper Gallatin elk winter range, southwest Montana. An exponential function (dashed line) was fitted to all tree frequencies in **(a)** and only to frequencies when wolves were present in **(b)**. "Missing age classes" in **(b)** indicates the difference between expected (exponential function) and observed tree frequencies (bars) after the 1920s, when wolves were absent.

many decades of exclosure studies in Yellowstone's northern range, where, in the absence of elk herbivory, recruitment of woody browse species continued to occur regardless of weather and climatic fluctuations (Singer 1996; Kay 1990, 2001; Barmore 2003). Since wolf reintroduction, young aspen in the Gallatin have been growing taller at some sites perceived as high risk by elk. Such high-risk sites benefit from reduced elk foraging and more plant growth (Ripple and Beschta 2004a; Halofsky and Ripple 2008a; see Chapter 14, this volume).

Cascading effects of predator exclusion may propagate widely through an ecosystem and result in some surprising consequences. In Yellowstone, senescing stands of aspen, unreplenished by recruitment, are replete with snags that support an enhanced abundance and diversity of cavity-nesting birds (Hollenbeck and Ripple 2008). However, as these stands continue to age, currently standing snags will fall and will not be replaced by regenerating stems for many decades. Even if aspen stem regeneration significantly increases after wolf reintroduction (Ripple and Beschta 2007b; Halofsky et al. 2008), cavity-nesting birds will probably decline until new snags arise from currently recruiting young aspen saplings.

Research in Yellowstone National Park is yielding new insights into the ecosystem-wide implications of tritrophic cascades. Moreover, the findings from Yellowstone are consistent with those from other sites in western North America, including several additional national parks: Grand Teton (Berger et al. 2001a), Banff (Hebblewhite et al. 2005), Jasper (Beschta and Ripple 2007b), Wind Cave (Ripple and Beschta 2007a), Rocky Mountain (Binkley 2008), and Olympic (Beschta and Ripple 2008).

Zion's Cougar and Mule Deer

Investigation of a cougar–mule deer system in Zion National Park (Utah, USA) provides additional insights into large predator trophic cascades in terrestrial ecosystems. Areas in the park were designated as "cougars common" or "cougars rare" (Ripple and Beschta 2006a). Spatial proximity of study sites helps minimize potentially confounding variation in climate or other factors.

Cougar have remained common in the roadless North Creek drainage, but since the mid-1930s they have largely avoided areas of high tourist visitation in Zion Canyon, only about 15 kilometers east of the North Creek drainage. Visitor numbers increased dramatically in Zion National Park in the 1920s and 1930s when park biologists noted a reduced frequency of cougar sign in Zion Canyon. Mule deer densities increased from less than four per square kilometer in 1930 to a peak of about thirty per square kilometer in 1942. Since then, the canyon deer population has declined to about ten per square kilometer, still considered high.

The relative strength of top-down trophic cascades in Zion Canyon and North Creek were assessed by evaluating the age structure of Fremont cottonwood (*Populus fremontii*) trees along watercourses (Ripple and Beschta 2006a). In North Creek, with uninterrupted cougar presence, cottonwood recruitment had occurred continuously over time, with more young trees than old (Figure 9.4a), a normal feature of a healthy riparian forest. In contrast, little cottonwood

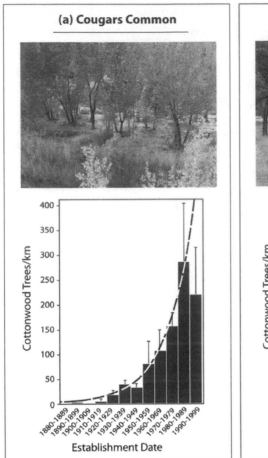
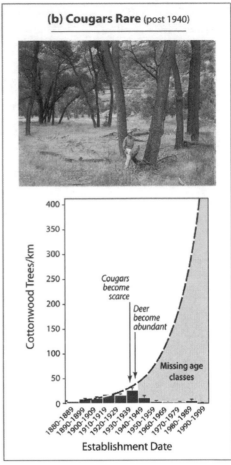

Figure 9.4. Photographs of cottonwoods and corresponding cottonwood age structure for riparian areas in Zion National Park: **(a)** North Creek, where cougars were common (control), and **(b)** Zion Canyon, where they were rare (treatment). The exponential function (dashed line) for tree recruitment cohorts in the control reaches **(a)** was also plotted in **(b)** for comparison after cougar became scarce. "Missing age classes" in **(b)** indicate the difference between expected (exponential function) and observed tree frequencies (bars) after the 1930s, when cougar were scarce. Error bars represent standard error of the means.

recruitment had occurred in Zion Canyon since the 1930s (Figure 9.4b). Post-1940 cottonwoods were forty-seven times more abundant along North Creek than in Zion Canyon.

To assess indirect effects of predator scarcity on biodiversity, Ripple and Beschta (2006a) conducted visual encounter surveys of indicator species on streambanks along North Creek and in Zion Canyon. Markedly higher numbers of wildflowers, amphibians, lizards, and butterflies were tallied along streams in the North Creek area than in Zion Canyon: Amphibians were more than 100 times, lizards three times, and butterflies five times more abundant (Figure 9.5). Although none of the censused wildflower species occurred within belt transects in Zion Canyon, they occurred outside the belt transects

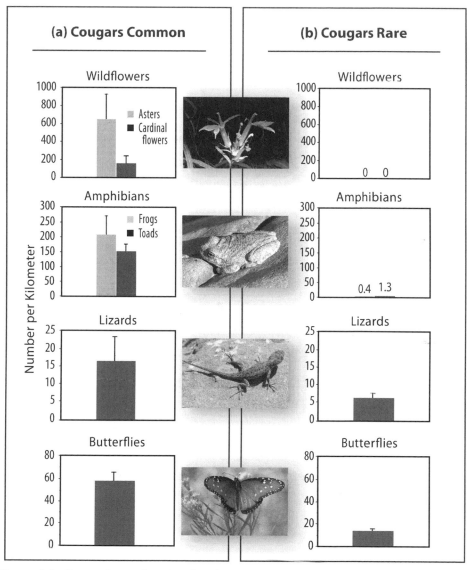

Figure 9.5. Comparison of species abundances in areas where cougars were common **(a)** and rare **(b)** for wildflowers, amphibians, lizards, and butterflies. Both locations are in Zion National Park.

in areas physically protected from deer browsing (i.e., plant refugia). Changes in cottonwood age structure and native species abundance indicate a shift to an alternative, low-diversity ecosystem state in Zion Canyon. This regime shift probably has also altered food web linkages between the aquatic and terrestrial components of the riverine system (Baxter et al. 2005).

Results from Zion National Park generally affirm Leopold's (1943) interpretation of the mule deer irruption on the Kaibab Plateau of Arizona and its effect on plant communities after cougar and wolf eradication. Aspen recruitment on the Kaibab subsequently plummeted, just as cottonwood recruitment nearly ceased in Zion Canyon after cougars vacated the area (Binkley et al. 2006). Findings from Yosemite National Park, another cougar–mule deer system, are consistent with those of both the Kaibab and Zion (Ripple and Beschta 2008).

From Predators to River Beds: Cascades Extend beyond the Producer Level

Investigations of tritrophic cascades normally conclude at the producer level. Yet in the western United States, as we have seen, highly productive and biodiverse riparian corridors are especially targeted by large herbivores when predators are scarce or absent. Destruction of riparian vegetation then cascades to the abiotic environment by exposing denuded streambanks to erosion. Channel widths expand, with consequent reduction of stream depth and altered fish habitat. Such chain reactions have been documented for streams in the upper Gallatin winter range, Wyoming (Beschta and Ripple 2006; Figure 9.6), Zion Canyon, Utah (Ripple and Beschta 2006a), and Olympic National Park, Washington (Beschta and Ripple 2008). Broadening of channel widths or channel incision reduces overbank flooding during spring snowmelt. Floodplains consequently fail to store floodwater during periods of high flow, to the detriment of riparian flora and fauna.

RECOVERY OF ECOSYSTEMS

We have summarized some of the ecological consequences of introducing deer into predator-free environments and compared the effects of predator presence and absence on several native species of deer. In each of these cases, large mammalian herbivores in the absence of strongly interacting predators (Soulé et al. 2005) have been found to have major impacts on vegetation, reducing and skewing tree recruitment in forests, transforming understory herbaceous communities, and denuding riparian vegetation. These profound and potentially devastating effects of large herbivore overabundance are little understood or appreciated by much of the public and many practicing biologists (Pyare and Berger 2003). For example, state cougar management plans in the United States typically overlook the role cougars play in regulating trophic cascades. Similarly, state wolf recovery plans base recovery goals on population size alone and generally ignore the potential biodiversity benefits provided by this keystone predator.

Can ecosystem decline initiated by the loss of a large carnivore be simply reversed if the carnivore is reintroduced to a part of its former range? Where large carnivores are reintroduced into fragmented ecosystems, what role does

Figure 9.6. Streamside plant communities and streambank conditions for riverine systems with low **(a)** and high **(b)** deer densities. The upper two photos are for elk summer range (left) and elk winter range (right) along the upper Gallatin River in Yellowstone National Park; the lower two photos show areas of low (left) and high (right) mule deer densities in Zion National Park.

management of predator and deer populations have on the structure and function of ecosystems? How does continuing human modification of ecosystems affect trophic cascades? And is it possible to apply trophic cascade science effectively to areas outside parks and reserves? These have emerged as key questions to be answered in future research. At present, we can draw only preliminary conclusions based on fragmentary evidence.

Whether ecosystem impacts caused by large carnivore removal can predictably be mitigated or reversed with large carnivore reintroductions remains to be proven, although initial responses to wolf reintroductions in Yellowstone are promising. For example, the diminished stature and density of willows after the extirpation of wolves in the upper Gallatin winter range of Wyoming (Patten 1968; Lovaas 1970; Figure 9.7) began to change when wolves recolonized the area in 1996, after nearly 70 years of absence. Increased heights of young willows began to occur as early as 1999 at sites associated with high predation risk (Ripple and Beschta 2004b; also see Chapter 14, this volume). By 2003, it was difficult to discern a difference between willow height inside and outside an exclosure (Ripple and Beschta 2004b) (Figure 9.7).

Recent studies in Yellowstone's northern range have similarly documented the recovery of woody browse species after wolf reintroduction. Willow, cottonwood, and aspen have begun growing taller in areas of high predation risk. However, heavy browsing by an expanding bison population may be a major factor limiting recovery, especially along main valley bottoms. To date, the release of woody browse species has been spatially patchy and has occurred mostly in riparian zones (Ripple and Beschta 2003, 2004a, 2006b, 2007b; Beyer et al. 2007; Beschta and Ripple 2007a). The extent to which the improving recruitment of these species will continue and perhaps usher in a broader recovery of riparian and upland plant communities is yet to be determined (Fortin et al. 2005).

Earlier we indicated that the removal of carnivores appears to have resulted in hyperabundant deer populations with subsequent impacts on biodiversity and ecosystems. Policymakers can use an understanding of trophic cascades to analyze the pros and cons of carnivore reintroductions, recolonizations, and the management of large carnivores and deer in boreal and temperate ecosystems around the globe. Where large carnivore reintroductions are impractical, forest managers and others will be able to use trophic cascades science within an adaptive management framework. Deer hyperabundance is only one, but perhaps the most obvious, consequence of large predator loss. Others include deer behavioral changes, mesopredator release, negative indirect effects on small vertebrates, changes in the composition or structure of plant communities, altered

Figure 9.7. Willow suppression outside the Snowflake exclosure in 1956 (upper photo) during the mid-1920s to mid-1990s period of wolf absence and in 2003 (lower photo) after the 1995–1996 reintroduction of wolves. The exclosure was constructed in 1945 along the Gallatin River in the upper Gallatin elk winter range. The graph shows decreased browsing (dashed line) and increased mean willow height (bars) outside the exclosure from 1998 to 2002, after wolves were reintroduced.

hydrologic interactions in riverine systems, and facilitation of the invasion of exotic plants. In the face of so many deleterious consequences of predator loss, there is no simple solution. Simply liberalizing hunting regulations will not suffice to mitigate or reverse so many adverse trends, including the loss of natural predator–prey behavioral dynamics (Berger 2005; Chapter 14, this volume).

CONCLUSIONS

Strong top-down forcing by apex predators appears to be widespread, if not universal, in intact boreal and temperate ecosystems. This is the consistent conclusion of numerous accounts from Asia, Oceania, Europe, and North America. Like Aldo Leopold more than 60 years ago, we found no evidence that deer populations ever irrupt in the presence of intact wolf and bear populations (i.e., a tritrophic cascade). Wherever deer have been released from historic predation pressure (truncated trophic cascade), we observe the same general pattern: high population densities and loss of predation risk accompanied by intensive browsing that begins to shift plant community composition toward domination by browse-tolerant or -resistant plants (Figure 9.8a). Even after an irruptive peak, persistently high foraging pressure continues to drive changes in plant community composition. Over time, the ecological integrity of terrestrial eco-

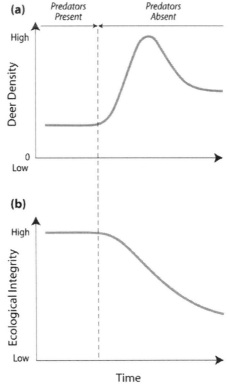

Figure 9.8. Numerical response of deer after the removal of large carnivores **(a)** and the consequent loss of ecological integrity with increased browsing pressure **(b)**.

systems (Woodley et al. 1993), including native plant communities and wildlife species dependent on them, is invariably affected (Figure 9.8b).

In forested ecosystems, declining species diversity first becomes apparent in understory vegetation, as less palatable ferns or grasses crowd out more palatable herbaceous dicots and inhibit the establishment of tree seedlings. Suppressed tree recruitment opens gaps in the age structure of palatable shrub and tree species, possibly hindering recovery, even when a natural predation regime is restored. With continued intensive browsing, plant communities shift toward an alternative stable state composed of species resistant or resilient to herbivore pressure (see Chapter 10, this volume). Decreased standing biomass and decreased diversity of native species, in turn, may trigger additional ecosystem effects, such as increased soil and streambank erosion, altered terrestrial and aquatic food webs, and changes in nutrient cycling (see Chapter 12, this volume). Thus, the removal or significant reduction of large predators sets in motion a chain of events that initiates a downward spiral toward ecosystem simplification. Although complete loss of forest ecosystems is a possible though still unproven endpoint (Chapter 8, this volume), preventing or reversing the impacts of large herbivores so that diverse native plant communities and dependent wildlife can thrive represents a daunting but pressing challenge for the scientific community and society as a whole.

ACKNOWLEDGMENTS

We thank C. Eisenberg, J. Hollenbeck, J. Terborgh, and two anonymous reviewers for comments and suggestions on an earlier draft of this manuscript.

CHAPTER 10

Islands as Tests of the Green World Hypothesis

*Lauri Oksanen, Tarja Oksanen, Jonas Dahlgren, Peter Hambäck,
Per Ekerholm, Åsa Lindgren, and Johan Olofsson*

The green world hypothesis (GWH) of Hairston et al. (1960) proposed that the collective regulatory action of predators prevents herbivores from depleting terrestrial vegetation, that is, that terrestrial ecosystems form three-level trophic cascades. In principle, GWH could be tested by maintaining predator–proof exclosures until a system reached asymptotic dynamics (Slobodkin et al. 1967). So far, however, such experiments have been almost limited to invertebrates (Schmitz 2006). Experiments testing GWH in vertebrate food webs are technically challenging (Chapters 18 and 19, this volume). For large vertebrates, minimum replicate sizes are measured in square kilometers. For small vertebrates, the problems of excluding all predators are formidable (Ekerholm et al. 2004). In most experiments on top-down regulation in vertebrate food webs, the focus has thus been on predation per se (Sinclair et al. 2000; Sundell 2006), not on the indirect effects of predation on vegetation. The few exceptions consist of experiments conducted in semiagricultural habitats (Klemola et al. 2000; Norrdahl et al. 2002), making inferences to species–rich natural communities unclear (Polis and Strong 1996; Halaj and Wise 2001).

Problems of temporal scale are equally formidable. An expanding herbivore population will start by consuming preferred forage plants. Unfortunately, the time frames typical for grant-financed experiments are normally insufficient for getting beyond such trivial transient dynamics. The results thus obtained do not tell what happens when the tastiest plants have been depleted. Do the herbivores then decline, obtaining asymptotic dynamics in a system where inedible plants prevail and sparse herbivore populations interact with equally sparse populations of palatable plants? Or do the herbivore populations keep on growing until the original vegetation has been replaced by a grazing lawn of palatable species (cf. Chapter 16, this volume)? Or does the outcome depend on the characteristics of the herbivore or the physical environment? Because of the time scale problem, exclosure experiments conducted so far leave these questions open.

Being isolated by natural barriers, islands provide opportunities for long-term, large-scale experimental tests of GWH and may also generate spontaneous natural experiments. Because of their low population densities, predators are especially prone to random extinctions (MacArthur and Wilson 1967), but on most marine coasts this is compensated for by abundant subsidies provided by marine-based resources (Polis and Hurd 1996). Before the invasion of the alien American mink, the land-lift archipelagos flanking the coasts of Finland and Sweden provided an exception as an area with no native predators capable of crossing water barriers and switching between terrestrial and marine resources. These archipelagos have supported active colonization–extinction dynamics, where many islands have at least periodically harbored small herbivorous mammals but not their predators. Consequently *Microtus* voles have had dramatic outbreaks on the outer skerries, influencing the vegetation, and these dynamics have recently been restored in mink removal areas (Pokki 1981; Fey et al. 2009). On larger islands, dense populations of snow hares (*Lepus timidus*) have imposed dramatic changes on the vegetation (Ottoson 1971). Unfortunately, these natural experiments have been short lived, because the limited degree of isolation has not completely blocked between-island movements of herbivores (Pokki 1981) and invasions of predators (Häkkinen and Jokinen 1974).

Corresponding experimental situations on larger spatial and longer temporal scales have arisen as consequences of herbivore introductions to oceanic islands (Carlquist 1974; Chapter 9, this volume), but complicating the interpretation of these natural experiments as tests of the GWH is that plants of oceanic islands have evolved in the absence of herbivorous vertebrates (Skottsberg 1922; Carlquist 1974). Introductions of deer to islands off the coasts of Canada (Chapter 9, this volume) share the instructive features of the spontaneous dynamics on the Finnish and Swedish land-lift archipelagos and herbivore introductions to oceanic islands. Many Canadian islands are isolated enough to generate stable long-term experimental conditions but nevertheless share the flora of the adjacent mainland. The introduction performed on Anticosti Island, Quebec provides an especially instructive case, because of its long history and the detailed studies conducted on this island. We shall expand on this case in this chapter.

Predation-free island populations of herbivorous vertebrates have also been generated during glacial–interglacial cycles, when the sea surface periodically regressed more than 100 meters below its current level. During interglacial periods with high sea levels, many landbridge islands thus emerged, especially in the Mediterranean basin and in the East Indian Sea between Asia and Australia (Kurtén 1969, 1971). On these landbridge islands, big herbivorous mammals rapidly evolved toward smaller size (Kurtén 1969, 1971; Lister 1989, 1993), indicating resource limitation (Damuth 1993). In the absence of more detailed data, we could only speculate that this development might be a consequence of lack of predation, in concordance with equivalent modern processes (Chapter 8, this volume; Terborgh et al. 2001, 2006; Anouk Simard et al. 2008a). Corresponding changes in food web dynamics have also occurred in the western Indian Ocean, where big islands existed during glacial maxima in areas currently harboring only submerged banks and tiny keys (Stoddart 1971; Masson 1984). During low sea level stages, dispersal distances were thus reduced, and tortoises were able to colonize a number of islands. Predators sometimes followed but went extinct when the islands became smaller and more isolated. The absence of predators persists on the Aldabra Atoll, and its consequences have been thoroughly studied, as will be detailed in this chapter.

Aware of these cases, we realized that the interactions between vertebrate herbivores and vegetation normally play out on extended time scales of decades or centuries. Recognizing this reality, our research group chose to study gray-sided voles (*Myodes rufocanus*) in low arctic scrubland within the freshwater archipelago of Iešjávri, Norwegian Lapland. Gray-sided voles are the dominant herbivore of the Fennoscandian inland tundra (Ekerholm et al. 2001) and are regulated by predators in corresponding mainland habitats (Ekerholm et al. 2004). Voles reproduce rapidly, and the foliage of shrubs cannot escape in height, as can the foliage of forest trees. It was thus reasonable to expect vole–vegetation dynamics to reach an asymptotic state within a feasible time interval. Moreover, gray-sided voles are reluctant to cross water barriers or stretches of open ice. Being in northernmost Fennoscandia, our study area was outside the migration routes of avian predators, and the low, treeless islands were practically invisible for mammalian predators. The system was thus suitable for conducting controlled long-term experiments.

We shall now examine these three cases in depth. Each involves a generalist herbivore isolated on an island or islands in the absence of predators. The Aldabra Atoll and its giant tortoises (tropical) will be first; Anticosti Island,

Quebec (temperate) will follow; and the tiny, experimentally manipulated islands of Iešjávri in arctic Norway will be last. These three cases, in our judgment, allow at least a tentative assessment of asymptotic dynamics of herbivorous vertebrates and plants in a predator-free world.

NATURAL EXPERIMENTS: WESTERN INDIAN OCEAN ISLANDS

During the latest ice age, all major islands of the western Indian Ocean were colonized by herbivorous giant tortoises (*Dipsochelys dussumieri*) originating from Madagascar (Braithwaite et al. 1973; Arnold 1976, 1979; Stoddart and Peake 1979; Palkovacs et al. 2002; Austin et al. 2003; Gerlach 2004). On the isolated Mascarenes, these tortoises evolved new morphological traits that suggested adaptations to resource limitation in a predator-free environment (Arnold 1979), and trees developed an odd variant of heterophylly, having narrow or finely lobed leaves as seedlings but broad, entire leaves when reaching full size. This pattern is opposite from the one expected on the basis of optimal light use (Horn 1971) and would thus be detrimental for seedlings growing in a closed forest. On the other hand, this seedling morphology reduces the risk of severe browsing by giant tortoises (Eskildsen et al. 2004), implying that, before their extinction, giant tortoises maintained savanna-like, semiopen vegetation on the Mascarenes (Terborgh 2009).

On the big islands of the Comores–Seychelles chain, giant tortoises co-existed with crocodiles, which still survive on the Comores and were only recently extirpated from the granite Seychelles (Taylor et al. 1979; Gerlach and Canning 1993; Brochu 2007). On Aldabra, the largest atoll between the Comores and the granite Seychelles, crocodiles had gone extinct by the time the island was discovered (Braithwaite et al. 1973; Arnold 1976; Gerlach and Canning 1993). Nevertheless, the giant tortoises had retained their antipredator adaptations (carapaces with small openings, thick carapace plates; Arnold 1979), remaining so similar to their Madagascan ancestors and to the probably extirpated tortoises of the granite Seychelles that Arnold (1979; see also Palkovacs et al. 2003) regards them as conspecifics. No heterophyllous trees of the Mascarenean type have been reported from the Comores–Seychelles chain. However, Aldabra harbors a set of dwarfish "tortoise turf" herbs, adapted to chronically intense tortoise herbivory (Renvoize 1971; Merton et al. 1976; Gibson and Phillipson 1983).

For us, the most plausible interpretation of these facts is as follows. By preying on juvenile tortoises, crocodiles regulated tortoises to the extent that they did not deplete the vegetation. The extinction of crocodiles from Aldabra triggered an explosion of the tortoise population, resulting in intense herbivory and a shift from woody to herbaceous vegetation. This change was recent, as compared with the long generation times of trees (more than 100 years) and giant tortoises (more than 30 years; see Grubb 1971; Swingland and Coe 1979), but ancient when measured in generations of herbaceous plants. The dynamics probably changed about 4,000 years ago, when the central basin of Aldabra was flooded and the former island of 400 square kilometers was reduced to an atoll archipelago with total land area of 155 square kilometers (Braithwaite et al. 1973; Arnold 1976; Taylor et al. 1979).

The discovery of Indian Ocean islands by Europeans was followed by the decimation of giant tortoises on Aldabra and by the extirpation of other tortoise populations (Arnold 1979; Stoddart and Peake 1979). The recovery of giant tortoises on Aldabra, thanks to fortunate circumstances (Stoddart 1971), provided an opportunity to recapitulate the switch from a predation-controlled state to a food-limited state and to observe the difference between transient and asymptotic dynamics. In 1967–1968, when the first thorough studies on Aldabra were conducted, the impact of giant tortoises was still selective. Grubb (1971) reported that 53 percent of woody plants and tall herbs of the island were seldom or never eaten by tortoises, because of their low palatability or unsuitable habit. Stoddart (1968) found that in coastal grasslands, tortoises ate the palatable, rhizomatous *Sporobolus virginicus* but avoided the spiny, tussock-forming *Sclerodactylon macrostachym*. He thus predicted that the former species would be replaced by the latter.

During and after the 1970s, the tortoise population stabilized at about 130,000 (eight tortoises per hectare) and even slightly declined (Grubb 1971; Bourn et al. 1999). At the same time, the impacts of herbivory increased and changed character. Inland, woody vegetation was replaced by the prostrate "tortoise turf" (Merton et al. 1976). In coastal habitats, shrubs became replaced by grasses, especially by the palatable *Sporobolus virginicus*, whereas the spiny *S. macrostachym* declined (Hnatiuk et al. 1976). Experimental exclusion of tortoises promptly resulted in reinvasion by taller plants (Gibson et al. 1983), and on the smaller islands, where tortoises were scarce or absent due to history or lack of freshwater, woodlands and scrublands persisted, excluding other potential explanations for the observed changes.

In his famous article "Are Trophic Cascades All Wet?" Strong (1992: 751) denies the significance of this natural experiment by describing it as "the exception that confirms the rule." Aldabra is characterized as "the most distant island on earth." The evolution of its giant tortoises is described as a local

process in a predation-free environment: "Neither natural predators of adult giant tortoises, like large cats and canids, nor those of tortoise eggs and young, like
snakes, lizards and rodents, occur on distant islands where giant tortoises have
evolved." As for the potential impact of predators, Strong (1992: 751) argues as
follows: "Rather than creating a whole trophic cascade, complete with predators, introduction of carnivores typical for terra firma would probably cause the
extinction of the giant tortoises. These keystone herbivores and their potential
trophic cascades are long gone from all but isolated areas; they have been eliminated by potent predators."

In reality, the giant tortoises represent an ancient Gondwana lineage (Gerlach 2004). Their big size increased buoyancy and thus the likelihood of surviving oceanic transits (Arnold 1976; Pritchard 1996). Before the evolution and
dispersal of hominids, whose hunting techniques turned out to be fatally efficient against the passive defenses of these lumbering behemoths, giant tortoises
coexisted with predators in the southern hemisphere and successfully invaded
Eurasia and North America when these continents came in contact with pieces
of Gondwana (Hooijer 1974; Caccone et al. 1999; Martin and Steadman 1999;
Edmeades 2006). On Madagascar, the ancestors of *D. dussumieri* coexisted with
giant crocodiles (Brochu 2007) and other currently extinct big predators and
omnivores (Burney et al. 2003). Even today Aldabra harbors predators of tortoise eggs and juveniles (land crabs, flightless rails, herons, ibises, feral cats, and
rats; see Stoddart 1971; Bourn and Coe 1979; Swingland and Coe 1979; White
and Wiseman 2002), in contrast with Strong's (1992) arguments.

Strong (1992: 751) also claims that "the case of Aldabran giant tortoises
gives hints of the historical developments at the opposite end of the food web,
the plants, that probably also thwart and hinder runaway consumption characteristic of trophic cascades. I would argue that were Aldabra to receive much of
the rich mainland flora, colonists would be likely to include unpalatable plants,
so defended, protected or just so different, that the tortoises could not graze
them down. Competition from these plants would then spell the demise of the
sward and of these aberrantly effective herbivores." This idea might be sound
if Aldabra were extremely distant and species poor (see Skottsberg 1922;
Carlquist 1974; Coblentz 1978), but botanists working on Aldabra have all
along been impressed by the richness of its flora, which has primarily Afro-
Madagascan affinities (Renvoize 1971, 1975). The majority of the woody species recorded from similar raised coral habitats on the Kenyan coast are also
present on Aldabra (Gibson and Phillipson 1983), including *Euphorbia pyrifolia*,
whose fresh leaves are so noxious that even its specialized insect pest shows a
strong preference for the less toxic senescent leaves (Newbery 1980). The rich

mainland flora with its inedible plants is thus well represented on Aldabra, but this has not prevented runaway consumption. Moreover, giant tortoises are anything but aberrantly efficient. They have inefficient digestive systems (Hnatiuk 1978) and are therefore vulnerable to mammalian competitors (Cherfas 1973; Campbell et al. 2004; Hamann 2004), providing a conservative estimate of the strength of trophic cascades in vertebrate food chains.

DEER–VEGETATION DYNAMICS ON THE ANTICOSTI ISLAND, QUEBEC, CANADA

Another thoroughly studied island system lacking natural predators and approaching the asymptotic state is Anticosti Island (area 7,950 square kilometers) in the Gulf of St. Lawrence. White-tailed deer (*Odocoileus virginianus*) were introduced to the previously deer-free island about 100 years ago. The island belongs to the southern subzone of the North American taiga, where disturbed habitats grow birch and aspen. Balsam fir is the dominating mid-successional tree, and white spruce is the regional climax species. Deciduous shrubs and trees form the normal winter diet of the white-tailed deer. Balsam fir and white spruce are at the bottom of the preference ladder, the latter being characterized as the "best of a bad choice" (Sauvé and Côté 2007).

The white-tailed deer population probably peaked around 1930 (Potvin et al. 2003; Anouk Simard et al. 2008a) and thereafter stabilized at a high level, currently numbering approximately 125,000 (sixteen deer per square kilometer on average, thirty deer per square kilometer in preferred habitats; see Cabascon and Pothier 2007). The sustained high deer densities have changed the vegetation. In clear-cuts, the previously dominant herbs and shrubs have been replaced by a scanty grass cover (Tremblay et al. 2005; Cabascon and Pothier 2008). Preferred deciduous trees have been almost eliminated. Recently, the regeneration of balsam fir has been failing because of high browsing pressure. Consequently, the share of white spruce in the winter diet of the deer has increased from 10 percent to 20 percent in the latest two decades (Potvin et al. 1997; Anouk Simard et al. 2008b).

The selective elimination of preferred forage plants has been accompanied by changes in characteristics of the deer. During one century of insular existence, mean body weight has declined by 50 percent (Anouk Simard et al. 2008a), the age at first reproduction has increased, and twinning rates have declined (Anouk Simard et al. 2008b). However, the winter survival rate has remained stable (.77 for adult does, .60 for fawns).

Potvin et al. (2003) argued that the current high deer densities would lead to replacement of balsam fir by white spruce, with the endpoint being an almost pure white spruce forest where deer have crashed and stabilized at low density. This scenario is supported by the indications of resource limitation in deer, summarized earlier. However, Taillon et al. (2006) showed that the deer of Anticosti can increase the share of white spruce in their winter diet to 40 percent (the highest level used in the experiment) without survival costs. Sauvé and Côté (2007) found only small differences between the forage value of balsam fir and white spruce. The ability of the Anticosti deer to survive while exploiting white spruce thus remains to be seen.

When projecting current preferences into the future, it is useful to know that fir and spruce react to recurrent browsing by becoming shrubbier and more palatable because of the breakdown of apical dominance and the consequent development of secondary leader shoots (Welch et al. 1991; Chouinard and Filion 2005). With increasing browsing pressure on white spruce, an increasing fraction of spruce saplings will probably be maintained as palatable "bonsais," as has happened to Sitka spruce, which forms the staple winter forage for red and roe deer in reforested parts of Scotland (Welch et al. 1991). Also, evidence from deer–vegetation dynamics in other predation-free areas (Chapter 9, this volume) helps in making educated guesses about the future of Anticosti. In the western United States, the absence of big predators has led to the thinning of forests at the arid timberline by cervid browsing and to the expansion of steppes. Even the hydrology of rivers has been influenced: Deer have denuded riverbanks and exposed them to erosion (Beschta 2003; Ripple and Beschta 2003, 2006a, 2006b, 2007a, 2007b; Ripple et al. Chapter 9, this volume). In the subalpine forests of Japan, dense sika deer populations are stopping the regeneration of the taiga and converting it to grassland (Ando et al. 2006). In Scotland, wolf extirpation and the consequent increase in red deer density has probably contributed to the large-scale replacement of Scots pine forests by heathlands (Palmer et al. 2007). The capacity of many cervids to reduce the community-level plant biomass and to survive on low-quality forage, though at the cost of reduced body size and reproductive output, is also illustrated by wild and semi-domesticated reindeer (Skogland 1990; Crête and Doucet 1998; Olofsson et al. 2001; den Herder et al. 2006; Bråthen et al. 2007). On the basis of the evidence summarized earlier, we regard it as likely that the asymptotic state of Anticosti will be a bonsai spruce chaparral, exploited by a dense but slowly reproducing population of dwarfish deer, but the scenario proposed by Potvin et al. (2003) also remains plausible.

VOLE VEGETATION DYNAMICS ON EXPERIMENTAL ISLANDS OF IEŠJÁVRI, NORWEGIAN LAPLAND

Our own studies have been carried out in low arctic scrublands on the freshwater archipelago of Iešjávri—"The Lake Itself"—in Norwegian Lapland. This system provided an opportunity for controlled experiments, because the most isolated islands are too far to be exploited by breeding avian predators and the area is far from main migration routes. Moreover, the low, treeless islands of Iešjávri are difficult to distinguish from the opposite shoreline even in summer. In winter, they look like another heap of windblown snow.

In 1991, we introduced gray-sided voles to previously vole-free islands in the middle of the lake and compared the development of the vegetation on these "two-trophic-level islands" to that in mainland areas and in vole exclosures. In the first 4 years, the introduced voles dramatically reduced the cover of the dominant shrub (dwarf birch, *Betula nana*), the dominant edible dwarf shrub (bilberry, *Vaccinium myrtillus*), and the dominant unpalatable ericoid (northern crowberry, *Empetrum nigrum* ssp. *hermaphroditum*) (Hambäck et al. 2004).

Our initial design had its weaknesses. Absence of browsing in the past might have influenced the quality of the plants. (The most isolated islands had only occasionally been invaded by lemmings; Oksanen and Oksanen 1981). Moreover, mainland–island comparisons might have been biased by the scanty snow cover of the wind-exposed islands, resulting in premature snowmelt and stressful spring conditions. Also, fences could influence physical conditions (wind velocity, snow accumulation). To control for these potential sources of bias, we repeated the experiment in 2000, working exclusively on islands and using blocks of scrubland vegetation transplanted from a big "three-trophic-level island" (where stoats had been trapped and breeding avian predators had been observed). Blocks of intact vegetation were translocated to four "two-trophic-level islands" with introduced gray-sided voles but without resident or breeding predators, to four vole-free "one-trophic-level islands," and back to the three-trophic-level island, thus establishing initially homogeneous vegetation that had suffered similar initial disturbances (removal, boat transport, and replanting) and was then exposed to the three treatments.

Impacts on plant biomass were estimated from point frequency versus biomass regressions obtained from similar mainland habitats (Aunapuu et al. 2008; Dahlgren et al. 2009a). (Note that this method is conservative because it ignores the impact of herbivory on stem biomass.) We computed indices of treatment effects on each species using the natural logarithm of response ratio (lnRR) approach (Hedges et al. 1999) as follows. We first computed the relative changes of each species on each island i as $C_i = F_{i,e}/F_{i,s}$, where $F_{i,s}$ is the point frequency score of the species i at the start (2000) and $F_{i,e}$ is the point frequency score at the end (2003). We then computed the island-specific treatment responses of each species as $\ln RR = \ln(C_i/C_{3\text{-tr}})$, where $C_{3\text{-tr}}$ is the relative change of the species on the three-trophic-level island.

On vegetation blocks transferred to two-trophic-level islands, community-level plant biomass declined and the composition of the vegetation changed, whereas transfers from the three-trophic-level island to one-trophic-level islands resulted in increases in community-level plant biomass (Aunapuu et al. 2008; Dahlgren et al. 2009a). Next, we shall delve into details of the responses by dividing the plants into three functional groups: palatable herbs, palatable deciduous woody plants, and unpalatable (or less preferred) evergreen ericaceous plants (Kalela 1957; Aleksandrova et al. 1964; Moen et al. 1993). We therefore averaged the island-specific responses of each species in each treatment and used these as input variables for estimating the mean responses for each functional plant group.

Our results show that the decimation of unpalatable evergreen ericoids largely accounted for declines in community-level plant biomass on two-trophic-level islands (Figure 10.1). Responses of palatable deciduous dwarf shrubs were more variable, and palatable herbaceous plants expanded as a consequence of intense vole herbivory. Notice that this functional group also responded positively to the transfer to one-trophic-level islands, indicating that for palatable herbs, the selective vole herbivory on the three-trophic-level island represented the worst of both worlds.

To extend the time horizon, we computed logarithms of ratios of point frequency scores in 2003 and in 1992 for each species in our long-term plots on the two-trophic-level island. (For this experiment, genuine treatment effects could no longer be presented because most vole exclosures, established in 1991, had been breached.) Moreover, we included two additional functional groups, which appeared and increased on vole-impacted plots: silica-accumulating plants (graminoids and *Equisetum* spp.), that is, herbaceous plants inedible for gray-sided voles (Kalela 1957; Moen et al. 1993), and trailing woody plants.

Results from the long-term plots (Figure 10.2) plausibly reflect continuations of the trends observed in the transplant experiment. The negative impacts of food-limited voles on unpalatable evergreen ericoids have persisted, but also the majority of deciduous woody species have been decimated. In both groups of woody plants, interspecific differences reflect differences in habit. Species

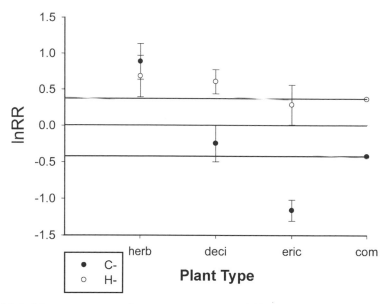

Figure 10.1. Mean responses of species representing different functional groups to transplantation from the three-trophic-level island to two-trophic-level islands (C-, black dots) and to one-trophic-level islands (H-, white circles); com = community-level plant biomass, denoted also by horizontal lines; deci = deciduous dwarf shrubs; eric = evergreen ericoids; herb = herbaceous dicotyledons. The responses are expressed as natural logarithms of response ratios. Error bars refer to standard errors.

with erect shoots have crashed, whereas species capable of taking a semiprostrate habit have survived better or increased (see Dahlgren et al. 2009a), as have species with an obligatorily trailing habit. The positive responses of palatable, herbaceous dicots and unpalatable silica accumulators are equally pronounced. The bottom line of these experiments is that differences in edibility matter little for plants interacting with food-limited gray-sided voles. The outcome depends on the accessibility of plants in winter. Less palatable species do have higher rates of shoot survival (Hambäck and Ekerholm 1997), but because these plants keep their nutrient and carbohydrate reserves in their evergreen shoots, they suffer higher costs per clipped shoot. In the end, recurrent, heavy winter herbivory drains the resources of all vertically growing woody plants, creating a shift from a scrubland to vegetation dominated by herbs and trailing woody plants (compare to Crête and Doucet 1998; Rammul et al. 2007; Olofsson et al. 2009).

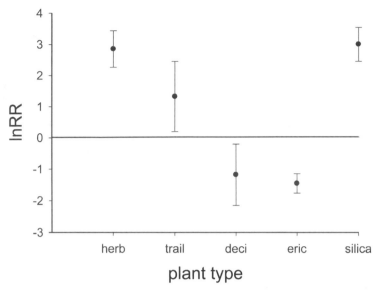

Figure 10.2. Mean responses of species representing different functional groups to exposure to 12 years of sustained impacts of food limited gray-sided voles; deci = deciduous dwarf shrubs; eric = evergreen ericoids; herb = herbaceous dicotyledons; silica = silica-accumulating plants (grasses, sedges, wood rushes, horsetails); trail = trailing woody plants. The responses are expressed as natural logarithms of ratios of final (2003) to initial (1992) biomass. Error bars refer to standard errors.

CONCLUSIONS

Despite the differences in habitat, climate, and herbivore characteristics, the evidence from Aldabra and Iešjávri yields a consistent picture. When plants are interacting with food-limited herbivorous vertebrates with broad diets, high tolerance is more important than maximal resistance due to low palatability (Hambäck 1998). If low palatability is associated with high vulnerability, it is a handicap rather than an asset. Therefore, intense herbivory can create and maintain communities dominated by palatable plants (McNaughton 1979b), especially if herbivory also speeds up nutrient circulation (Zimov et al. 1995; Olofsson et al. 2001).

The secondary role of plant defenses in the interaction between plants and food-limited herbivorous vertebrates with broad diets emerges from the principles of apparent competition (Holt 1977). The success of a plant subjected to intense herbivory depends on the ratio of its expansion rate to its loss rate. A woody plant can be killed by girdling (Hansson 1985a), which imposes a minuscule loss of biomass but destroys the transport system from the shoot to the roots. Large herbivores can also kill trees by removing the ground vegetation, thus triggering erosion (Chapter 9, this volume) and making tree roots vulnerable to trampling (Merton et al. 1976). Even unpalatable plants can suffer heavy losses when interacting with food-limited herbivores. If investments in defensive compounds carry a substantial cost, such costs will reduce the ratio of expansion to loss and therefore be maladaptive. Conversely, prostrate growth form and herbaceous habit reduce loss rates, being thus adaptive in interactions with food-limited herbivores. This point is illustrated by the open landscapes created by traditional, large-scale grazing systems throughout Eurasia, where scattered trees may grow in poorly accessible spots but prostrate plants prevail (Walter 1964; Oksanen and Olofsson 2009).

Although food-limited herbivores with broad diets appear invariably to create plant communities dominated by prostrate or semiprostrate plants, the chemical characteristics of the dominant plants vary, leading to big differences in palatability. A tradeoff between tolerance and resistance is therefore not inevitable (Leimu and Koricheva 2006; Dahlgren et al. 2009b). The cost of reduced carbon depends on nutrient supply (Bryant et al. 1983). Moreover, secondary chemicals can act as weapons in plant–plant chemical warfare (Nilsson et al. 2000), and they can protect plants against physical stresses (Close and McArthur 2002). In nutrient-poor environments, where the costs of reduced carbon are low, these advantages seem to make the production of secondary chemicals profitable. This was illustrated in our experiment, where evergreen ericoids, loaded with secondary chemicals, flourished when transferred to herbivore-free one-trophic-level islands (Figure 10.1). The role of nutrient supply is demonstrated by the seminatural communities created throughout the Old World by predator extirpation and large-scale outback grazing with mixed livestock (Tansley 1926, 1939; Godwin and Tansley 1941; Walter 1964, 1968; Gimingham 1972; Ellenberg 1988; Rosén 1982; Zobel and Kont 1992; Baines et al. 1994; Oksanen and Olofsson 2009; Chapter 9, this volume). In nutrient-rich areas, such seminatural vegetation is dominated by palatable grasses, whereas in nutrient-poor areas, the same treatment has created various kinds of heaths, dominated by grazing-tolerant but unpalatable plants.

The evidence for strong community-level trophic cascades in terrestrial nature, summarized in this chapter, does not negate the existence of differences between ecosystems. The case of Anticosti Island is still open to different interpretations but might indicate that the capacities of herbivorous vertebrates to exploit low-quality forage vary (Hansson 1985b). Selective species, such as white-tailed deer, might act like the invertebrates in the experiments of Schmitz (2006), changing the composition of the vegetation but not overall plant biomass. If there is a tradeoff between the ability to exploit low-quality forage and the ability to elude predators (L. Oksanen 1992), white-tailed deer represent the elusive and selective end of this spectrum. Their ability to survive while exploiting low-quality forage such as white spruce remains to be seen. Whether or not this is the case, all three examples discussed in this chapter represent community-level rather than species-specific trophic cascades. Contrary to the arguments of Polis and Strong (1996), one food-limited herbivore species can change the vegetation beyond recognition. The profound changes imposed by a single herbivore exceeded our a priori expectations. Tolerating the silica-based defenses of graminoids and the multifarious defenses of woody dicots requires such different adaptations that few animals can do both. Therefore, all balanced Old World grazing systems contain both grazers and browsers. Specialization on graminoid-eating cattle results in the expansion of "inedible" woody plants, whereas specialization on browsing sheep triggers the expansion of coarse-leaved graminoids (Walter 1964, 1968).

The strength of trophic cascades involving endotherms can also be expected to depend on primary productivity. To be able to survive and reproduce, predators need herbivores at densities exceeding a threshold level, $H^\star$, and this critical density is much higher for endotherms than for ectotherms (Shurin et al. 2006). Moreover, for each area, there must be some maximum herbivore density, H_{mx}, which the vegetation can barely sustain. For productive areas such as Aldabra and Anticosti, it is reasonable to assume that $H^\star << H_{mx}$, that is, that the rapidly growing vegetation can readily absorb the impacts of herbivore stocking dense enough to make the habitat attractive to predators. But the less productive the area, the lower H_{mx} must also be. At some point along a gradient of decreasing primary productivity, we must encounter areas where $H_{mx} < H^\star$, that is, the minimum density of herbivores that predators would need exceeds the maximum density of herbivores that the vegetation could sustain.

In the vicinity of this threshold, predators should still prevent runaway consumption of plant biomass, but herbivory pressure should be substantial, even in the presence of predators, as illustrated by the case of low arctic scrublands at Iešjávri. In still less productive areas, where $H_{mx} < H^\star$, three-trophic-level cascades should be replaced by two-trophic-level dynamics, that is, by strong herbivore–plant interactions, where plant biomass is chronically depressed or the vegetation is periodically devastated, depending on the dynamics of the system (L. Oksanen et al. 1981; T. Oksanen 1990; L. Oksanen and T. Oksanen 2000; Crête 1999; Crête and Manseau 1996; Turchin et al. 2000; Hansen et al. 2007; Aunapuu et al. 2008; Oksanen et al. 2008).

The high energy demands of endotherms also influence herbivore–plant interactions. The tundra, exploited by lemmings, voles, and reindeer or caribou, and grassland–heath landscapes, exploited by domestic herbivores, are much "greener" than the benthic habitats exploited by food-limited sea urchin populations (Chapters 2 and 3, this volume). However, there is also a big difference between the aboveground plant biomass of a forest and a grass sward or a heath. At least in areas with polyphagous browsing vertebrates, the cascading effects of predation seem to account for this difference, in concordance to the GWH of Hairston et al. (1960) and Chapter 16, this volume.

ACKNOWLEDGMENTS

Sincerest thanks to John Terborgh for organizing the meeting and assisting with manuscript preparation, and to Mary E. Power and Ilkka Hanski for constructive comments on earlier drafts. We thank the Swedish Natural Science Research Council, the Swedish Council for Forestry and Agricultural Research, and the European Commission (projects: DART, grant ENV4-CT97-0586, and LACOPE, grant EVK-2-CT_2002-00150) for their financial support and Norwegian authorities and Guovdageainu Ovttasgasguohtun (Kautokeino Fellesbeite) for the necessary permits. We also thank Nils Rolf "Molles Røffe" Johnsen and Oskar Eriksen for technical assistance, Helge and Britta Romsdal for their hospitality, and numerous field assistants for their devoted work.

CHAPTER 11

Trophic Cascades on Islands

Thomas W. Schoener and David A. Spiller

As MacArthur and Wilson (1967) pointed out in *The Theory of Island Biogeography*, Darwin was among the first to promote islands as ideal for the study of evolution. Islands are also ideal for ecological research (Schoener 2008), and the study of trophic cascades is no exception. Islands are simple, with relatively few species and habitats, so they are a good starting point for investigating ecological complexity. Islands are discrete and often small and therefore manageable. Islands are isolated, so that they are freer from colonization and possible disruption from outside species. Islands are natural, so unlike human-constructed experimental containers, islands have biotas at least in part adapted to them. Islands are combinatorial, so that species occur in various combinations, allowing comparative studies. Islands are numerous, so that often a number of them are available for replicated experimentation.

Unlike some of the subjects of other chapters, islands are not a biome type per se but rather range over a variety of biome types, from tropical rainforest to desert. Despite the recognition that bodies of water or patches of terrestrial habitat can be conceptualized for many purposes as islands, we restrict our definition for purposes of this chapter to areas of land surrounded by water.

Especially since MacArthur and Wilson (1967), the two properties of islands that have received the most attention are area and isolation (generally measured as distance from a larger source area). These two variables could be used to refine a definition of islands. If a body of land is too large, it is considered a mainland or even a continent. If too little water separates a body of land from other such bodies or from the mainland, it is considered part of a larger area rather than isolated. Therefore, one might select arbitrary area and distance cutoffs to designate "islandness," although there might be disagreement as to the exact such values, because what is large to some organisms is small to others, and what is near to some organisms is far to others. For example, the threshold between islands and mainlands for lizard species will be at a larger area than for the relatively small spider species. Likewise, for lizards the threshold between isolated and connected in these relatively poor dispersers will be at a smaller distance than for many birds, some of which include both mainland areas and fringing islands in their home ranges. Such considerations suggest that the characterization of "islandness" may be profitably viewed as continua along the two axes of area and distance, with smaller or more isolated designated as more insular. However, the traits area and isolation in themselves are not those most contributory toward our understanding of cascades on islands; instead, as seen below, other traits correlated with one or both of those axes are more explanatory.

EVIDENCE FOR ISLAND CASCADES

We now review existing evidence for trophic cascades on islands. In so doing, we define cascades as top-down effects, and we include both effects that are direct (one link in the pathway from the affecting to the affected food web element) and indirect (more than one link in the pathway), even though some investigators include only several-link pathways in their definition of *cascade*. Most of the evidence reviewed comes from our work and that of others on Caribbean islands, but we also summarize the effects of predators on islands in the Gulf of California and the spectacular cascade on remote Christmas Island, which harbors a generalist herbivore in the absence of any natural predator.

Observations and Experiments on Small Caribbean Islands

In the 1970s, one of us (T.W.S.) visited more than 500 Bahamian islands to survey distributions of vertebrates, with special emphasis on lizards and birds. We traveled in a 34-foot sailing vessel with retractable keels that allowed travel over shallow banks. A key objective was to determine the threshold island area on which vertebrate populations could just survive. We were astonished to find lizards, particularly *Anolis sagrei* (Figure 11.1), on some tiny islands, a discovery that multiplied by at least two orders of magnitude the list of Bahamian islands surmised or known to have resident populations of vertebrates. We realized that

Figure 11.1. *Anolis sagrei* of the Bahamas.

we had to check many quite small islands to determine such thresholds, and in the course of that endeavor we came upon a large number of islands without lizards. This led to a second, even more exciting discovery: Such islands sometimes had extraordinarily high densities of spiders, the omnipresent webbing giving them the appearance of the proverbial grandmother's attic.

In 1981, we had time to investigate this phenomenon systematically for the many small islands in the central Bahamas near the relatively large island of Staniel Cay, a major stopover in our earlier survey. Our first study (Schoener and Toft 1983) found that spiders (Figure 11.2) were about an order of magnitude denser on no-lizard than lizard islands (adjusted for the positive and negative correlations with area and distance, respectively). A second observational

Figure 11.2. The four most common web spider species at the Staniel Cay site in the Central Bahamas: **(a)** *Argiope argentata*, **(b)** *Eustala cazieri*, **(c)** *Metepeira datona*, and **(d)** *Gasteracantha cancriformis*.

study in 1982 (Toft and Schoener 1983) found much the same relation for spider density as for 1981. In addition, the second study examined numbers of spider species, finding that no-lizard islands had 1.5–2 times the number of species as had lizard islands (again adjusted for area and distance, and for the maximum height attained by the vegetation on the island, which correlated positively with number of spider species). This result was quite different from Paine's (1966) famous one in the rocky intertidal, in which diversity increased with increasing predation, and it presaged other such results for terrestrial arthropods in our system and in others also (Spiller and Schoener 1998).

Figure 11.3. Experimental enclosure on Staniel Cay. Dave Spiller is nearby.

Such comparative data pointed to a strong negative effect of lizards on spiders, but as is true of all comparative studies, the observations did not suffice to eliminate alternative hypotheses about why islands with and without lizards might differ. A more reliable investigation would be experimental, and toward that end we staked out nine approximately 83-square-meter plots on Staniel Cay in 1985 (Figure 11.3). Three of the plots were unenclosed, and the others had wood-framed fences made of hardware cloth topped with smooth plastic to impede lizard locomotion in and out (modeled on Pacala and Roughgarden's 1984 design). The substantial materials for this endeavor were flown from Florida by small plane to an airstrip near our study site or shipped to the island by mailboat. Three of the enclosed plots were randomly chosen to maintain lizards at natural densities, whereas the other three had lizards removed. Thus we had three treatments: The two types of enclosed plots tested the lizard effect, and the unenclosed plots were a cage control, to be compared with the enclosed lizard plots. The 18-month experiment showed that lizard removal enclosures had spider densities three times higher than those in control enclosures and the unenclosed (also having lizards) plots. Numbers of spider species were higher without lizards as well, in parallel to the comparative studies. Numbers and biomasses of insects caught in sticky traps were also higher in lizard removal enclosures than in control enclosures; therefore, an increase in spiders did not completely compensate for the absence of lizards. There was some effect of the enclosures: Sticky traps in enclosed plots caught about 20 percent fewer arthropod individuals than those in open plots.

What was the mechanism of the (now firmly established) lizard effect on spiders? The obvious one is predation. However, a second is competition for food: Spiders consume prey large in relation to their own size, so lizards and spiders might overlap in prey size well beyond expectations from their relative body sizes alone. Detailed observation and experiment (Spiller and Schoener 1990a) showed that both predation and competition were important. Short-term spider survival in lizard enclosures was about one fourth that in no–lizard enclosures, supporting the predation mechanism. Five lines of evidence supported the competition mechanism. First, lizards and spiders overlapped in diet, both preferring large to small prey. Second, lizards reduced large prey consumed by spiders but not small prey; total biomass consumed by spiders was reduced by lizards. Third, lizards reduced the number of large but not small arthropods caught in sticky traps. Fourth, lizards reduced the body size and fecundity of spiders. Fifth, food supplementation increased spider fecundity. Predation appeared stronger than competition in this study, and only one of two subsequent studies gave evidence of competition. Therefore, we believe predation is in general the dominant effect, but the discovery of competition was exciting, especially because we had evidence of an indirect food web effect; lizards affect spiders negatively via their common prey. The double whammy of predation and competition explains why the lizard effect on spiders was so strong.

Given that lizards affected spiders so much, both directly and indirectly, it might seem likely that their effects would permeate farther down the food web, perhaps even to plants. In fact, relevant observational data from seventy-four islands near Staniel Cay had been reported a few years earlier (Schoener 1988): Leaf damage in buttonwood (*Conocarpus erectus*) was about 1.5 times higher on islands without lizards. That lizards would have a positive effect on plants was not obvious, because lizards eat arthropod predators of herbivores as well as eating the herbivores themselves. Does the dominant pathway from lizards to plants have two links (effect on plants positive) or three links (effect on plants negative)? Figure 11.4 shows the two possibilities. A three-link (four-level) system had been published for lakes (Carpenter and Kitchell 1988) and was about to be published for rivers (Power 1990). On the other hand, the venerable Hairston, Smith, and Slobodkin (HSS 1960) proposition about terrestrial systems assumes three effective levels.

Again, an experiment was called for, and we used the same enclosure setup as before (Spiller and Schoener 1990b). The most common plant species at

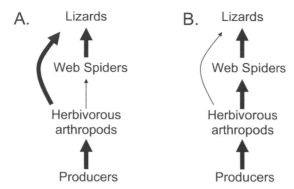

Figure 11.4. Alternative cascade models for (partial) food webs on small Bahamian islands. (a) Lizards affect plants positively (three effective levels). (b) Lizards affect plants negatively (four effective levels). For elaboration, see Spiller and Schoener (1996).

Figure 11.5. Sea grape (*Coccoloba uvifera*) showing herbivore damage.

the experimental site was not buttonwood, however, but sea grape (*Coccoloba uvifera*; Figure 11.5), so we measured leaf damage in the latter. Despite this difference in plant subject, the direction of the lizard effect was the same as that observed on the islands with buttonwood: Lizards decreased damage overall, so the system was effectively three-level (two links) rather than four-level (three links).

An interesting twist involved an herbivore not found on buttonwood, the gall midge (*Ctenodactylomyia watsoni*). Leaf damage from this species increased in the presence of lizards, not decreased, although gall damage was sparse enough to make the overall effect of lizards a decrease in damage. Because the adults of this herbivore are a small fly, we hypothesized that spiders were probably its major predator, and we had already shown that lizards reduce spiders, so for this part of the food web three links would be involved. An elaborate attempt to demonstrate three links rigorously, using a 2 × 2 removal of lizards with spiders (Spiller and Schoener 1994), was unsuccessful, however, because gall midges nearly disappeared everywhere from the Staniel study site, in the seemingly capricious way that arthropods sometimes do. Finally, given that we did an experiment showing a significant lizard effect on plant damage in sea grape, we reversed our observation-followed-by-experiment protocol, performing an 8-year monitoring of natural sea grape leaf damage on islands near Staniel Cay, including many in the buttonwood study. Only eighteen islands could be used, however, because sea grape is less common than buttonwood on small islands. The result went in the same direction as in the just-described experiments, and even quantitatively it was rather similar (Spiller and Schoener 1997).

So far our experiments had involved only removal of a predator. Moreover, they had been done in enclosed plots on a large island rather than on smaller islands that served as the basis for most of our observations. Would introductions of lizards onto lizard-free islands give significantly novel information? The project seemed timely for the 1990s, because attention was becoming increasingly focused on invasive species, of which *Anolis sagrei* is a prime example. To do this experiment we moved northward to the Great Abaco area, another region first visited in the 1970s surveys and found then to have numerous tiny islands embedded in its creek waterways (Figure 11.6). The islands chosen for the experiment have other arthropod predators, particularly parasitoids, but no breeding populations of vertebrates other than *Anolis sagrei* (although they are visited by itinerant birds).

The experiment had three treatments: islands naturally supporting lizard populations, lizard-free islands onto which lizards were experimentally introduced, and islands without lizards (the areas of all these islands were near the threshold island size at which lizards can just maintain populations, so we could get fairly comparable islands for the three treatments). In addition to the usual food web questions about the effects of lizards, the design enabled us to ask whether lizard presence was sufficient to generate the huge natural difference between lizard and no-lizard islands (Schoener and Spiller 1999a).

Figure 11.6. A tiny island in the Abaco area. Tom Schoener is on it. (Photo courtesy of Jonathan Losos)

Lizard introductions had an even more devastating effect on spiders than what we found in our enclosure experiments farther south. Both spider abundance and diversity plummeted on the lizard introduction islands, and within 2 years both variables had reached the levels found on natural lizard islands (Figure 11.7, top). Common and rare spider species were both reduced by lizards, but local extinction of the latter was much more extreme: The proportion of species becoming extinct was 12.6 times higher on islands with introduced lizards than on those without lizards (and higher than on natural lizard islands, whose spider populations were presumably filtered in favor of species less likely to be affected strongly by lizard predation). Thus predators caused a huge reduction in the species diversity of prey, exactly as had been found farther south.

The negative effect of lizards on spiders was so pronounced that it prompted another experiment. This one was better conducted in the more controlled setting of the enclosures on Staniel Cay (Spiller and Schoener 2001); the latter site, like the Abaco islets, showed a differential exclusion of rare spider species by lizards (Spiller and Schoener 1998). Under the "Paine effect"

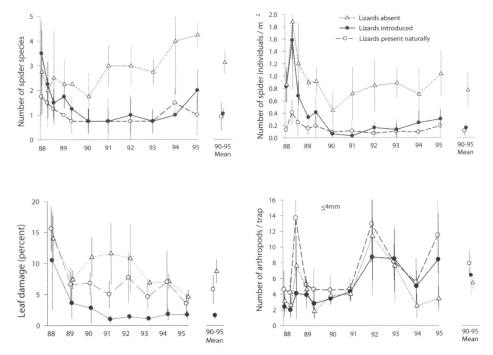

Figure 11.7. Results of experimental introductions of lizards onto Bahamian islets. Top: Effect of lizards on spider species (left) and spider individuals (right). Bottom: Effect of lizards on total leaf damage (left) and number of small aerial arthropods (right). Bars represent ±1 standard deviation.

described earlier, predation increases prey diversity by reducing competition between prey species. What if in our system predators actually increased competition between prey, contributing further to a reduction in prey diversity? To test this, we did another 2 × 2 experiment, with lizard presence or absence crossed with the presence or absence of the common spider *Metepeira datona*; the other web spiders at the site served as species potentially affected by competition from *Metepeira*. We found neither an increase nor a decrease in competition but rather no statistically significant difference; indeed, there was no significant spider competition with or without lizards. Perhaps physical factors or birds (banaquits, *Coereba flaveola*; Bahama mockingbirds, *Mimus gundlachii*; and thick-billed vireos, *Vireo crassirostris* were common at the Staniel site) keep spider populations down in the absence of lizards.

Returning to the introduction of lizards onto islands near Abaco, we also measured effects on buttonwood leaf damage and on arthropods caught in sticky traps. Again, lizards reduced leaf damage (Figure 11.7, bottom left) but with a novel quirk: In the middle years of the experiment, damage was actually less on lizard introduction islands than on islands having lizards naturally. This overshoot of the protective effect of lizards may have resulted from especially vulnerable herbivore species not accustomed to lizard presence being reduced or eliminated by the invading lizards (similar to the spiders on the same islands, as discussed earlier). An intriguing positive indirect effect of lizards was also found: an increase in the number of small aerial arthropods (Figure 11.7, bottom right). We surmised that this occurred via a reduction in the density of spiders (and showed this more rigorously later on; Schoener et al. 2002). All these indirect effects were weaker than the mostly direct effect of lizards on spiders, in accordance with theory (Schoener 1993; Abrams et al. 1996).

Despite the relatively weak indirect effects, we were sufficiently encouraged by the results of the *A. sagrei* introduction to attempt an investigation of an even longer cascade: that involving a predator of *A. sagrei*, the curly-tail lizard (*Leiocephalus carinatus*) (Figure 11.8). *Leiocephalus* is found erratically on islands near the Abaco site and is common on Great Abaco itself. We used islands averaging somewhat larger in area for this experiment; we introduced *Leiocephalus* to islands having *A. sagrei* naturally, using other such (*A. sagrei* only) islands as controls (Schoener et al. 2002). We also monitored islands without lizards to ask whether the effect of *Leiocephalus* on *A. sagrei* is sufficient to reverse the latter's pervasive and sometimes great effects on lower levels of the food web.

Figure 11.8. *Leiocephalus carinatus* consumes *Anolis sagrei*.

A complete answer was foiled, however, when Floyd, a monstrous hurricane with winds 140 miles per hour and a 10-foot storm surge, terminated the experiment only two and a half years after inception. Nonetheless, we were able to show huge effects of *Leiocephalus* on various traits of *A. sagrei* and major indirect effects on spiders. In addition, *A. sagrei* in the absence of *Leiocephalus* had a major effect on large ground arthropods. One small lizardless island teemed with roaches, sometimes filling our pitfall traps to overflow in a 3-day period. The reversal of *A. sagrei*'s effects on spiders was significant but not quite complete, whereas no reversal was evident for large ground arthropods, probably because *Leiocephalus* also consumed them, thereby compensating for the reduction of *A. sagrei*. Note that this is unlike the result mentioned earlier for web spiders because these were mostly out of reach to the large terrestrial *Leiocephalus*. The experiment failed to yield any longer-chained effects (on plants, parasitoids, or other aerial arthropods) resulting from the introduction of the higher predator. Of course, these results might have been very different had it not been for Hurricane Floyd.

The untimely end of the experiment with *Leiocephalus* marked the beginning for us of the current upsurge in Atlantic hurricane activity, and perhaps we have not seen the whole of it yet. Several of our recent experiments on trophic cascades have been turned into studies of the effects of hurricanes. Such hybrid studies combining cascades and disturbances have led to many new insights, however, and this has been the silver lining.

One of the more interesting such hybrid studies is a recent one (Spiller and Schoener 2007) on how hurricanes Floyd and Michelle affected the cascade on small Abaco islands. With marvelous serendipity, we had laid the groundwork before Floyd hit the site: For 3 years before the hurricane, we had accumulated data on key food web features, including buttonwood leaf damage, lizard density, web spider density, and hymenopteran parasitoid abundance. Herbivory increased dramatically after each hurricane, and the lizard effect on herbivory did as well. Predator densities declined over the same period. Spiders were especially pounded: Not a single spider was found on half the islands shortly after Floyd flooded them, and mean spider density was reduced to 3 percent of that of the previous census. We attributed the increase in herbivory primarily to this reduction of predators that eat herbivores (although an increase in plant palatability, caused in part by an increase in nitrogen content, could also have been involved; see Spiller and Agrawal 2003). The increased lizard effect on herbivores would occur if arthropod predators that compensated for the absence of lizard predation on no-lizard islands were sufficiently reduced.

Observations and Experiments on Large Caribbean Islands

Pacala and Roughgarden (1984) pioneered the experimental study of the food web effects of lizards, using plots on St. Eustatius, an island substantially larger than Staniel Cay, Bahamas, and located at the other end of the Caribbean, the Lesser Antilles. Six large enclosures were constructed, and both native species of *Anolis* lizards were removed from three of them. After 6 months, forest floor arthropods (measured with sticky paper plates) were two to three times as abundant in the enclosures without lizards, and web spiders were twenty to thirty times as abundant. Arthropods caught in sticky traps between the forest floor and the canopy decreased in abundance, a result Pacala and Roughgarden interpreted as resulting from the increase in spiders. Herbivory on two unspecified plants was unaffected. The positive effect of lizards on small aerial arthropods is similar to ours, whereas the negative effect on spiders is similar in direction if not in magnitude. The St. Eustatius study site consisted of much higher vegetation (6- to 8-meter trees rather than 0.15- to 0.53-meter shrubs on average), a fact that necessitated higher fences, thereby entailing a behemoth logistical effort on Pacala's part (according to Roughgarden) to organize transport of the necessary materials.

Another logistical tour de force was involved in the experimental *Anolis* removal by Dial and Roughgarden (1995) on the huge island of Puerto Rico. These investigators took advantage of the forest fragmentation resulting from Hurricane Hugo to study fourteen isolated tabonuco trees (*Dacryodes excelsa*) up to 29 meters in height. Dial squirreled over this vegetation using what he called modestly and perhaps euphemistically "noninjurious, modified arborist methods"—one can only imagine!—to remove lizards and perform the subsequent monitorings. Thirty-five arthropod groups were followed through the 6-month duration of the experiment, and leaf damage was sampled at the end. Lizard removal had strong positive effects on large arthropods, including orb spiders and many other large-bodied groups such as roaches and large hymenopteran parasitoids. Lizard removal had negative effects on small arthropods. Leaf damage was greater in lizard removals than controls, and this was correlated with an increase in Orthoptera and roaches. All their results are remarkably similar to ours in the Bahamas, with the possible exception of parasitoids, which in the Puerto Rican study either increased upon lizard removal (large species) or showed no change (small species). This similarity holds despite the fact that Puerto Rico has many more kinds of predators that may eat anoles, including certain birds and snakes, than do the islands we study in the Bahamas.

Islands of the Gulf of California, Mexico

Polis and colleagues (1998) studied twenty desert islands in the Gulf of California that are chronically so dry that much of the usable productivity is allochthonous, consisting of detritus washed ashore and seabird guano (Figure 11.9). Spiders in particular were increased by an increased prey base fueled by coastal
detritus and seabird guano. Some top-down control also occurred: Spiders were
strongly affected by the predatory scorpions. Unlike in the Bahamas, lizards had
little effect in the Gulf of California, perhaps because the species (e.g., *Uta stans-
buriana*) are so much more terrestrial, rarely climbing into the vegetation where
spiders abound.

An intriguing sequence of ecological effects was observed during the
1992–1993 El Niño rainfall. Spider density at first doubled in apparent response
to the increase in productivity of their prey, then crashed with an eruption of
pompilid (spider-hunting) wasps in 1993; populations of the latter were appar-

Figure 11.9. Gemelos West, a desert island in the Gulf of California. Reproduced with
permission from Wendy B. Anderson.

ently greatly increased by a pulse of nectar and pollen resulting from a great flowering of annual plants brought on by heavy rains.

Christmas Island

An impressive trophic cascade has been documented on Christmas Island, located 360 kilometers south of Java in the Indian Ocean. The island supports a generalist herbivore, the red crab (*Gecarcoidea natalis*) at the astonishing density of more than a million per square kilometer (O'Dowd et al. 1990, 2003; Green et al. 1997). However, in 1989 multiqueened supercolonies of the yellow crazy ant (*Anoplolepis gracilipes*) were noticed to be spreading across the island, reaching densities in the thousands per square meter. These ants wipe out the defenseless crabs and in so doing have caused a massive top-down effect on the vegetation (Figure 11.10). Areas whose crab density is yet unaffected are largely open, with few seedlings and little leaf litter. The ants caused a spectacular increase in plant biomass, because of the saplings that are now able to survive, and plant species diversity increased threefold. Leaf litter is now at the state expected on mainlands. These effects on herbivory are substantially stronger than those of arthropod folivores reported earlier for various Caribbean localities. John Terborgh (personal communication, 2009) suggests that the lower selectivity of major generalist herbivores such as red crabs may be responsible for the difference.

Figure 11.10. Understory of forest on Christmas Island, Indian Ocean. Left: Natural state with red crabs (*Gecarcoidea natalis*). Right: Without red crabs after invasion of the yellow crazy ant (*Anoplolepis gracilipes*). (From O'Dowd et al. 2003: 815)

THE STRENGTH OF TROPHIC CASCADES ON ISLANDS

The preceding review has shown that in some cases trophic cascades can be very strong on islands. In other cases, however, the cascades appear weak. Why should island cascades differ in strength? A related, intriguing question is the following: Is there any tendency for trophic cascades to be stronger or weaker on islands than on mainlands? Our strategy in examining such issues is first to ask what key traits might vary on islands that would affect their trophic cascades. As appropriate, we will briefly review both the theory and evidence for such traits. We shall see that such traits are correlated with island area and sometimes with island distance also. We will form hypotheses about how such variation might affect the strength of trophic cascades. Next, we will survey a particular system whose islands vary in area to see whether this basic island feature is related to a stronger or weaker cascade. Finally, we will comment on the likelihood that each of the hypotheses contributes toward understanding the survey results for that system.

Key Island Traits and Their Hypothetical Impact on the Strength of Trophic Cascades

We distinguish six such traits, each generating one or two directional hypotheses about how the trait relates to the strength of trophic cascades on islands. Table 11.1 briefly characterizes these hypotheses.

Table 11.1. Hypotheses for the strength of trophic cascades on islands.

Hypothesis	Prediction for Those Islands*
1. Species diversity is lower on smaller or more isolated islands.	Stronger cascade
2. Intermediate predators are at higher density on smaller or more isolated islands.	Stronger cascade
3a. Disturbance is more severe on smaller or more exposed (isolated) islands.	Weaker cascade
3b. Disturbance selectively reduces compensating lower-level predators on smaller islands.	Stronger cascade
4a. Predators switch away from local prey to consume subsidy on smaller islands.	Weaker cascade
4b. Predator numerical response to subsidy is greater on smaller islands.	Stronger cascade
5. Plant defenses are lower on smaller or more isolated islands.	Stronger cascade
6. Refuges from which affected species can repopulate are fewer on smaller islands.	Stronger cascade

*Comparison also applies to islands versus mainlands.

1. LOWER NUMBER OF SPECIES

The MacArthur–Wilson Species Equilibrium Model is the most influential theoretical construct predicting how a given island's area and distance determine its number of species. The model stipulates that the immigration rate of species not already present on the island and extinction rate of species present will through time become equal, at which point the number of species on the island is at dynamic equilibrium. Islands with smaller areas have larger extinction rates for the same number of species present (because of their smaller average population sizes), giving fewer species at equilibrium. Because distance serves as a filter, islands farther from a source of immigrants will have smaller immigration rates for the same numbers of species present, giving fewer species at equilibrium. Theoretical explanations other than MacArthur–Wilson exist for both area and distance effects. For area, these include random sampling, larger interception rate of dispersing organisms on larger islands, higher habitat diversity on larger islands, greater in situ speciation rate on larger islands, and lower effective abiotic disturbance on larger islands (see Trait 3) (Spiller and Schoener 2009). For distance, these include the nonequilibrial explanation that, given enough time, far islands will eventually attain the same number of species as near islands, and that far islands have lower habitat diversity (Schoener 2009). The evidence for area and distance effects is extensive, and so it will not be reviewed here.

A lower species diversity, implying a less reticulate food web, should strengthen the trophic cascade (Strong 1992; Polis and Strong 1996). Manipulation of the predator level may have less far-reaching effects on a single species or other element farther down the food web (Strong 1992; Polis and Strong 1996). Moreover, if there are several kinds of predators, changes in abundance of one may have less effect if there is compensation among them. For example, the effect of removing a given predator is less if other predators compensate (Spiller and Schoener 1998; Schoener and Spiller 1999b; see also Hypothesis 3b). The expectation is a stronger cascade on smaller or more isolated islands (Hypothesis 1).

2. DIFFERENTIAL ABSENCE OF TOP LEVELS AND SPECIES

Independent treatments of similar continuous-time Markov models (Schoener et al. 1995; Holt 1996) predicted that smaller islands and farther islands are more likely to lack a top level. The tendency for islands with smaller areas to have shorter food chains was shown for "real" (i.e., terrestrial) islands in the Bahamas (Schoener 1989) and Lago Guri (Chapter 8, this volume). Schoener et al.

(1995) also showed that the fraction of parasitoid individuals (and, by implication, species) was less for smaller as well as farther islands in the central Bahamas. Random reduction in species number (see Trait 1) would also by chance result in fewer food webs with a top level, but the tendency seems more regular and stronger than such a null model would predict (although this has never been tested).

The presence of a top predator level implies that the predator at the next level down—the intermediate predator—has a relatively low density. Therefore, removal of the latter will have little effect. Moreover, introductions of the intermediate predator will be largely absorbed into the next highest trophic level (Fretwell 1977; Chapter 13, this volume). The expectation is a stronger cascade on smaller or more isolated islands (Hypothesis 2).

3. Higher Vulnerability to Disturbance

Smaller islands have more shoreline per area and therefore are more vulnerable per area to the effects of coastal disturbance such as waves and wind. Not only do they have a greater ratio of perimeter to area, but they often have a lower maximum altitude, also increasing vulnerability to high water (Schoener et al. 2001). Moreover, islands farther from a mainland are often more exposed and are thus less protected from storm surges (Spiller et al. 1998) and from other coastal disturbance. Noncoastal disturbance, such as heavy rain whose distribution is uniform over the entire island, seems less likely to vary with island size or distance. Indeed, we found no statistical evidence that island area modulates the strength of the indirect effect of lizards on buttonwood (Spiller and Schoener 2007) because it varied before and after the passage of two hurricanes.

A greater amount of disturbance can have opposite effects. According to the Menge–Sutherland (1987) hypothesis, more disturbance reduces the top-down effect of predators on their prey, through a reduction in predator density or predator per capita effect, or both. For example, the rocky intertidal gastropod predator *Thais* (*Nucella*) was unable to reduce the abundance of its prey at sites exposed to severe wave action, whereas at protected sites it had a very strong effect, particularly on the preferred prey type (mussels). In the northern Bahamas, we found that the effect of lizards on spiders diminished with increasing rainfall, some associated with hurricanes (rainfall is measured as days of rain; Spiller and Schoener 2008). The expectation is a weaker cascade on smaller and more exposed (often more isolated) islands (Hypothesis 3a). However, the opposite can occur if disturbance intensifies the effect of a particular predator by reducing the ability of other predators to compensate (Spiller and Schoener 2007). For example, larger predators on a given prey may be less vulnerable to the effects of disturbance than smaller predators on that same prey, being more resistant to wind or desiccation. Therefore, when the latter are reduced by a disturbance such as a hurricane, they will fail to compensate upon experimental removal of the larger predator, increasing the effect of the removal. (Note that this hypothesis can be viewed as a more specific version of Hypothesis 1: Disturbance selectively reduces certain species on small islands, contributing to the latter's lower species diversity.) The expectation is a stronger cascade on smaller or more exposed (often more isolated) islands (Hypothesis 3b).

4. Higher Subsidy from Marine Systems per Area

As with shoreline-dependent abiotic disturbance, bottom-up marine subsidies should be related to an island's perimeter, so that the amount of such subsidy per area is greater on smaller islands. For desert islands, this subsidy can be overwhelming in importance: In non–El Niño years, in situ productivity on islands in the Gulf of California is less than 100 grams dry mass per year, but the islands receive about 3,090 grams dry mass per year per meter of shoreline in the form of marine detritus (Anderson and Polis 1998). This subsidy may or may not be important to the energy budget of consumers, depending on its utility. In fact, carbon stable isotope analysis showed for a species of *Anolis* and two species of spiders in the Bahamas that the greater the perimeter/area ratio, the greater the marine contribution (Sears and Spiller, unpublished data).

A greater amount of marine subsidy can have opposite effects (depending on time after the subsidy). The initial effect of an allochthonous pulse of resources will often be to divert predators away from local prey to prey generated by the subsidy (e.g., flies breeding in seaweed detritus). The consequent effect on local prey, such as herbivores of in situ plants, should be positive. The expectation for this earlier phase is a weaker cascade on smaller islands (Hypothesis 4a). However, as time goes on the predators will show a numerical response, and if the subsidy is not renewed, they will be driven back to eating the local prey, but with a greater effect than before the subsidy because of their elevated populations. The expectation for this later phase is a stronger cascade on smaller islands (Hypothesis 4b).

5. Lower Defenses in Plants

Plant defenses may relax on smaller and more isolated islands because they are less likely to have herbivores, or certain species of herbivores (see Traits 1 and 2). Data from islands in the central Bahamas directly support this idea: The smaller and farther the island, the lower the proportion of buttonwood plants that are defended by trichomes (Schoener 1987). Even when these defenses are controlled for, leaf damage is greater for larger islands but not for nearer islands (Schoener 1988). However, on another set of islands less than 100 kilometers south, exposed islands had greater buttonwood herbivory than protected islands; the former were on average more distant. Moreover, after Hurricane Lili, an increase in herbivory was observed only on the exposed islands. Spiller and Agrawal (2003) suggest a variety of hypotheses for the latter result, including the extermination and subsequent relatively slow recolonization of predators on the more distant exposed islands.

Lower defenses in plants means that the plant level should show more response to variation in herbivores, which in turn may be caused by variation in populations of their predators (Strong 1992). The expectation is a stronger cascade on smaller or more isolated islands (Hypothesis 5).

6. Greater Ability of Populations Affected by Predators to Repopulate from Elsewhere

Because of their heterogeneity and greater area, larger islands are more likely to have prey refuges that would enable repopulation of predator-depleted sites (Spiller and Schoener 1998; Schoener and Spiller 1999b). The same is true for less isolated islands.

If the diminishing effects of predators on their prey at a particular site are muffled by the prey reimmigrating from elsewhere, the trophic cascade is weakened. The prediction is a stronger cascade on smaller and more isolated islands (Hypothesis 6).

Procedures for Determining How the Strength of Trophic Cascades Varies with Island Area and Isolation, Including Whether Cascades Are Stronger on Islands or Mainlands

These hypotheses for the strength of trophic cascades on islands do not run uniformly in one direction, even sometimes for the same key trait. But mostly they point to a stronger cascade on smaller islands, with a subset of those hypotheses predicting a stronger cascade on more isolated islands. How do the trends come out in reality?

One approach to answering this question is to perform a meta-analysis of literature experiments on terrestrial cascades. The simplest version of this approach would be to separate experimental arenas into two groups, those on islands and those on mainlands, where the latter are operationally defined as exceeding some area threshold; this approach is similar to the ones comparing terrestrial and aquatic cascades (Schmitz et al. 2000; Halaj and Wise 2001; Shurin et al. 2002). The specific question here would be whether cascades are stronger on islands or mainlands. A more complicated version of this approach is to regress experimental effect size against (log) area of the island on which the experiment was performed. A complication could be that the strength of the cascade may depend on the level of the predator being manipulated. For example, the longer the effect chain, the weaker on average it should be (Schoener 1993). Moreover, if a manipulated predator is low in abundance because it has a predator at the next highest level, its effect should again be weakened. Certainly longer chains are more likely on larger islands (see Trait 1), but this complication is not an issue if the island is small enough to have no secondary predator; it then alternates in likelihood as the length of the chain above the focal predator increases with island area.

It may be that such effects could be controlled for in an ideal meta-analysis. Here we will try a simpler approach by comparing sites with the same predator on islands of various sizes. We shall consider top-down effects resulting only from perturbations of that predator. The obvious candidate for our Bahamas system is the lizard *Anolis sagrei*. This common and widespread species occurs on islands of very different areas, down to very small sizes (roughly 10–100 square meters, depending on the exact location), although not the smallest sizes, where only arthropod predators such as spiders occur. The species also can occur over a range of distances from the nearest source population, but because it is sensitive to high water resulting from hurricane surges and is unable to disperse readily (Spiller et al. 1998; Schoener et al. 2001), it is often absent from the more remote islands in an area. Therefore, focusing on island area as our measure of "islandness" is the more practical strategy. In fact, we have already published such a survey of *Anolis sagrei* on Bahamian islands (Schoener and Spiller 1999b). Here we summarize results that bear on the current topic.

We use as our measure of effect magnitude the ratio of the treatment to control value or vice versa, whichever is larger. This varies monotonically with the absolute value of the log of this ratio, the measure recommended by Hedges et al. (1999). For traits of spider populations, including abundance, the overwhelming tendency is for the effect magnitude of the predator to be greater on smaller islands. Herbivory is also affected on average more on small islands, but variation in effect magnitude with island area is less for this longer-chained effect. Aerial arthropods on average are also more affected on small islands, but the direction of the effect can vary with size of the arthropods, and the effect is often weak. We have shown (Schoener et al. 2002) that the positive effect of lizards on small aerial arthropods (including parasitoids) can be relatively long-chained, proceeding from lizards through spiders to aerial arthropods and thus involving a second predator level.

One possible explanation for part of the results involves the fact that the experiments on the large island were conducted in enclosures, whereas observations and experiments on medium to small islands encompass the entire island. As elaborated elsewhere (Schoener and Spiller 1999b), on the large islands dispersal from removal to control treatments may have diluted the lizard effect on lower trophic level variables. Note that this is an artifact of the experimental procedure, but it incorporates the same mechanism as for Hypothesis 6.

Taking the results as not resulting primarily from the study design but rather relating directly to one axis of islandness, island area, we see that they jibe with most of the reasonable hypotheses we were able to suggest. Specifically, Hypotheses 1, 2, 3b, 4b, 5, and 6 are all at the most qualitative level consistent with our result. Thus our procedure so far doesn't eliminate many possibilities. Perhaps we could narrow them down, eliminating certain hypotheses that for other reasons (i.e., by assumption of the given hypothesis) are unlikely to apply to our particular test system. To perform such an analysis, we need to have a closer look at specific properties of that system, small Bahamian islands.

Are the Conditions for the Predictions Likely to Occur in the Test System: Bahamian Islands Having the Lizard Anolis sagrei?

To examine this question, we proceed through the list of hypotheses consistent with the main result: stronger cascades on smaller islands.

Hypothesis 1: Lower Species Diversity on Smaller Islands

Wherever studied in detail, the taxa of our Bahamian system show a species–area relation: spiders (Schoener and Toft 1983; Toft and Schoener 1983; Spiller and Schoener 2009), plants (Morrison 1997), and ants (Morrison 2002), but with a very low log–log slope.

Hypothesis 2: Intermediate Predators (Such as Anolis Lizards) at Higher Density on Smaller Islands

Larger islands in the Bahamas contain more kinds of organisms that would probably eat *Anolis sagrei*. These can include *Ameiva* (a teiid lizard), *Leiocephalus* (a larger iguanid lizard), birds (e.g., the Bahamas mockingbird, *Mimus gund-*

lachii), and certain snakes. In fact, comparing two of the three classes of island areas in the survey summarized earlier, we find that densities of *A. sagrei* vary in the direction expected from the hypothesis: On very tiny Abaco islands, where no higher predators are present, densities were about 0.2–0.3 per square meter (before Hurricane Floyd), whereas on the large island of Staniel Cay, they were about 0.1 per square meter. Further support is found in the experimental introduction of *Leiocephalus* (Schoener et al. 2002), in which *A. sagrei* loses some of its effect on spiders on islands where the larger predatory lizard was introduced.

HYPOTHESIS 3B: DISTURBANCE SELECTIVELY REDUCES COMPENSATING PREDATORS LOWER IN THE FOOD WEB ON SMALLER ISLANDS

On islands of the northern Bahamas (Spiller and Schoener 2007), herbivory markedly increased after each of two hurricanes (3.2 times higher after Floyd; 1.7 times higher after Michelle). The indirect effect of lizards on herbivory was substantially higher as well (2.4 times higher during the post-hurricane period than before). Although all kinds of predators studied diminished in abundance after the hurricane, lizards fell substantially less (30 percent) than arthropod predators (spiders 66 percent, hymenopteran parasitoids 59 percent). The greater vulnerability of in situ spider populations compared with lizard populations was also shown after Hurricane Lili at a site farther south (Spiller et al. 1998). Thus, on islands without lizards, the arthropod predators that might have compensated were greatly reduced, magnifying the lizard effect (difference between lizard and no-lizard islands). These considerations support the second version of this hypothesis, a stronger cascade on smaller islands.

HYPOTHESIS 4B: INCREASED PREDATOR NUMERICAL RESPONSE TO A SUBSIDY ON SMALLER ISLANDS

Contrary to the survey results, preliminary results of an ongoing field experiment support Hypothesis 4a; addition of seaweed to shoreline plots increases herbivory, even though lizard density also increases. Therefore, the positive effect of lizards on leaf damage via increased predation on in situ herbivorous arthropods (as in Hypothesis 4b) does not appear to occur, at least in the short term, disfavoring the second version of this hypothesis, a stronger cascade on smaller islands. However, in addition to increasing subsidized prey (detritivorous arthropods), seaweed might also be fertilizing the plants, providing nutrients that could increase the plants' palatability to herbivores.

HYPOTHESIS 5: REDUCED PLANT DEFENSES ON SMALLER ISLANDS

For the islands in the central Bahamas, lizard islands show less buttonwood leaf damage than no-lizard islands. However, relatively defended buttonwood (called "silver" because of its trichomes) shows the same difference in leaf damage between lizard and no-lizard islands as does undefended (called "green") buttonwood (Spiller and Schoener 1996; Figure 11.7). Thus, in this example the lizard effect on herbivory did not vary with degree of defense.

HYPOTHESIS 6: FEWER REFUGES FROM WHICH AFFECTED SPECIES CAN REPOPULATE ON SMALLER ISLANDS

On islands of the central Bahamas having lizards, spiders are often abundant (and sometimes very abundant) on peninsulas or points of land where lizards rarely occur (Schoener and Spiller, unpublished data). Thus a type of refuge seems to exist (another is relatively high vegetation), and substantial such refuges are more likely for larger islands.

CONCLUSION: WHICH HYPOTHESES CONTAIN LIKELY MECHANISMS TO EXPLAIN STRONGER CASCADES ON SMALLER BAHAMIAN ISLANDS?

Hypotheses 1, 2, 3b, and 6 are supported, and Hypotheses 4b and 5 are contradicted. Thus it appears that the three most likely mechanisms for why trophic cascades are stronger on small islands in the Bahamian system are their lower species diversity (perhaps resulting from disturbance), the relative absence of top predators, and the paucity of prey refuges that can replenish predator-depleted populations by immigration from elsewhere. In short, although the possibilities are still multiple and somewhat complex, we have narrowed them down, and it will be interesting to see whether the same set of hypotheses would be supported for other insular systems.

CHAPTER 12

Trophic Cascades, Aboveground–Belowground Linkages, and Ecosystem Functioning

David A. Wardle

Over the past decade, there has been a greatly increasing recognition of the interdependence between aboveground and belowground communities and the importance of interactions between these communities in controlling ecosystem processes and properties (Van der Putten et al. 2001; Wardle 2002; Wardle et al. 2004). Plants integrate these communities (Figure 12.1). They generate the carbon that drives the aboveground (foliage-based) food web and the root-based and detritus-based food webs in the soil. Consumption of foliage by aboveground consumers can influence the quantity and quality of resources that plants provide to belowground organisms, both as live plant roots and as plant litter (Bardgett and Wardle 2003). Furthermore, they return processed plant material to the soil as fecal material (Lovett and Ruesink 1995). Similarly, soil organisms that are directly associated with plants such as mutualists and antagonists (direct pathway in Figure 12.1) may indirectly influence the provision by plants of resources to aboveground consumers (Masters et al. 2001). Furthermore, soil organisms in the detritus food web that are indirectly associated with plants (indirect pathway in Figure 12.1) can exert important indirect effects on the consumers of both foliage and roots by regulating the decomposition of plant organic matter and thus the availability of nutrients to plants and their consumers (Laakso and Setälä 1999a; Scheu et al. 1999). Aboveground and belowground food webs thus do not function in isolation from each other.

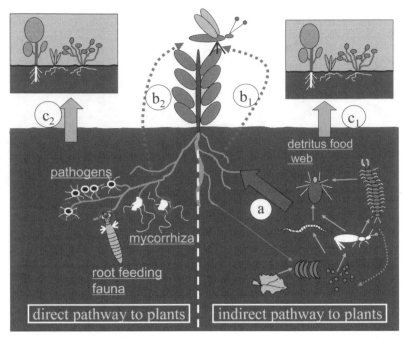

Figure 12.1. Aboveground communities as affected by both direct and indirect consequences of food web organisms. Right: Feeding activities in the detritus food web (slender
arrows) stimulate nutrient turnover, plant nutrient acquisition (a), and plant performance,
and thereby indirectly influence aboveground herbivores (b_1). Left: Soil biota exert direct
effects on plants by feeding on roots and forming antagonistic or mutualistic relationships
with their host plants. Such direct interactions with plants influence not just the performance of the host plants themselves but also that of the herbivores (b_2) and potentially their
predators. Furthermore, the soil food web can control the successional development of
plant communities both directly (c_2) and indirectly (c_1), and these plant community
changes can in turn influence soil biota. (Reproduced from Wardle et al. 2004 with permission from the American Association of the Advancement of Science. Drawing by Heikki
Setälä)

Most studies recognizing that aboveground and belowground biota can indirectly affect one another have emphasized the influence of aboveground consumers on the provisioning by plants of resources for belowground consumers,
and vice versa (Wardle 2002). However, lower-level consumers in both the
aboveground and belowground food webs are in turn eaten by their predators.
Therefore, there is increasing evidence that regulation by top predators of their
prey can exert cascading effects on the lower trophic levels in both aboveground and belowground food webs (Figure 12.2). The issue of whether
aboveground trophic cascades can indirectly influence belowground food webs

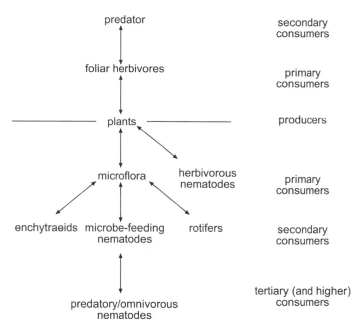

Figure 12.2. Linkages of aboveground and belowground food chains, focusing on the groups investigated in the study by Wardle et al. (2005). Here, an aboveground trophic cascade induced by top predators altered the plant community and also had trickle-through effects on several components of the soil biota. (From Wardle et al. (2005). Reproduced with permission from Wiley–Blackwell Publishing)

has seldom been explored, despite several studies pointing to aboveground cascades affecting production by plants of resources needed by soil organisms (Pace et al. 1999). Similarly, the indirect consequences of trophic cascades in soil food webs for aboveground organisms has been little studied, despite several studies pointing to belowground cascades affecting decomposition and nutrient mineralization processes known to influence nutrition and growth of plants and the animals that eat their foliage (Laakso and Setälä 1999a; Lawrence and Wise 2004).

In this chapter, I aim to draw together our knowledge on the role of two separate drivers of ecological processes: trophic cascades driven by predators and aboveground–belowground feedbacks. In doing this I will first consider the aboveground consequences of trophic cascades that occur in the soil. Second, I will address the belowground consequences of trophic cascades that occur aboveground. Third, I will focus on a case study that demonstrates how an invasive aboveground predator can exert cascading effects at an ecosystem-level scale both aboveground and belowground. Most terrestrial studies on trophic cascades are focused solely on one or the other side of the aboveground–belowground interface, and the ultimate goal of this chapter will therefore be to show how upper-level consumers may drive feedbacks between the aboveground and belowground components of ecosystems, and the consequences of these relationships for the functioning of terrestrial ecosystems.

DO TROPHIC CASCADES IN SOIL AFFECT WHAT WE SEE ABOVEGROUND?

The presence of trophic cascades in the decomposer food web has been long established, and the potential of these cascades to alter ecosystem properties was recognized for soil food webs long before it was recognized for aboveground food webs. Santos et al. (1981) showed that the experimental removal of predatory tydeid mites promoted bacterial-feeding nematode prey, reduced the bacterial prey that the nematodes feed on, and in turn reduced microbial decomposition of plant litter (indirect pathway in Figure 12.1). Many subsequent studies have shown that top predators in the soil food web can exert cascading effects on multiple trophic levels, ultimately affecting litter decomposition and mineralization of carbon and nitrogen, either positively (Allen-Morley and Coleman 1989; Kajak et al. 1993; Lensing and Wise 2006) or negatively (Mikola and Setälä 1998; Wyman 1998). Most studies that have identified these effects have involved manipulations of specific groups of top predators that eat other consumer organisms, notably predatory mites that eat microbe-feeding nematodes and springtails, predatory nematodes that eat microbe-feeding and herbivorous nematodes, and spiders and other large predators that eat springtails and other saprophagous animals. These effects occur even when biomasses of organisms in lower trophic levels in the food web are unaffected by the presence or absence of the predator, as long as the predator promotes production and turnover of organisms at lower levels of the food web (Mikola and Setälä 1998).

These effects of trophic cascades on decomposition and nutrient mineralization could conceivably influence the supply from the soil of plant-available nutrients and thereby influence plants and their consumers. Although it is well known that soil fauna that consume microbes and plant litter can indirectly promote plant nutrition and growth, there is also some evidence that they can indirectly affect aboveground (foliar) herbivores (Scheu et al. 1999; Wurst et al. 2003). However, the effect of trophic cascades in the decomposer food web on plant nutrition and growth is little explored. Two studies (Laakso and Setälä 1999a, 1999b) used experimental microcosms containing individual seedlings of *Betula pendula* and soil containing food webs of different structures, including with and without top predatory mites. In both cases, there were no effects of top predators on plant growth, although in one of those studies (Laakso and Setälä 1999a) they also found no effect of top predators on the basal consumer level of the soil food web (i.e., decomposer microbes). However, it is conceivable that for the cases in which top predators have strong cascading effects on soil processes, plant growth should also be responsive. Belowground influences could thus extend to the aboveground food web, although no study to date has explicitly explored this.

How widespread are trophic cascades in soil food webs? Knowing the answer would help us predict their importance in affecting aboveground biota. First, whether or not trophic cascades occur in decomposer communities depends on environmental context. For example, Lenoir et al. (2007) used microcosms to show that experimental addition of predators (notably gamasid mites) induced trophic cascades in nutrient-poor but not nutrient-rich substrates. Furthermore, Lensing and Wise (2006) found trophic cascades driven by spiders to be indirectly by influenced decomposition in a high-rainfall site when moisture was reduced but not when moisture was ambient or at a low-rainfall site. Second, the occurrence and importance of trophic cascades are strongly driven by the types of organisms present. Decomposer food webs consist of both fungal-based and bacterial-based energy channels (Moore and Hunt 1988); lower trophic levels in the fungal-based channel are often regulated by bottom-up forces (i.e., resources), whereas those in the bacterial-based channel are regulated by top-down forces (i.e., predation) (Wardle and Yeates 1993). Several lines of evidence suggest that trophic cascades should be more important when the bacterial-based rather than the fungal-based energy channel dominates (Wardle 2002). Many predators that cause belowground trophic cascades have small body sizes (e.g., predatory mites and nematodes). Many animals with larger body sizes are unlikely to induce cascades because they enter mutualistic rather than antagonistic relationships with lower trophic levels (e.g., millipedes, earthworms, termites, many collembolans) (Lavelle 1997). Exceptions involve large-bodied soil organisms that are entirely predatory, such as spiders (Lawrence and Wise 2004). Trophic cascades in decomposer food webs, and potentially their indirect effects on aboveground biota, are therefore highly dependent on the relative abundances of certain types of belowground predators. However, much remains unknown about what sorts of environmental conditions are conducive to allowing belowground predators to reach sufficient densities to induce trophic cascades or in which types of ecosystems they should be the most important.

Belowground biota can influence plant growth and aboveground consumers not just indirectly through mediating decomposition processes, as described earlier, but also through more intimate associations with plant roots via the direct pathway (Figure 12.1). Mycorrhizal fungi, root herbivores, and root pathogens can greatly influence the growth, foliar nutrient concentrations, and secondary defense chemistry of plants aboveground. This has in turn been shown to exert important effects not just on foliar herbivores (Masters et al. 1993; Gange and West 1994) but also on pollinators (Poveda et al. 2005) and higher trophic level groups such as parasitoids (Gange et al. 2003; Soler et al. 2007). Such root-associated biota can in turn be consumed by soil animals, indirectly influencing plant growth (Klironomos and Kendrick 1995; Setälä 1995; Strong et al. 1999; Denno et al. 2008). All such studies to date have focused on predatory invertebrates, and little is known about how soil-dwelling predatory mammals that feed on root herbivores such as shrews and moles affect the aboveground subsystem. Multitrophic belowground interactions that involve root associates may also affect aboveground consumers, although this issue has not been directly tested experimentally. However, it could be hypothesized that aboveground consequences of belowground trophic cascades involving root-associated biota may be stronger than cascades involving only decomposers because interactions between root-associated biota and plants are much more direct (Figure 12.1) (Wardle et al. 2004).

DO ABOVEGROUND TROPHIC CASCADES TRICKLE BELOWGROUND?

Just as belowground biota influence what we see aboveground, there is ample evidence that aboveground consumers affect what we see in the soil. There is accumulating evidence that foliar herbivores can indirectly influence the soil biota by altering the quantity and quality of resources that plants return to the belowground subsystem. These effects occur through herbivores either inducing physiological changes at the whole plant level or causing replacement of some plant species with others and thus altering vegetation composition (Pastor et al. 1988; Bardgett and Wardle 2003). Herbivore-induced changes in vegetation composition can simultaneously influence multiple trophic groups in the soil food web via alterations in their abundance, community composition, and species diversity (Mikola at al. 2001; Wardle et al. 2001). Effects of herbivores on decomposers can be either negative or positive depending on context (Bardgett and Wardle 2003). The effects of herbivory on decomposer communities in turn influence the ecosystem-level processes they drive, such as decomposition of plant litter and mineralization of soil carbon and nitrogen (Pastor et al. 1988; Suominen 1999). Foliar herbivores also affect the allocation by plants of resources to the soil biota intimately associated with the root system, such as mycorrhizal fungi (usually negatively) (Gehring and Whitham 1994), root-feeding fauna (Masters et al. 1993), and root pathogens (Van der Putten et al. 2001).

Although many studies have considered the effects of plants and herbivores on belowground biota and the processes that they drive, remarkably few studies have considered the belowground consequences of predation on herbivores or whether aboveground trophic cascades have effects that trickle belowground. Aboveground trophic cascades could plausibly exert belowground effects, given that aboveground cascades can greatly affect production by plants of resources known to drive the decomposer subsystem (Letourneau and Dyer 1998; Post et al. 1999; Feeley and Terborgh 2005; Chapter 8, this volume). In an experimental study involving an understory forest shrub, Dyer and Letourneau (2003) showed that species diversity within each of three consumer trophic levels in a decomposer food web was unaffected by manipulation of a top predatory beetle that feeds on herbivorous and possibly detritivorous insects. Diversity within each trophic level was instead driven mainly by the manipulation of basal resources. Furthermore, Dunham (2008) performed an experiment that showed that exclusion of insectivorous birds and mammals caused enhanced seedling herbivory and that cascading effects of the insectivores indirectly enhanced several groups of soil invertebrates. Other recent experimental studies (Schmitz 2006; Frank 2008) have provided evidence that aboveground trophic cascades can indirectly affect the rates of soil nitrogen mineralization. Because nitrogen mineralization is a microbially driven process, the trophic cascade identified in this system must have also influenced the activity of the soil microbial community.

The trickle-through effect of aboveground trophic cascades on the soil food web was experimentally investigated in model communities by Wardle et al. (2005); treatments included soil only, soil + plants, soil + plants + aphids, and soil + plants + aphids + predators. In this study, the plant and aphid treatments consisted of mixtures of three and four species, respectively, and the predator treatments consisted of the lacewing (*Micromus tasmaniae*) and the ladybird beetle (*Coccinella undecimpunctata*), both of which consume aphids. The indirect effects of aphid and predator treatments on the plant community and several trophic levels of the soil food web were then monitored (Figure 12.2). Aphids had negative effects on plant biomass that were reversed by *Micromus*, and both predator species caused large changes in the relative abundances of dominant plant species. Predators of aphids also indirectly affected several components of the belowground subsystem. *Micromus* had positive indirect effects on the primary consumer of the soil decomposer food web (microflora), probably by indirectly promoting greater input of plant-derived resources to the decomposer subsystem. Predator treatments also indirectly influenced densities of belowground tertiary consumers (top predatory nematodes). However, the secondary consumers of the soil food web (microbe-feeding nematodes) were unresponsive. The fact that some belowground trophic groups but not others responded to aboveground manipulations is probably due to top-down and bottom-up forces differentially regulating different trophic levels. Aphids also affected microbial community structure, promoted the ratio of bacteria to fungi, and increased diversity of herbivorous nematodes; these effects were always at least partially reversed by *Micromus*. The results of this study provide evidence that aboveground predator-induced trophic cascades are capable of simultaneously influencing multiple trophic levels of soil food webs.

At a larger scale, large herbivorous mammals exert important effects on the quantity and quality of resources that enter the belowground subsystem, and this can have either positive or negative effects on soil processes and organisms through a variety of mechanisms (Bardgett and Wardle 2003). Different mechanisms dominate under different environmental contexts (Bardgett and Wardle 2003) and under different densities of herbivores within the same habitat (Olofsson et al. 2004). For example, grazing ungulates in African savannas can accelerate soil process rates by returning organic matter to the soil as labile fecal material (McNaughton et al. 1988), whereas moose in boreal forests can retard soil processes by promoting unpalatable plant species that produce poor-quality litter (Pastor et al. 1988). Trophic cascades induced by predators of these herbivores are likely to reverse these effects. Therefore, there are many instances in which predators of mammalian herbivores are well known to have important indirect effects on the plant community (Chapter 10, this volume) and the likely return of resources to the belowground subsystem (Chapter 9, this volume). For example, reintroduction of wolves into Yellowstone National Park reduces densities of ungulates, and this in turn indirectly decreases the rate of soil nitrogen mineralization (Frank 2008) and presumably the soil organisms that drive this mineralization.

Humans can also function as important predators of megaherbivores and thereby cause trophic cascades that are likely to affect soil processes and ecosystem functioning. This is apparent through human-induced extinctions of megafaunal populations that have occurred in many parts of the world. For example, it has been proposed that megaherbivore extinctions in Alaska and Russia more than 10,000 years ago have led to the change of vegetation from grazed steppe to moss-dominated tundra (Zimov et al. 1995). This has involved increasing domination by plant species that produce poor-quality litter and probably led to the retardation of soil processes such as nitrogen mineralization, leading to a feedback that maintains dominance by the tundra. Furthermore, human reductions of megaherbivores may greatly alter fire regimes at the landscape scale (Flannery 1994; Chapter 16, this volume). This will inevitably have important consequences for the functioning of the belowground subsystem and feedbacks aboveground, given the wide-ranging effects of fire on soil organisms and the processes that they drive (Wardle 2002; Certini 2005).

INVASIVE PREDATORS AND TROPHIC CASCADES: THE RATS AND SEABIRDS SYSTEM

Some of the strongest documented evidence for the effects of trophic cascades transcending the aboveground–belowground interface and affecting ecosystem processes involves invasions by top predators into new ecosystems. For example, invasion of Christmas Island in the Indian Ocean by the predatory yellow crazy ant locally extirpated the red land crab, which otherwise serves as the island's dominant primary consumer (O'Dowd et al. 2003). By eliminating the crab, a predator of seeds and seedlings and a consumer of leaf litter, the ants indirectly promote tree seedling establishment but retard litter decomposition. Introduction of predatory foxes to islands in the Aleutian Island chain has resulted in substantial reductions of seabird colonies and thwarted nutrient transfer by these birds from the ocean to the land (Croll et al. 2005; Maron et al. 2006). This has cascading effects by reducing soil fertility, altering plant community structure, and changing the flow of nitrogen through consumer trophic levels. In a belowground example, the accidental introduction of the predatory New Zealand flatworm to the British Isles and Faroe Islands has caused the widespread reduction of earthworms, resulting in reduced soil porosity and waterlogging, greater domination by rushes, and reduced populations of moles (Boag 2000).

The case study I will explore here is the Rats and Seabirds Project, an ongoing study initiated in 2003 (Fukami et al. 2006; Wardle et al. 2007, 2009; Mulder et al. 2009; Towns et al. 2009) (see also www.landcareresearch.co.nz/research/ecosystems/rasp/index.asp). This project focuses on forested oceanic islands off the coast of northern New Zealand. When seabirds are present, they exert two major effects on the ecosystem. First, they transport significant quantities of nutrients from the ocean to the land. Second, during nesting they create extensive networks of burrows in the soil, and by doing so they serve as an agent of physical disturbance. Several of these islands have been invaded by rats spread by European contact (i.e., *Rattus norvegicus* and *R. rattus*) or the Pacific rat (*R. exulans*), whereas other islands have never been invaded. These rats serve as top predators, and when present they feed on seabird chicks and eggs and therefore severely reduce seabird densities. They therefore potentially induce cascading effects through the ecosystem, by mitigating the ecological impacts of their prey. The Rats and Seabirds Project focuses on nine islands that are dominated by seabirds and lack rats and nine that have been invaded by European rat species between 150 and 50 years ago, with each island serving as an independent replicate ecosystem (Figure 12.3).

It was shown that soils from the rat-invaded islands had on average much lower concentrations of nutrients than soils from uninvaded islands, and this in turn exerted a strong effect on many components of the belowground food web (Fukami et al. 2006; Towns et al. 2009). Primary consumers in the litter or soil layers, such as herbivorous nematodes and land snails, and secondary consumers, such as enchytraeids, microbe-feeding nematodes, rotifers, collembola, and chilopods, all showed a statistically significant negative response to rat invasion. Soil microbes and tertiary consumers (predatory nematodes) were also less on invaded islands, although this response was not statistically significant. The simplest explanation for this outcome is that predation by rats on seabirds has reduced nutrient inputs to the soil, which has resulted in strong bottom-up effects that have cascaded through the soil food web. Rat invasion was also found to influence the rates of key processes that are driven by the soil biota, such as total respiration rates in the litter layer and decomposition of plant litter.

The effects of rat invasion are also manifested aboveground, and this occurs through two main mechanisms. First, by eliminating seabirds, rats induce nutrient limitation of plants. Plants grown in soil collected from invaded islands on average show less growth and uptake of nutrients than those grown in soil from uninvaded islands (Fukami et al. 2006). Leaves and litter collected from plants on invaded islands also have lower concentrations of nutrients (notably nitrogen) than these tissues collected from the same species on uninvaded islands

Figure 12.3. Study system used for the Rats and Seabirds Project. **(a)** Aorangaia (5.6 hectares), a typical island used in this study. **(b)** Forest floor on Tawhiti Rahi, a rat-free island. **(c)** Forest floor on Aiguilles, a rat-invaded island. Rat-free islands are characterized by dense seabird burrows on forest floor, such as those of Buller's shearwater (*Puffinus bulleri*), shown in **(b)**. Burrow entrances are up to 50 centimeters wide, some of which are indicated by arrows in **(b)**. Rat-free islands are in sharp contrast to rat-invaded islands, where seabird burrows are virtually nonexistent because of rat predation of seabirds **(c)**. (From Fukami et al. 2006. Reproduced with permission from Wiley–Blackwell Publishing)

(Mulder et al. 2009; Wardle et al. 2009). Fresh foliage and litter collected from uninvaded islands from each of twelve perennial plant species, ranging from ground-dwelling species to emergent trees, showed that resorption of nitrogen (but not of phosphorus) was consistently greater for plants growing on invaded islands (Wardle et al. 2009). Furthermore, release of nitrogen (but not phosphorus) from each of these species during litter decomposition was much less for plants collected from invaded islands. These lines of evidence point to greater nitrogen limitation and tighter cycling of nitrogen on rat-invaded islands deprived of the inputs provided by seabirds (Wardle et al. 2009).

Rats exert a second effect by reducing disturbances to plants caused by seabirds. Islands with rats support a greater density of establishing tree seedlings (Fukami et al. 2006; Mulder et al. 2009), and a higher tree standing biomass (Wardle et al. 2007), through reduced seabird trampling and burrowing activity. Therefore, inhibition of seabirds by rats has two opposing effects on vegetation: reducing the negative effects of disturbance and inducing nutrient limitation.

This study also highlights the effects that invasive predators can exert on ecosystem carbon storage. Although there has been much recent interest in the effects of human-induced global climate change on carbon sequestration, very little attention has been given to the effects of invasive biota relative to the attention given to other global change drivers. Ecosystem carbon storage (per unit land area) has been quantified both aboveground and belowground for islands with and without rats (Wardle et al. 2007). Overall, rat invasion caused carbon storage in plant biomass to increase by 104 percent. This was probably because seabirds damage tree roots during burrowing and nesting and suppress seedling establishment, thus reducing forest standing biomass; predation by rats of seabirds reverses this effect. These negative effects of seabirds on plant biomass more than offset their positive effects on plant growth through increasing nutrient inputs (Fukami et al. 2006). In contrast, invasion by rats caused a 26 percent decline in carbon storage in nonliving (mainly soil) pools, probably by thwarting the transfer by seabirds of carbon from ocean to the land. The net effect of this increase in aboveground carbon storage and decreased belowground carbon storage is that rat invasion has caused total ecosystem carbon sequestration to be increased by 37 percent. If coastal and island forests were to be managed solely for the purpose of enhancing carbon sequestration, then invasion of rats in new ecosystems would be perceived as beneficial. This highlights the problems that can be inherent in managing natural ecosystems for goals other than biodiversity conservation.

Removal of invasive predators from islands is appropriately seen as beneficial, and numerous rat eradication programs have therefore been conducted in many regions worldwide, including in New Zealand. Mulder et al. (2009) studied the islands featured here, plus additional islands in the same region from which rats had been controlled or eradicated. Eradication efforts on these islands began between 22 and 11 years ago. Measurement of several variables, including plant community properties, leaf characteristics, and soil properties, revealed that rat eradication efforts had not led to any detectable recovery of these islands. Recovery did not occur presumably because seabirds did not recolonize these islands and restore the transfer of nutrients from sea to land. It appears likely that without the return of seabirds, islands that have had rats eradicated may differ in community and ecosystem attributes from both invaded and uninvaded islands, potentially leading to a new type of ecosystem (Mulder et al. 2009). Restoration of these islands may therefore require not only rat removal but also deliberate efforts to reintroduce seabirds.

The Rats and Seabirds Project focuses on offshore islands, but the types of effects that we found are likely to be important elsewhere, given the widespread occurrence of introduced predatory mammals on seabird-dominated islands worldwide (Courchamp et al. 2003b). In this light, the disappearance of extensive nesting seabird communities from much of the mainland of New Zealand over the past few centuries due to the introduction of predators (notably rats) (Worthy and Holdaway 2002) has probably exerted wide-ranging effects in many forests in New Zealand, similar to those described earlier. Before the introduction of predators to New Zealand, seabirds nested 50 kilometers or more inland from the coast (Worthy and Holdaway 2002), so the types of results obtained in this study may not necessarily be restricted to coastal forests. Ultimately, this case study highlights the wide-ranging cascading effects at the community and ecosystem level resulting from predator introduction to predator-free ecosystems.

CONCLUSIONS

Although the importance of trophic cascades in terrestrial ecosystems has been debated, there are a growing number of documented examples in which top predators can have important cascading effects on lower trophic levels for both aboveground (Pace et al. 1999) and belowground (Wardle 2002) food webs. The concepts and examples presented in this chapter highlight the potential significance of trophic cascades crossing the aboveground–belowground interface. Although few studies have explicitly explored this issue, evidence is accumulating that important multitrophic linkages mediated by plants extend across the aboveground–belowground interface. An improved understanding of the linkages between aboveground and belowground trophic relationships is essential for improving our knowledge about how the two subsystems interact.

Trophic cascades influence not just densities or biomasses of organisms in lower trophic levels, such as primary producers (plants) and primary decomposers (bacteria and fungi), but also the key ecosystem-level functions that they perform, such as ecosystem production, decomposition, and mineralization of carbon and nutrients. Soil biologists have long recognized that belowground top predators and the cascades they induce can indirectly influence decomposition and nutrient release from plant litter (Santos et al. 1981); this has obvious implications for plants and their consumers. Similarly, studies in the past half-decade have begun to reveal that aboveground trophic cascades can influence the quality and quantity of resources that plants return to the soil, potentially influencing belowground consumers and the processes they regulate. There is also evidence that trophic cascades can strongly influence ecosystem processes even when the effect of the cascade on the biomass of organisms in lower trophic levels is weak or nonexistent, by influencing the turnover of these organisms (Mikola and Setälä 1998). There is increasing evidence that strong interactions between aboveground and belowground food webs can influence ecosystem functioning, and understanding the interactions and feedbacks between them should greatly enhance our understanding of the biotic controls of ecosystem functioning.

The literature on trophic cascades from an aboveground–belowground perspective remains somewhat disparate, and developing general principles about the conditions under which trophic cascades should be important remains a major challenge. In the belowground subsystem, some studies have found evidence of strong trophic cascades in the soil food web and effects on soil processes, whereas others have not (Wardle 2002). Although these differences may be driven by environmental conditions such as climate (Lensing and Wise 2006) or soil fertility (Lenoir et al. 2007), we are still not in a strong position to predict the ecosystems for which belowground trophic cascades may be important as ecosystem drivers. However, for the aboveground food web, there is evidence that large mammal effects on decomposer processes are often positive in fertile conditions and negative in infertile conditions (Bardgett and Wardle 2003); these effects should be reduced by predators and enhanced by the elimination of predators. An improved understanding of these types of effects both aboveground and belowground would be useful in predicting whether the gain or loss of predator species is likely to be reversible or propel the ecosystem to an alternative stable state (Chapter 17, this volume); however, our knowledge of most systems is insufficient to make this type of prediction (but see Mulder et al. 2009).

The strongest documented examples of top predators and trophic cascades affecting aboveground–belowground linkages arguably involve human-induced alterations of densities of top predators and the effect of humans themselves operating as predators (Wardle and Bardgett 2004). There are many situations in which top predators invading new localities can alter community and ecosystem properties both aboveground and belowground, especially whenever the prey species is naive of predators. This includes the historic widespread extirpation of megaherbivores by humans over much of the earth's terrestrial surface (Zimov et al. 1995; Chapter 16, this volume). Furthermore, humans have directly or indirectly reduced or eliminated populations of top predators throughout the terrestrial realm (see Chapter 14, this volume); this is likely to have important cascading effects on both sides of the aboveground–belowground interface. Understanding these sorts of effects is essential for recognizing how human-induced changes in top predator densities may influence ecosystem functioning and the consequences of restoration efforts that involve reversing these changes.

ACKNOWLEDGMENTS

I thank my colleagues on the Rats and Seabirds Project for development of data sets and ideas explored in this chapter, notably Peter Bellingham, Karen Boot, Tadashi Fukami, Christa Mulder, Dave Towns, and Gregor Yeates. Two anonymous reviewers and John Terborgh gave very useful feedback on this chapter.

PART III

Predation and Ecosystem Processes

The chapters in Parts I and II of this book have considered top-down forcing and trophic cascades in various kinds of ecosystems, or in several cases in particular places. By implication, the regulatory processes occurred through what are often called density-mediated effects, or the flux of energy and materials between consumer and prey. Almost everyone recognizes that food web dynamics are more complex than this. For instance, consumer–prey interactions are often mediated by behavior; equilibrial states in the interactions between consumers and their prey can occur at strikingly different abundances in otherwise similar environments because of nonlinear processes; prey vulnerability to consumers and consumer access to prey often vary with body size; and plant communities across much of the terrestrial world are regulated not only by herbivores but by water and fire. How do these processes modify the dynamics of species interactions and the consequences of those interactions for the structure and organization of communities and ecosystems? The chapters in this part provide some of the answers to these questions.

Classical Hairston, Smith, and Slobodkin (HSS) trophic cascades involve a simple three-trophic-level predator–herbivore–plant food chain. In Chapter 13 Brashares and colleagues investigate some of the consequences of top-down forcing processes in food chains that either are longer than those envisioned by HSS or do not end at the bottom with autotrophs. The loss of consumers at higher trophic levels in these less traditionally viewed food chains often results in what is commonly known as mesopredator release. The materials in Chapter 13 establish that with the selective loss of large apex predators, mesopredator release is rampant in nature. Just as the removal of apex predators from simple three-trophic-level cascades causes uncontrolled herbivory, mesopredator release has caused widespread predation-induced declines in many other lower trophic forms, including numerous small mammals, birds, and herps. The chapter concludes with a fascinating but unsettling account of the causes and consequences of baboon population irruptions across sub-Saharan Africa. Chapter 14 (Berger) treats what has been referred to in popular parlance as the ecology of fear and more recently in ecological circles as trait-mediated effects. In fitness terms, the cost of being eaten is nearly always substantial; therefore, it is not surprising that prey have adaptations to reduce the risk of that happening. For mobile species this includes such behavioral modifications as vigilance, habitat selection, migration, and grouping. Chapter 14 provides an overview of the diversity, lability, and ecosystem-level consequences of these trait-mediated effects. Body size is yet another attribute that influences the ease with which predators can subdue their prey, the ability of prey to avoid being eaten, and thus the strength of a consumer–prey interaction. These various effects of body size are explored in Chapter 15 (Sinclair et al.) for the large mammal predator–prey systems in the Serengeti, one of the few remaining terrestrial ecosystems with a largely intact fauna. This elegant and important body of research establishes that increased body size confers an advantage to both predators (through their ability to consume a greater diversity of prey) and prey (through their ability to resist predators), which when coupled with the influences of herding, migration, weather, and fire helps explain long-term changes in the landscapes of east African savannas. The interplay between predation, body size, herbivory, and fire is further explored in Chapter 16 (Bond). As a prelude to his overall conclusion that bottom-up forcing processes alone do not adequately explain vegetation dynamics, Bond reminds us that the distribution of forests and grasslands across mid- to lower-latitude terrestrial ecosystems is not well predicted by temperature, rainfall, and soil alone. He then goes on to argue that intense herbivory (by predation-resistant megaherbivores or in predator-free ecosystems) or fire tends to drive what would otherwise be woodlands to savannas and grasslands but that grazing also limits the spread of fire. Chapter 17 (Scheffer) briefly discusses the dynamic properties of entire ecosystems, focusing in particular on how abrupt shifts can occur between phase states and the conditions under which such phase state shifts are characterized by hysteresis. These phenomena are especially important in the management and conservation of predators because they imply that abrupt transitions between phase states may depend not only on predator population density but on whether that density is achieved from a population that is trending upward or downward.

CHAPTER 13

Ecological and Conservation Implications of Mesopredator Release

Justin S. Brashares, Laura R. Prugh, Chantal J. Stoner, and Clinton W. Epps

Human knowledge of the ability of predators to regulate prey goes back at least 10,000 years, when cats were domesticated to control rodent pests in Syria and Turkey. However, our appreciation of the ability of predators to regulate other predators and the importance of these interactions for community structure is much more recent. Although ecological narratives (e.g., Leopold 1949) and studies of pest control in agricultural systems have long considered interactions between predators to be important to the dynamics of communities, the impact of top predator removal on intermediate predators was seldom considered empirically before the 1980s (see Terborgh and Winter 1980; Wilcove 1985). During this period, growing interest and concern about habitat fragmentation, the troubled status of large carnivores, and the apparent increase and spread of smaller predators led to a growth in studies examining the ecological consequences of top predator decline that continues today. A review of recent literature shows that the release of intermediate predators from top-down control is a critical component of many, though not all, trophic cascades (e.g., Polis et al. 2000; Table 13.1). In fact, over the last 15 years a multitude of studies from both terrestrial and aquatic systems around the world have examined the causes and consequences of what Hollywood might call *The Rise of the Mesopredator.*

Table 13.1. Characteristics of studies that have examined mesopredator release. Apex predator, mesopredator, and prey species of interest are shown, in parentheses if the species was not studied directly.

Reference	Location	Continent	Apex	Mesopredator	Prey	Method	Scale	Shape
Ball et al. (1995)	Montana	North America	Coyote	Red fox	Ducks	o	m	l
Barton (2005)	Florida	North America	Raccoon	Crab	Sea turtle eggs	e	s	t
Berger et al. (2008)	Wyoming	North America	Wolf	Coyote	Pronghorn	o	m	l
Brashares, this chapter	Africa	Africa	Lion	Baboon	Ungulates	o	l	t
Burkepile and Hay (2007)	Florida Keys	Atlantic Ocean	Fish	Gastropod	Coral	e	s	l
Catling and Burt (1995)★	Australia	Australia	Dingo	Red fox	(Unspecified)	o	l	
Crooks and Soulé (1999)	California	North America	Coyote	Cat	Birds and lizards	o	m	l
Elmhagen and Rushton (2007)	Sweden	Europe	Wolf, lynx	Red fox	(Hares, grouse)	o	l	t
Estes et al. (1998)	British Columbia and Alaska coast	Pacific Ocean	(Killer whale)	Sea otter	Sea urchin	o	l	l
Frank et al. (2005)	Scotian Shelf	Atlantic Ocean	Cod	Shrimp, crab	Zooplankton	o	l	t
Gehrt and Prange (2007)★	Illinois	North America	Coyote	Raccoon	(Unspecified)	o	s	
Helldin et al. (2006)	Sweden	Europe	Lynx	Red fox	(Unspecified but probably shared)	o	m	t
Henke and Bryant (1999)	Texas	North America	Coyote	Badger, bobcat, gray fox	Rodents	o	m	t
Johnson et al. (2007)	Australia	Australia	Dingo	Red fox	Marsupials	o	l	t
Lloyd (2007)★	South Africa	Africa	(Jackal, caracal)	(Cat, badger, fox, mongoose)	Birds	o	l	l
Maezono et al. (2005)	Japan	Asia	Bass, bluegill	Crayfish	Invertebrates	e	s	l
Maina and Jackson (2003)	Kenya	Africa	Dog, leopard, lion	(Mongoose, civet, others)	Songbirds	o	m	l
Myers et al. (2007)	Eastern U.S. coast	Atlantic Ocean	Sharks	Cownose ray	Bay scallop	o	l	l

Pacala and Roughgarden

Study	Region	Location	Apex predator	Mesopredator	Prey	Method	Scale	Shape
Pacala and Roughgarden (1984)	North America	Caribbean island	Lizard	Spider	Insects	e	s	t
Palomares et al. (1995)	Europe	Spain	Lynx	Mongoose	Rabbit	o	m	t
Polis and McCormick (1986)	North America	California	Scorpions	Spiders and solpugids	Arthropods	e	s	t
Rayner et al. (2007)	Australia	New Zealand	(Cat)	(Rat)	Petrel	o	s	t
Rogers and Caro (1998)	North America	Michigan	Coyote	Raccoon	Song sparrow	o	s	1
Salo et al. (2008)	Europe	Finland	Sea eagle	American mink	(Birds, voles, amphibians)	o	m	1
Schmidt (2003)	North America	Illinois	(Coyote)	Raccoon	Birds	o	1	1
Schoener and Spiller (1987)	North America	Bahamas	Lizard	Spider	Insects	e	s	t
Sieving (1992)	North America	Panama	(Jaguar, eagle)	Monkeys, pigs, others	Birds	o	s	1
Soulé et al. (1988)	North America	California	Coyote	Gray fox	Birds	o	m	1
Sovada et al. (1995)	North America	North and South Dakota	Coyote	Red fox	Ducks	o	1	1
Stallings (2008)	North America	Bahamas	Nassau grouper	Smaller groupers	Other reef fish	e	s	1
Terborgh et al. (1997a)	South America	Venezuela	(Unspecified)	Monkeys	Birds	o	1	1
Wangchuk (2004)	Asia	Bhutan	Wild dog	Wild boar	(Unspecified)	o	1	1
Wilcove (1985)	North America	Eastern U.S.	(Mountain lion, bobcat)	(Cat, raccoon, opossum, others)	Birds	o	1	1
Wright et al. (1994)*	North America	Panama	Puma, jaguar	Mid-sized mammals	Small to mid-sized mammals	o	1	t

*Results from these studies are not interpreted by their authors as supporting mesopredator release (all other studies are presented as supporting the release hypothesis).

Method: e = experimental; o = observational. Studies were considered to be experimental if apex predators were reduced or excluded from replicated areas, and mesopredator and prey populations were monitored on experimental and control plots. Scale: s = small (0–100 km²), m = medium (100–1,000 km²), l = large (>1,000 km²). The shape of the apex–mesopredator–prey interaction is categorized as triangular (t) if the apex and mesopredator both rely on the focal prey and linear (l) if the apex predator relies on different prey than the mesopredator (listed focal prey are those eaten by the mesopredator in the study).

MESOPREDATOR RELEASE DEFINED

The term *mesopredator release* first appeared in the scientific literature in 1988 (Soulé et al. 1988), but the concepts behind it can be traced back several decades (e.g., Paine 1969a; Terborgh and Winter 1980; Pacala and Roughgarden 1984; Wilcove 1985). Like many somewhat new and popular terms in the lexicon of science, *mesopredator release* has seen alterations and, perhaps, misapplications since it was coined 20 years ago. In the context of Soulé et al.'s (1988) article, the term was created to describe a process in which predators of intermediate body size (foxes and domestic cats) were more prevalent in the absence of a larger predator (coyote) and showed an increased effect on prey species (birds). The term is now often defined as any case in which a predator weighing approximately 3–20 kilograms plays an increasingly important role in regulating prey (Gehrt and Prange 2007). This common usage is problematic for two reasons: The somewhat arbitrary body size threshold has made the term unnecessarily restrictive and disjointed from ecological principles, and it places an emphasis on the impacts of mesopredators on prey rather than on the interactions of apex predators and mesopredators, where it belongs. In light of this confusion, we suggest the term *mesopredator release* be recast as a narrowing of Crowell's (1961) concept of ecological release (see also Terborgh and Faaborg 1973) to describe scenarios in which the absence or negative change in the density or distribution of an apex predator results in an expansion in density, distribution, or behavior of a middle-rank predator in a trophic web, irrespective of their relative or absolute body sizes. In most cases in which the term has been applied, mesopredator release is demonstrated or hypothesized to have a negative effect on the survival or productivity of predators and prey at lower trophic levels (popularly known as the mesopredator release hypothesis). However, such cascading effects should not be viewed as a required component of the term's definition.

CHAPTER OVERVIEW

In this chapter, we examine the history, significance, and practical implications of more than two decades of research on mesopredator release. We also briefly recount our own work on the causes and many consequences of the release of a primate mesopredator, the olive baboon (*Papio anubis*), in West Africa. An important manifestation of trophic cascades, the observation of mesopredator release provides ecologists with rare opportunities to illuminate and isolate complex ecological relationships that otherwise are difficult to detect, much less measure. However, to the conservation biologist, the cascading effects or aftershocks that often accompany mesopredator release are cause for alarm and suggest that communities are easily pushed over thresholds to alternative ecosystem states. Such state changes may lead to dramatic shifts in community structure or, potentially, irreversible ecosystem meltdown (Chapter 4, this volume). In this chapter, we consider these and other aspects of mesopredator release as we set out to review the mesopredator release literature and, in the process, search for patterns among studies; discuss challenges, both conceptual and practical, that are faced in studies of mesopredator release; and examine the conservation implications of mesopredator release for management and policy.

REVIEWING THE LITERATURE

Likely cases of mesopredator release have been documented throughout the world in terrestrial, marine, and freshwater systems. Using Web of Science and Google Scholar searches with the terms *mesopredator**, *carnivore release*, and *mesocarnivore**, we identified thirty-four studies that investigated mesopredator release in thirty-two independent landscapes (Table 13.1). Because the term *mesopredator release* was not coined until 1988, our search probably missed earlier studies that used different terminology. Nonetheless, examination of more recent studies illuminates several interesting patterns. We begin by summarizing the systems, spatial scales, and methods that have been used to study mesopredator release. Based on results of these studies, we then highlight key factors that can affect the outcome of apex predator losses and mesopredator effects.

Mesopredator Studies

Studies of mesopredator release have been conducted most frequently in North America or in waters off the North American coast (twenty-one of thirty-four studies), commonly in systems where coyotes (*Canis latrans*) are the apex predator; cats, raccoons, skunks, or opossums are the mesopredators; and birds or rodents are the affected prey (eight studies, Table 13.1). However, studies focusing on mesopredator overabundance have been conducted in all continents except Antarctica, in a wide variety of systems, and at large spatial scales (*n* = nine state, country, or continent-wide studies). Studies often focus on ground-nesting birds as the prey of concern (fourteen studies). This is because nest depredation is the leading cause of reproductive failure for birds. Many bird populations are declining worldwide, and mesopredators are often important nest predators

(Chalfoun et al. 2002). Most of these studies used artificial nests to examine rates of nest depredation with respect to apex and mesopredator presence or abundance (e.g., Sieving 1992; Maina and Jackson 2003; Schmidt 2003). Other prey of concern include reptiles, amphibians, marsupials, rabbits, rodents, scallops, and ungulates. The weight of evidence suggests that mesopredator release is a common result of apex predator loss throughout the world.

Interestingly, the focal apex predators and mesopredators in published studies are often exotic or "undesirable" species. For example, on island systems cats are an exotic apex predator, and rats are an exotic mesopredator (Courchamp et al. 1999; Rayner et al. 2007). Similarly, in Japanese ponds, exotic bass and bluegill are apex predators and exotic crayfish are mesopredators (Maezono et al. 2005). Even when apex predators are native, such as wild dogs in Asia or coyotes in North America, they are usually considered to be undesirable because they prey on domestic livestock. Numerous studies have highlighted the perverse consequences that can result from such "nuisance control": the unleashing of even worse nuisances (i.e., mesopredators). For example, control of raccoons in Florida to protect sea turtle eggs paradoxically resulted in increased predation on the eggs because another egg predator, the ghost crab, was released from control by raccoons (Barton 2005). In contrast, one study has shown that suppression of beneficial mesopredators can damage ecological systems. In the Pacific Ocean, killer whales are an apex predator that have recently begun preying heavily on sea otters, a mesopredator that controls sea urchins. Without sea otters, urchin numbers explode, destroying kelp forests (Estes et al. 1998).

Most mesopredator studies have been observational, in which existing patterns of apex predator, mesopredator, or prey abundance have been examined over time or space, sometimes long after the apex predator has been eradicated (Wilcove 1985). Many of these studies acknowledge that land use changes are intimately connected with the loss of top predators and that both factors can lead to increased mesopredator populations. Two studies have successfully used modeling techniques to separate the contributions of land use change and apex predator removal on mesopredator abundance (Crooks and Soulé 1999; Elmhagen and Rushton 2007). Seven experimental studies have also been conducted (Table 13.1), in which apex predator removals were replicated and the abundance of mesopredators and prey was monitored on control and removal plots. Six of the seven experiments resulted in significant cascades, wherein mesopredators increased and prey numbers decreased in the absence of apex predators. Cascading effects have also been shown by observational studies, but many of these studies monitored only two of the three levels relevant to mesopredator release, and cascading effects were therefore difficult to demonstrate. Only four studies failed to find evidence supporting the mesopredator release hypothesis (Table 13.1), but this paucity of negative results could reflect publication bias. Because mesopredator release has been studied in so many different ways (e.g., behavioral changes, correlations in densities, presence or absence trends, artificial nest predation rates), a formal meta-analysis examining the hypothesis is not possible at this time.

Factors That Influence Mesopredator Release

The loss of apex predators may or may not cause mesopredator numbers to increase, and increased numbers of mesopredators may or may not cause prey populations to decline. Two factors should strongly influence mesopredator release: the productivity of the system and the strength of interactions between apex predators, mesopredators, and prey.

Theoretical ecology provides insights into the effect of productivity on mesopredator release. Theory predicts domination by mesopredators in low-productivity systems, domination by apex predators in high-productivity systems, and the greatest chances for coexistence in systems of intermediate productivity (Chapter 18, this volume; Oksanen et al. 1981; Holt and Polis 1997). This prediction has been supported by several empirical studies in laboratory and agricultural systems (e.g., Morin 1999; Borer et al. 2003). The recent study by Elmhagen and Rushton (2007) also demonstrates how the productivity of a system can moderate the strength of top-down effects such as mesopredator release. Using a spatially explicit, 90-year data set on wolf, lynx, and red fox numbers throughout Sweden, they show that top-down effects are strongest in the most productive regions. Put simply, mesopredators may be regulated by the limited abundance of prey rather than by predation in less productive areas. In productive areas, abundant food should allow mesopredator numbers to increase when control by apex predators is removed.

Regardless of system productivity, mesopredator release should be most dramatic when links between the species of interest are strong. Thus, high species diversity and wide diet niche breadths should weaken the strength of mesopredator release. Removal of an apex predator from a system with many apex predators, many mesopredators, and many prey species should not have a strong effect compared with a system dominated by a few species (Chapter 8, this volume). This was the case in one of the studies that failed to support mesopredator release (Lloyd 2007). The dramatic responses to predator loss on depauperate islands also support this idea (Terborgh et al. 1997a; Rayner et al. 2007). Likewise, if apex predators and mesopredators consume a variety of prey, the cascading impact of apex predator removal on a particular prey species should be lower than in a system with prey specialists (Hanski et al. 1991).

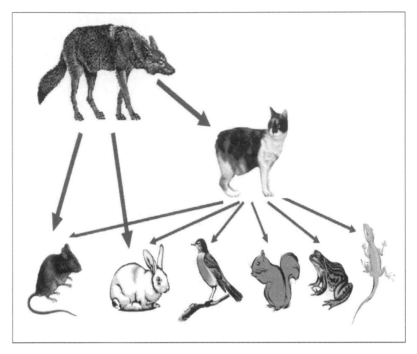

Figure 13.1. Example of a mesopredator interaction involving coyotes, domestic cats, and small vertebrates. Note the greater dietary breadth of cats compared with coyotes.

Differences in diet breadth and trophic position may explain why removal of apex predators often leads to increased predation pressure on lower trophic levels. Apex predators typically have a more restricted and carnivorous diet than mesopredators (Figure 13.1). Because mesopredators are positioned lower in the food web and tend to have a wider resource base than apex predators, they are poised to reach high densities and suppress a variety of prey species when released from top-down control (Roemer et al. 2001). Omnivorous mesopredators that switch easily between multiple food sources are demographically insulated from the rise and fall of individual prey species and thus maintain high densities where specialized predators could not. This is particularly evident in mesopredators that effectively use agricultural plants, livestock, or human refuse and reach and maintain strikingly high densities in modified environments (e.g., baboons, raccoons, and wild boars; Table 13.1).

Differences in diet breadth and population density are two reasons why mesopredator populations that increase after apex predator removal should not be regarded as simply ecological replacements. In 60 percent (nineteen of thirty-two) of communities we surveyed (Table 13.1), mesopredators relied on a distinctly different suite of prey than sympatric apex predators. A second important difference lies in the foraging efficiency of apex predators and mesopredators. Many mesopredators such as mongooses, snakes, civets, lizards, raccoons, and opossums can exploit their environment more thoroughly than the apex predators they replace because they are smaller and able to forage belowground and aboveground. This ability to scour the landscape for prey of various shapes and sizes, combined with the potential for occurring at high densities, explains how mesopredators can wreak havoc on a large portion of their community.

CHALLENGES IN DETECTING MESOPREDATOR RELEASE

Despite significant research interest, unambiguous demonstration of mesopredator release has often been difficult. The typical constraints of many ecological studies, such as short time scales and uncontrolled environmental variation, are exacerbated by the complexity of interactions at multiple trophic levels that must be examined in a test of mesopredator release. Terborgh et al. (1999) describe many of the fundamental challenges of this area of research, including the difficulty of studying apex predators that are rare or cryptic, the risk of inappropriate generalizations from studies relying solely on analysis of correlation, and the fact that mesopredator release is often invoked when the apex predator has already disappeared.

Each of the research approaches typically used has inherent limitations. For instance, meta-analyses of changes in apex predator, mesopredator, and prey populations (e.g., Gehrt and Clark 2003) may be limited by lack of correspondence in scale, habitat, and other factors in individual studies not necessarily designed to test for mesopredator release. Studies that do not track populations over time but merely examine the spatial distribution of predators, mesopredators, and prey may be confounded by uncontrolled variation in habitat (Gehrt and Clark 2003). Attempts to assess the spatial distribution of predators or prey may often reflect use rather than abundance (e.g., Crooks and Soulé 1999). Studies that track temporal changes in mesopredator or prey abundance may be unable to provide spatial replication because of the intensive monitoring needed, making it difficult to control for environmental variation or other outside influences (Rayner et al. 2007). Anecdotal accounts of population changes consistent with mesopredator release could result from natural population variability, particularly when results are not replicated in space or tracked for long periods of time (Wright et al. 1994). Finally, testing mesopredator release involves detecting a response in abundance or behavior (Ale and Whelan 2008) of the mesopredator when the apex predator is removed, but what magnitude of response really demonstrates release? Presumably, an increase in mesopredator abundance that exceeds normal short-term variability should be detected. Ideally, studies would also demonstrate that predation by the apex predator is a limiting factor for the mesopredator population. A further complication is that the consequences of mesopredator release are realized only if an effect on the population of the prey species is also detected. In addition to changes in prey abundance, biologically significant changes in diversity, richness, or biomass of the prey species community could result from mesopredator release (Henke and Bryant 1999).

Despite these challenges, several studies have addressed problems inherent in teasing apart complex interactions between apex predators, mesopredators, and prey. For instance, Myers et al. (2007) compiled long-term data sets over a wide geographic extent to show that population trends for multiple species at all three trophic levels (apex predators, mesopredators, and selected prey species) have changed in a manner consistent with mesopredator release. The extensive anecdotal evidence for mesopredator release has been bolstered by controlled experimental approaches involving removal or reintroduction of apex predators with subsequent monitoring of mesopredator and prey communities (e.g., Henke and Bryant 1999). It is also increasingly recognized that variation in habitat quality must be controlled because mesopredator release may be weak in cases where trophic regulation is more bottom up than top down, as demonstrated by Elmhagen and Rushton (2007). The inconsistent evidence that apex predators control mesopredators (e.g., Gehrt and Prange 2007) could result from that variation or the weakening effect of increased food web complexity on mesopredator release. Techniques such as path analysis (e.g., Elmhagen and Rushton 2007) may be a particularly effective method for testing the strength of various trophic interactions and inferring mesopredator release. Although local variation in trophic control and food web complexity often confounds detailed predictions about the cascading impacts of removing or reintroducing apex predators, removal of top predators can be expected to have large and obvious consequences in most ecosystems.

An ideal study following experimental manipulation of a population of apex predators would examine both the mesopredator population or community and their prey through both time and space. In practice, that ideal will remain difficult to achieve. However, future studies should carefully consider the strength of interactions between apex and mesopredator in the context of the larger food web and the likelihood of bottom-up versus top-down control. More research is also needed on the often underestimated indirect effects of predation (such as changes in prey foraging behavior; Chapter 14, this volume) on mesopredators and their prey.

CONSERVATION AND MANAGEMENT IMPLICATIONS

Mesopredator release was originally introduced in the context of highly frag-mented landscapes (Diamond 1988; Crooks and Soulé 1999). Habitat fragmen-tation catalyzes mesopredator release in part because apex predators commonly have larger area needs than the mesopredators they control and thus are more sensitive to reductions in the size of remaining habitat. Wide-ranging species of large carnivores are particularly prone to population decline because they tend to spill out from small, isolated areas into regions of high conflict with humans (Woodroffe and Ginsberg 1998). Conversely, prey species driven to decline by mesopredator outbreaks may have small area needs and be tolerated, if not de-sired (e.g., game species; Palomares et al. 1995), by humans. Therefore, meso-predator release not only is intensified by habitat fragmentation but also ex-tends fragmentation effects to species that might otherwise appear resilient to human-induced landscape changes. If fragmented landscapes that fail to retain apex predators are exceptionally vulnerable to the influences of mesopredators, as case studies from across the world indicate, then current rates of large carni-vore declines and habitat loss are certain to set the stage for mesopredator out-breaks of increasing intensity.

In addition to predator removal, other characteristics of fragmented land-scapes probably increase mesopredator abundance. Mesopredators (e.g., feral cats; Soulé et al. 1988) may be subsidized by human food sources, for example, or otherwise directly benefit from habitat modifications (Litvaitis and Villa-fuerte 1996). Under such scenarios, mesopredators are likely to experience re-lease from bottom-up constraints, and the strength of top-down effects can be expected to simultaneously increase (Elmhagen and Rushton 2007). Therefore, apex predation can be expected to play an increasingly important regulatory role as mesopredator resource availability improves. The loss of large carnivores in fragmented landscapes is of great ecological consequence not only where mesopredators are naturally heavily regulated by top predators but particularly where other fragmentation effects directly benefit mesopredators and remove resource constraints.

Population Control of Mesopredators

In the absence of apex predators, efforts to mediate the ecological conse-quences of mesopredators could be directed toward controlling overabundant populations, but several factors indicate that such management can be problem-atic. Overabundant mesopredators are likely to be characterized by high densi-ties, high rates of recruitment, and high rates of dispersal, all of which may make them resilient to control programs (Palomares et al. 1995). In addition, density-dependent responses to control efforts, public outrage, and unintended consequences for other species additionally can significantly complicate control efforts. Goodrich and Buskirk (1995) argue that to be effective, such control efforts must be intensive (and probably expensive) and therefore are less desirable management options than addressing the ecodisturbances generating overabundance. This contention is borne out by Henke and Bryant (1999), who observe that lowering coyote density by half in a 10-square-kilometer area of Texas entailed the lethal removal of 354 coyotes over a 2-year period.

Emerging studies of behavior-mediated interactions between predators and prey additionally hint that replicating the full ecosystem effects of apex predation is likely to be exceptionally difficult (Chapter 14, this volume). In a review of intraguild predation, for example, Palomares and Caro (1999) note that interactions between predators result not only in direct killing but also in avoidance in the form of shifts in space use, temporal segregation (activity pattern), and group formation. Although intensive management may be successful in controlling mesopredator densities, mimicking the behaviorally mediated ecosystem functions of apex predators is likely to be much more challenging.

Mesopredator release has also provided management lessons for eradication efforts that target both an invasive apex predator and an invasive mesopredator. Using multispecies models that accounted for the presence of two invasive predators (cats and rats) on native island birds, for example, Courchamp et al. (1999) conclude that the eradication of cats alone could result in a release in the rat population and ultimately intensified bird declines. More sophisticated models (Fan et al. 2005) similarly predict that as an apex predator, cats offer birds some degree of protection from rats. Recent field observations of island rat–petrel dynamics across systems that varied in elevation and presence of cats (Rayner et al. 2007) hint that the order of cat versus rat eradications dramatically influences the breeding success of native birds. Together, these studies and those showing the consequences of rat overabundance on nesting seabirds (Chapter 12, this volume) indicate that ignoring mesopredator release effects in control efforts could hasten rather than slow prey declines.

Perceived Costs of Large Carnivore Conservation

More than two decades of publications focusing on mesopredator release confirm that conserving apex predators is critical in preventing rippling waves of secondary extinctions (Terborgh and Winter 1980) that stretch across trophic levels (Terborgh 1988) and thus trigger widespread faunal collapse (Soulé et al. 1988). Nonetheless, the species most likely to prevent mesopredator outbreaks are large carnivore species that are not only highly sensitive to habitat loss but often the focus of human–wildlife conflict. As noted by Palomares et al. (1995), large predators have particularly suffered from a public relations crisis where they are perceived to threaten prey species of value to humans.

Mesopredator release is of conservation significance not only because of its ecological and management implications but also because it highlights the need to revise long-standing perceptions of the social and economic costs of sustaining large carnivores. Whereas large carnivores have traditionally been viewed as clear competitors with humans for game species, for example, evidence of mesopredator release points to scenarios under which large carnivores can boost populations of desirable game species (Palomares et al. 1995; Rogers and Caro 1998) and thereby reduce conflict with game ranch objectives (Palomares and Caro 1999). As research on mesopredator release develops, suggestions that carnivore conservation can yield financial benefits (e.g., Rogers and Caro 1998) are being replaced by clear examples in which local extinctions of carnivores resulted in the collapse of prey populations that are economically valuable, such as fishery collapse in the absence of sharks (Myers et al. 2007) and waterfowl declines in the absence of coyotes (Sovada et al. 1995). Therefore, although mesopredator release has generated strong ecological arguments for ensuring carnivore persistence to best protect biodiversity (Soulé and Terborgh 1999), it also suggests that increased tolerance of carnivores may be beneficial in preventing large financial losses. Furthermore, the ecological consequences of recent carnivore reintroductions (e.g., wolves in Yellowstone; Berger and Gese 2007) imply that early state changes in ecosystems that result from the loss of apex predators can be reversed with their reintroduction, providing hope for the success of restoration efforts in remedying the progression of ecosystems to undesirable alternative states. Large carnivore conservation and reintroduction efforts might similarly provide a promising solution for reversing the social and economic costs of mesopredator release that are detailed in the following case study.

CASE STUDY: OLIVE BABOONS IN WEST AFRICA

Olive baboon outbreaks in Ghana, West Africa provide a striking example of mesopredator release and, more generally, illustrate the far-reaching effects of apex predator extinctions. In 1968, the Ghana Wildlife Division initiated a monitoring program in which staff at sixty-three posts spread throughout six protected areas (hereafter called parks) recorded the type and number of

forty-one species of larger mammals encountered on monthly walking transects (Brashares et al. 2001). Transects were 10–15 kilometers in length, started
and finished at a ranger post, and were repeated monthly almost continuously
through 2004. The six parks in which monitoring occurred are characterized
by savanna habitat and, at 58–4,840 square kilometers in size, are the largest
members of a protected area system in Ghana that today includes 321 sites, 95
percent of which are smaller than 10 square kilometers (Figure 13.2). When
monitoring started in the late 1960s, wildlife communities in these parks were
largely intact, harboring a high diversity of ungulates and primates (thirty-three

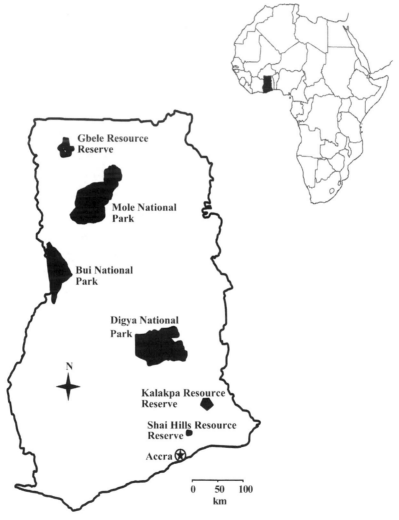

Figure 13.2. Approximate location and size of six protected areas in Ghana, West Africa,
discussed in this chapter.

species) and eight species of large carnivores, including several that today are exceedingly rare (e.g., lion, *Panthera leo*) or regionally extinct (e.g., wild dog, *Lycaon pictus*) (Brashares et al. 2001; Brashares 2003). Thus, this 37-year data set provides unparalleled insight into the dynamics of wildlife populations and communities undergoing change (Brashares and Sam 2005), and, as relates to this chapter, a rare glimpse of the process of mesopredator release.

Large mammals in Ghana are actively hunted as bushmeat for human consumption, and count data from 1969–2004 show a 68 percent decline in wildlife biomass across the six monitored parks (Brashares et al. 2004). As elsewhere (see Woodroffe and Ginsberg 1998), Ghana's large carnivores are particularly hard hit by human activities (Brashares et al. 2001), in part because their low densities and slow rates of reproduction make them incapable of sustaining heavy harvest from bushmeat and pelt hunters but also because they are viewed as a threat to people and livestock and widely persecuted. Count data from the six parks show that the four largest carnivores in Ghana (lion, leopard, spotted hyena, and wild dog) declined from an average of 348 combined detections per month in 1969 (*SD* ±66) to 31 (±7) in 2004 (Figure 13.3). In fact, all large carnivores became extinct in three of the six parks by 1986. In the three parks where lion, leopard, and spotted hyena remained extant in 2004, they showed significant range contractions away from areas frequented by humans (Brashares, unpublished data). Nevertheless, wildlife declines were not evenly distributed across parks, and core areas within Ghana's three largest protected areas (Mole, Bui, and Digya national parks; Figures 13.3 and 13.4) show counts of predators and prey that resemble observations made almost 40 years ago.

The extirpation of large carnivores in three of Ghana's six savanna parks provides a somewhat replicated natural experiment for examining the effect of apex predator removal on mesopredators. Moreover, the three parks currently without apex predators lost them at distinctly different times (approximately 1976, 1983, and 1986, respectively); thus, community responses to predator removal are expected to be staggered in time. A visual examination of counts for all forty-one species in the six parks shows only one species, olive baboon, was consistently observed more frequently and along more transects in 2004 than in 1969 (Figure 13.3). Specifically, olive baboons showed a 365 percent increase in observations over this period, and as quantified from an analysis of sightings over time at the sixty-three sampling sites, the species expanded its range within parks by more than 500 percent (Brashares, unpublished data). However, these increases in baboon abundance and distribution did not occur evenly across parks. Consistent with a hypothesis of mesopredator release, baboons showed the greatest increases in the three parks where apex predators became

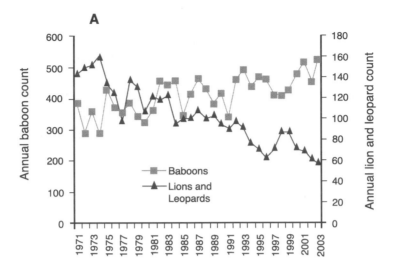

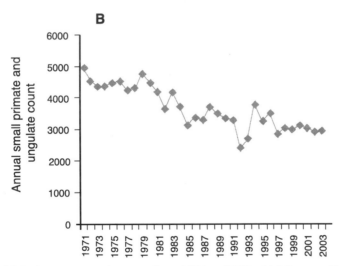

Figure 13.3. Average annual counts from 1971 to 2003 of **(a)** lions and leopards (triangles) and baboons (squares) and **(b)** 11 species of smaller (less than 10 kilograms) primates and ungulates at 20 largely intact sites within parks in Ghana where apex predators remain common.

extinct. The rate of baboon increases was correlated closely with apex predator declines ($r = 0.72$–0.88, using a 3-year lag). Even in parks that maintained apex predators, baboons were observed to spread and increase in density in areas where large predators were absent. These broad- and fine-scale temporal and spatial responses of this mesopredator to the removal of apex predators were quantified both through time series analyses and multiple regression models

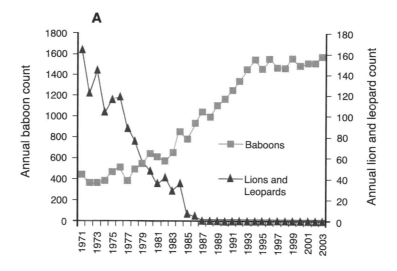

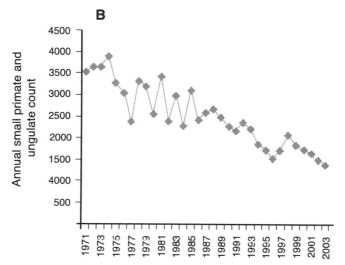

Figure 13.4. Average annual counts from 1971 to 2003 of **(a)** lions and leopards (triangles) and baboons (squares) and **(b)** 11 species of smaller (<10 kilograms) primates and ungulates at 20 altered sites within parks in Ghana where apex predators were extirpated before 1987.

that selected lion and leopard density, above seven other biotic and abiotic variables, as the single best predictor of baboon occupancy and growth rates within parks (Brashares, unpublished data). Finally, although apex predator extinctions occurred at different times in the six parks, increases in baboon density and range after predator extinctions followed similar trajectories and, amazingly, a consistent 3-year time lag (Brashares, unpublished data).

Like other savanna baboons, olive baboons are opportunistic omnivores that augment their primarily vegetarian diet with insects, eggs, birds, reptiles, mammals, and fish (Hamilton and Busse 1978; Norton et al. 1987; Barton and Whiten 1994). Olive baboons are extremely efficient foragers and capable of organized hunting (Strum 1975) and, like other dominant mesopredators, show greater dietary breadth than apex predators in their community. These features and the fact that in the absence of large carnivores baboons can sustain local densities orders of magnitude higher than apex predators (Brashares, unpublished data) suggest that these mesopredators, once released, can have greatly amplified impacts on their prey. Strum (1975) observed the development of a "tradition" of carnivory in olive baboon troops in Kenya, and many other authors comment on high rates of predation by savanna baboons on young antelope, large rodents, and small primates (e.g., Hamilton and Busse 1978). Observations in the field and analysis of count data suggest that olive baboons in Ghana have assumed the role of top predator where apex predators are removed. The impact of baboons on their prey is most easily quantified by comparing population trends of potential prey species in areas where baboons have become superabundant with those in areas where they have not (i.e., where apex predators remain extant). Such an analysis reveals that increases in baboon density and range are correlated closely ($r = 0.65$–0.92 with a 2-year lag) with accelerated declines of five species of smaller primates and nine species of antelope (Figure 13.4). In the most extreme cases, estimates of annual growth rates in populations of smaller monkeys (e.g., *Cercopithecus petaurista* and *Chlorocebus sabaeus*) and ungulates (e.g., *Ourebia ourebi*) shift from values above 1.1 to below 0.7 within 2 years of baboon outbreaks, with local extinction of these prey occurring within 5 years of outbreaks. In sum, although higher concentrations of prey in the presence of apex predators may seem counterintuitive, spatial patterns of wildlife communities in Ghana mirror other cases of mesopredator release in that primates and smaller ungulates persist in highest densities where apex predators remain most abundant.

One obvious alternative explanation for this pattern posits that apex carnivores, primates, and ungulates, but not baboons, are negatively affected by bushmeat hunting, and this pressure, rather than trophic cascades or mesopredator release, has created a community in which one species is hyperabundant and others rare. However, this hypothesis ignores several key details of the long-term data: Identical cascades are observed in areas of Ghana's parks where apex predators are absent but hunting does not occur or occurs at low rates; the staggered timing of apex predator extirpation, baboon release, and prey declines, replicated in three different parks, suggests causation more than correlation; and ungulate species too large to experience predation by baboons (e.g., buffalo, *Syncerus caffer*, roan antelope, *Hippotragus equinus*) but prized by bushmeat hunters do not show accelerated declines where apex predators are extirpated. Furthermore, baboons are hunted as intensively or more so than other species in their community, particularly when they occur near human settlements, a fact that stands in contrast to the alternative hypothesis outlined earlier. Preliminary results of stable isotope analysis of baboons in Ghana's parks also support the hypothesis that baboons released from predation increase their reliance on animal prey. Tissue samples from baboons living in high density (without apex predators) showed significantly higher nitrogen-15 isotope ratios (an indication of a more carnivorous diet) than those living in intact wildlife communities (8.6 ± 0.8 per mil vs. 5.5 ± 0.6 per mil; $n = 7$ and 5, respectively). However, broader sampling is needed before this result can be confirmed because current sample sizes are prohibitively small and the observed differences in nitrogen-15 isotope ratios could reflect only a transient shift in diet.

The impact of baboon release is observed not only in declines of primates and ungulates but also in reduced rates of nest success among ground- and tree-nesting birds. A 4-year study of nest success in areas of high and low baboon density showed that nests subject to heavy predation by baboons had average fledging rates of 19 percent (± 6 percent; $n = 132$ nests), as compared to average rates of 52 percent (± 11 percent; $n = 169$). It is likely that reptiles, insects, parasites, and perhaps vegetation structure will also show a response to high baboon densities in Ghana's parks, but data necessary to test the impact of this mesopredator release on these groups are yet to be collected.

CONCLUSIONS

Surprisingly few studies have attempted to quantify or have even considered the short- and long-term costs of mesopredator release for people (Chapter 20, this volume). Efforts to identify the economic, social, or public health implications of hyperabundant raccoons, rampaging red foxes, and egg-hungry rats may have no place in formal community ecology, but such pursuits may go far toward enlisting the help of the public in conserving apex predators. The case of baboon release in Ghana lends itself to such outreach for several reasons. First, savanna baboons become voracious predators of crops and livestock where they are released from predation (Butler 2000; Hill 2000). This conflict goes far beyond Ghana, and among large mammals baboons are the greatest threat to crops and livestock in twenty-seven countries in sub-Saharan Africa

(Brashares, unpublished data).Villages suffering loss of crops and livestock to baboons must enlist the services of school-age children as crop guards, and this reduces school attendance rates and exposes children to physical harm. Second, baboons compete directly with rural people in Ghana and elsewhere for wild sources of animal protein. Terrestrial wildlife, most often antelope and rodent species, is the primary source of animal protein and a major source of livelihoods for tens of millions of Africans (Brashares et al. 2004), and populations of these harvested species do not persist in areas of high baboon density. Last, baboons and humans share many parasites and pathogens, and baboon outbreaks, particularly near human settlements, create hotspots of infectious disease for wildlife, people, and livestock (Brashares, unpublished data). Taken together, the economic, social, and health costs of this mesopredator release dwarf the investment that would be necessary to restore, protect, and manage Africa's apex predators.

CHAPTER 14

Fear-Mediated Food Webs

Joel Berger

Modern humans feel a primordial dread when walking alone through habitat occupied by lions, tigers, or grizzly bears. Many prey species must similarly be gripped with fear in the presence of predators. Even today's megavertebrates—elephants, rhinos, and whales—who some might claim are immune to predation by virtue of their size, alter their movements and patterns of habitat use to reduce predation on neonates (Berger 1993; Reeves et al. 2006). In this chapter I consider the behavioral and ecological consequences of fear as manifested primarily by large mammals and as revealed in comparisons of prey species in contrasting situations of high and low risk of predation.

For behavior in general and fear in particular to affect trophic relationships requires that prey are cognizant of predators, that they adjust their behaviors in relation to predation risk, and that prey responses influence the distribution of herbivory on the landscape. In this chapter, I ask, To what extent does the threat of predation govern prey behavior? How do the indirect effects of predators affect ecological dynamics? And do such indirect effects matter?

I concentrate on large terrestrial mammals because ample evidence already indicates that big carnivores play key (top-down) roles (Terborgh et al. 2001; Smith et al. 2003a), although not all authors agree on this point (Ray et al. 2005). Furthermore, studies of known individuals reveal population-level variability in response to predation and indicate how this variability drives community dynamics.

BACKDROP: LABILE BEHAVIOR AND PREDATION THREAT

Tradeoffs between food acquisition and predator avoidance are widespread in nature and operate in both the terrestrial and aquatic realms (Sih 1992; Bowyer and Kie 2004; Estes et al. 2004; Polis et al. 2004). In marine settings, for example, dolphins, harbor seals, and dugongs use such tradeoffs, as do many fish species (Werner 1991; Wirsing et al. 2007), demonstrating that fear operates across diverse biomes (Kie et al. 2003; Heithaus et al. 2007). Predator-sensitive behaviors are those that differ under situations of high or low risk of predation. When predator-sensitive behavior is pronounced, predators may indirectly regulate food web interactions via shifts in group size, habitat use, or activity period.

To investigate these issues further, I compared a number of Holarctic localities that either did or did not support large carnivores. I then performed field experiments to simulate carnivore presence by exposing prey species to the sounds or odors of predators (Berger et al. 2001b; Berger 2007a). Associated conditions were also explored as covariates, including snow depth and habitat (proximity to dense cover), because each such covariate might either heighten or dampen behavioral responses by altering the perceived vulnerability to predation (Figure 14.1a).

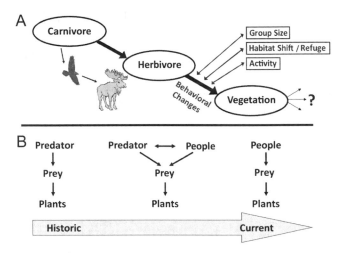

Figure 14.1. (a) Depiction of primary pathways by which apex carnivores might be expected to affect indirectly prey behavior and subsequent trophic relationships. (b) Depiction of changing temporal dynamics in which historic conditions reflect a simple food web with large apex large mammalian predators in the absence of predator, a transition into a time when people and predators experienced intraguild predation, and, finally, relationships today with people replacing predators.

Conservation Context and Changed Functional Responses

Predator-mediated selection on ungulates has been relaxed over much of the contiguous United States. Only small parts of Washington, Montana, and Wyoming have retained grizzly bears, and wolves survived only in Minnesota and have recently been restored or naturally spread to Michigan, Wisconsin, Montana, Idaho, and Wyoming. However, carnivore presence has been continuous at more northern latitudes, creating opportunities to compare predator-sensitive behavior from behavioral and ecological perspectives.

Fear responses may be general, so that prey such as mule deer or peccaries may still show a full repertoire of predator-sensitive responses in the absence of wolves or cougars, if coyotes or black bears fill the void. Therefore, prey behavior may remain unaltered, even where presumptive top predators are missing. Relationships between prey behavior and carnivore presence were evaluated under three treatments: predators absent (naturally or extirpated by humans); predators present, having coexisted continuously with target prey; and predators reintroduced (Figure 14.2).

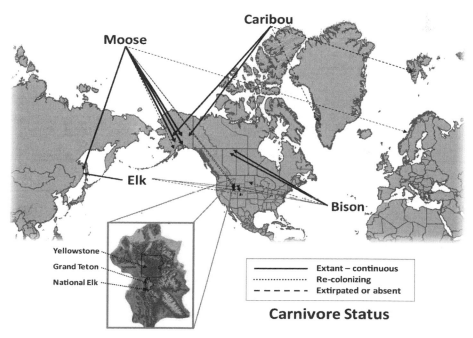

Figure 14.2. Overview of study areas and carnivore status by prey. Extant carnivore distribution is as follows. Russian Far East, tigers, brown bears, and wolves; Alaska, wolves and brown bears except Kalgin island (no carnivores); northern Alberta, wolves; Yellowstone Ecosystem has reintroduced wolves and expanding grizzly bears (see Berger 2007a for further details).

I conducted two types of experiments: exposure to auditory signals (all prey) and olfactory stimuli (moose only). Recordings of running water, wolf howls, tiger roars, and howler monkey hoots were broadcast for 25 seconds at fixed levels (90–100 decibels at 1 meter) under calm conditions. Prey response was gauged as the proportion of time diverted from feeding by vigilance. Additionally, whether the members of a group moved closer together or departed from their feeding sites was noted as responses to wolf playbacks except in the Russian Far East, where tigers were the key predator (Figure 14.3). Covariates

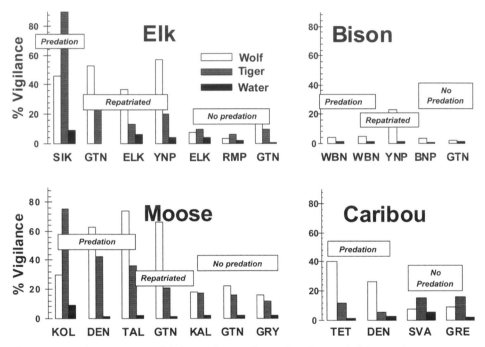

Figure 14.3. Comparison of vigilance by species at sites that varied in carnivore presence. Values reflect untransformed means to playbacks of wolves, tigers, and water. Sample sizes as follows: elk (1,438), bison (797), moose (1,334), and caribou (860). Upper left: SIK = Sikote Alin (Russian Far East); GTN = Grand Teton (Wyoming, post-wolf); ELK = National Elk Refuge (Wyoming, post-wolf); YNP = Yellowstone National Park (Wyoming, post-wolf); ELK = National Elk Refuge (Wyoming, pre-wolf); RMP = Rocky Mountain National Park (Colorado); GTN = Grand Teton (Wyoming, pre-wolf). Upper right: WBN = Wood Buffalo National Park (Alberta, Canada) and a different area of WBN about 60 kilometers distant; YNP = Yellowstone National Park; BNP = Badlands National Park (South Dakota); GTN = Grand Teton (Wyoming). Lower left: KOL = Kolumbe River Basin (Russian Far East); DEN = Denali National Park (Alaska); TAL = Talkeetna Mountains (Alaska); GTN = Grand Teton (Wyoming); KAL = Kalgin Island (Alaska); GRY = Grey's River (Wyoming). Lower right: TET = Tetlin Wildlife Refuge (Alaska); DEN = Denali National Park (Alaska); SVA = Svalbard (Norway); GRE = Greenland.

included distance between an animal and the speaker, distance to potential habitat cover (defined as vegetation sufficient for concealment), snow depth (playbacks were in mid- to late winter), and group size (Berger 1999, 2007a).

Predation-Induced Changes in Prey Behavior

In localities with wolves, elk and caribou responded to playbacks with heightened vigilance, a greater propensity for clustering, and more frequent abandonment of feeding sites than they did in wolf-free areas. Moose responded similarly except, being asocial, they rarely had opportunities to cluster. By contrast, bison were significantly less responsive to wolf howls (Figure 14.3).

Of the covariates, snow depth and distance to suitable vegetation cover significantly influenced vigilance levels in elk and moose but only at sites with wolves. In areas lacking top predators, group sizes were smaller; for caribou on wolf-free Arctic islands, females occurred in associations of two or one between 18 and 29 percent of the time but were never in such small groups in Alaska. In other words, when sympatric with large carnivores, all three species—caribou, elk, and moose—were more vigilant and more likely to cluster (elk and caribou) and to vacate feeding areas when confronted with cues indicating wolf or bear presence.

INDIRECT RESPONSES: SCAVENGERS AND PREY–PREDATOR INTERACTIONS

Scavengers live in mutualistic associations with carnivores. Vultures are the principal scavengers in much of the world, but ravens assume this role in the north, where they consume carcasses killed by wolves and grizzly bears. Ravens depend on carnivores to open carcasses because their bills are unable to penetrate thick hides, especially frozen ones (Heinrich 1991; Marzluff and Angell 2007). In Yellowstone, ravens often arrive at wolf-killed carcasses within 30 seconds (Stahler et al. 2002). Groups of up to 100 may accumulate and remove significant amounts of meat (Wilmers et al. 2003). Usurpation of spoils by ravens leads to increased frequency of wolf predation in a positive feedback (Vucetich et al. 2004).

The association of ravens with kills allows prey species to rely on ravens to assay the presence of predators, using their calls as proxies. Moose retreated from raven calls where they were sympatric with large carnivores in Alaska, but not in Wyoming (Berger 1999). The contrasting effects were not a consequence of geographic variation per se because moose at another Alaskan locality where they had been isolated on a predator-free island for four decades were similar to Wyoming moose in ignoring raven calls. However, once reintroduced wolves began consuming the calves of Wyoming moose, mothers became hypervigilant to wolves, not to ravens (Berger et al. 2001b), an indication that learning in moose proceeds first to the carnivores themselves and secondarily to the scavengers.

As shown by moose living in predator-free environments in Alaska and Wyoming, a widely distributed behavioral response of moose to ravens is readily lost where not reinforced by experience, suggesting that the response is a socially transmitted trait. One must wonder how many other behavioral responses of wild species around the world have been lost as a consequence of the widespread extirpation of large carnivores (Berger 1999).

FEAR-MEDIATED INDIRECT EFFECTS AT LOWER TROPHIC LEVELS

Fear of predators influences prey behavior in several ways. Three of the best-documented behavioral modifications are changes in group size, refuging behavior, and daily activity patterns (Figure 14.1a; Lima 1998; Ripple and Beschta 2004a; Stephens et al. 2007).

Group Size

Per capita predation risk decreases as a function of group size ("safety in numbers"; Hamilton 1971), although larger groups may be more likely to attract the attention of predators (Hebblewhite and Pletscher 2002). Prey aggregations can lead to an increase in the patchiness of the grazing regime, a modification that affects plant growth rates, recruitment, and nutrient cycling (Kie et al. 2003).

Frequent tensions between foraging success and group size can lead to reduced group size under reduced predation risk. For example, long-tailed macaques on leopard-free islands in the Simuele archipelago of Indonesia occur in smaller groups than on Sumatra, where leopards are present (van Schaik and van Noordjwick 1985). Similarly, coloniality of seabirds is often reduced on predator-free islands (Beauchamp 2004). Musk oxen show similar group responses: reductions on predator-free islands and larger groups when wolves are present (Caro 2005).

Movements and Refuge Use

In solitary species, predator-induced adjustments in group size are not an option, but other responses occur at coarse or fine scales. Moose mothers that have lost young to wolves or bears move to new areas immediately upon the detection of fresh predator sign, but those that have not remain more sedentary (Pyare and Berger 2003). By contrast, mothers whose offspring perish for reasons unrelated to predation do not move in response to the same stimuli (Berger et al. 2001b), suggesting that variation in parental experience with predators affects fine-scale movements.

Furthermore, even in the absence of experimental evidence, other changes are also notable. Moose mothers give birth in the same general region annually where calf survival is high but shift birth sites up to 10 kilometers the year after a neonate has perished (Bowyer et al. 1999; Testa et al. 2000). Choice of parturition and nursery sites is therefore governed not simply by evidence of carnivore presence but by carnivore-specific experiences (Berger 2008). Such predator-induced movements and habitat selection affect herbivore density and subsequently food webs (Kie et al. 2003).

Diet may also vary with predation risk and habitat (Chapter 9, this volume), as exemplified by elk use of open grasslands or closed-canopy forests (Winnie and Creel 2007), where grass, sedge, and forb layers are differentially affected (Christianson and Creel 2007).

Activity Patterns

Although predation-sensitive activity patterns are known in rodents and other small mammals (Kotler et al. 1991; Griffin et al. 2005), many of the more striking effects of predation on the activity budgets of larger mammals stem from hunting by humans. Both coyotes and elk become more nocturnal with intensified harvest, adjustments that may lead to diet changes (Berger 2005). The evidence that native predators maintain or govern activity levels is less clear, however.

Somewhat intermediate in this discussion is the endangered endemic island gray fox. Unlike nocturnal mainland foxes, this diurnal insular form has been severely reduced in numbers by golden eagles (Roemer et al. 2001), partly because of its inability to become nocturnal. Despite a few case-specific studies, little attention has focused on relationships between activity patterns, forcing by native predation, and trophic dynamics.

INTRAGUILD PREDATION: THE CONTEXT OF BEHAVIORAL AVOIDANCE

Predators routinely kill and sometimes consume other members of the carnivore guild, a behavior known as intraguild predation (Palomares and Caro

1999; Polis et al. 2004). Where intraguild predation is frequent, lesser carnivores may occur at lower densities than in the absence of larger guild members (e.g., coyotes, wolves, and smaller cats). Altered relative abundances of sympatric carnivores may affect prey species and other components of community structure (Chapters 13 and 20, this volume).

Spatial and Temporal Avoidance

Lesser carnivores may attempt to reduce encounters with larger carnivores by temporal or spatial partitioning of activities. One of the first demonstrations of this was obtained with simultaneously radio-tracked leopards and tigers in Chitwan National Park in Nepal. Leopards systematically avoided neighborhoods that contained tigers (Seidensticker 1976). Whether leopard diet was affected as a consequence is unclear (Woodroffe and Ginsberg 2005). Similarly, cheetahs avoid lions, as demonstrated by movements of targeted cheetahs to areas of lower prey density in response to broadcast lion roars.

Carnivore–carnivore avoidance can be deduced from other forms of spatial evidence. One particularly striking example comes from areas in and around national forests surrounding Yellowstone National Park. Grizzly bears are attracted to gut piles left by hunters outside the park. Cougars move away from these areas, but wolf distribution remains unaffected, suggesting that the presence of bears partially mediates cougar ranging patterns (Ruth et al. 2003; Smith et al. 2003a).

A final example involves coyotes. After wolf reintroduction into Yellowstone, coyotes did not necessarily avoid areas of high wolf density but adjusted their behavior, primarily by resting less and increasing vigilance at carcasses (Gese and Grothe 1995; Gese et al. 1996; Switalski 2003).

Habitat Selection and Diet

Densities of coyotes and wolves in the Yellowstone ecosystem are inversely related (Berger et al. 2008). Solitary coyotes wandering outside their natal territories were killed by wolves at greater rates than pack-living residents (Berger and Gese 2007). Benefiting from this interaction were pronghorn fawns, which experienced four times greater survival in areas with than without wolves, presumably because fawns are not sought by hunting wolves, which prefer larger prey (Berger et al. 2008). Hence, pronghorn recruitment was enhanced by the interaction between a mesopredator and an apex predator.

A similar dynamic has unfolded in the Banff region of Canada, where recolonizing wolves and resident cougars interact. With the recent arrival of wolves, elk have declined about 65 percent, cougars have been killed, and displacement of cougars from carcasses has become more common. As a consequence, the primary prey of cougars has switched from elk to deer. As elk declined, wolves also turned increasingly to deer, a change that lagged 1 year behind the decrease in cougars (Kortello et al. 2007).

INTRAGUILD PREDATION: THE CONTEXT OF HUMAN PERSECUTION

In many areas of the United States, humans have replaced carnivores as the primary hunters of big game. However, nowhere do human hunters play the same functional role in ecological dynamics as do the carnivores they replace (Berger 2005). Next, I present three cases in which prey redistribute themselves in response to humans as predators, and then I draw analogies about how these behavior-mediated changes may affect other trophic levels.

Fear and the Use of Humans as Shields

Given human domination over most terrestrial systems, the strong interaction that once involved only native predators and prey is either nonexistent or modulated by an even more complex, three-way interaction involving people, predators, and prey (Figure 14.1b). In some situations, animals appear to use humans as shields against predators (Terborgh 2000). Conservationist John Muir reported a case of this while traveling in Africa in 1910. "Most of the animals seen today were on the Athi Plains (Kenya) and have learned that the nearer the railroad the safer they are from the attack of either men or lions" (Branch 2001).

However, the presence of animals near humans or their infrastructure can be subject to more than one interpretation. The notion that prey have figured out how to use humans and associated infrastructure to shield against danger is a tricky one. For instance, coyotes may invade suburbs, not because of the danger presented by wolves or varmint hunters on the towns' outskirts but because culinary options such as poodles and unattended food bowls may be attractive. To detect whether distributional patterns change because of buffering against predation requires simultaneous information on both the pace of prey redistribution and the intensity of predation across changing landscapes.

Expanding populations of wolves and grizzly bears in the Greater Yellowstone Ecosystem enable a test of the idea that prey derive protection from predators by using humans as shields. During a 10-year period, grizzly bear numbers increased in and around Grand Teton National Park, whereas adjacent regions with moose remained essentially bear-free. If predation on juveniles was

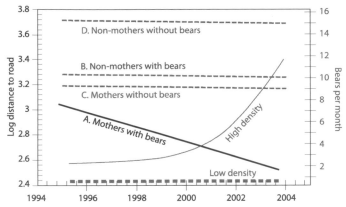

Figure 14.4. Relationships between log median distance to a paved road and expected birthdates by year for adult females at sites differing in bear densities. Dotted lines are not significant: (B) $Y = -0.0126x + 27.259$ ($r^2 = .211$; $p = .182$); (C) $Y = -0.003x + 9.722$ ($r^2 = .003$; $p < .894$); (D) $Y = 0.002x - 0.739$ ($r^2 = .02$; $p = .920$); the low–density site had 3 bears/10 years. (A) $Y = -0.056x + 114.97$ ($r^2 = .875$; $p < .001$) (details in Berger 2007b). For bears per month and time, $Y = 6E–214e^{0.2462}x$ ($r^2 = .614$; $p < .01$).

a major factor guiding female moose distribution, then females with calves should have been more likely to alter their patterns of distribution than non-parous females.

Over a 10-year period, the locations of birth sites in areas with and without bears were revealing. Moose mothers in bear-free areas showed no consistent patterns in birth sites, but in areas with grizzlies, the median distance between birth sites and paved highways decreased by about 120 meters per year during the 10-year period (Figure 14.4). Because grizzly bears avoid areas within about 500 meters of highways, moose mothers apparently shifted into these areas to avoid bears (Berger 2007b). Hence, grizzly bears appeared to determine the distribution of parous moose. A semianalogous situation occurs with wolves and elk in the Canadian Rockies. Elk congregate around or within towns, areas wolves avoid. As a consequence, high elk densities affected aspen tree growth and willows, as well as beavers and songbirds (Hebblewhite et al. 2005).

Warfare and Game Sinks

As in the preceding example, areas of extraordinary danger tend to be avoided, and prey may fill interstitial zones where the risk of death is diminished. More than 200 years ago Lewis and Clark described wildlife as shy or reclusive near human settlements, but in other areas "the whole face of the country was covered with herds of buffalo, elk and antelopes; deer are also abundant, but keep themselves more concealed in the woodland. The buffalo and elk and antelope are so gentle that we pass near them while feeding, without appearing to excite any alarm among them, and when we attract their attention, they frequently approach us more nearly to discover what we are, and in some instances pursue us a considerable distance" (Moulton 1986–1996: Vol. 4, p. 67). Similarly, boundary zones between mutually hostile wolf packs are places where large mammals are found at greater than expected densities (Mech 1977; Martin and Szuter 1999; Laliberte and Ripple 2003).

Altered Migration Routes and Predation

Mass migrations of mammals constitute another behavioral mechanism of predator avoidance (Fryxell et al. 1988; Hebblewhite and Merrill 2007). However, evidence that migration routes change as a consequence of natural predation is weak. There are only a few studied examples, such as the seasonal movements of elk into and out of Jackson Hole, Wyoming, in which data support the idea that predation (in this case by humans) modulates the temporal and spatial distribution of migration (Connor 2001; Smith et al. 2003a).

CONSIDERATIONS FOR THE FUTURE

I have explored some of the links between behavior and food web interactions by exploiting instances of loss, retention, or restoration of native large carnivores to gauge effects on sentient prey. Predation-sensitive changes in foraging, grouping, movement, and refuge use reveal a powerful mechanism that modifies the operation of top-down effects.

Human interventions into natural landscapes can disrupt historical predator–prey relationships. Intense forestry practices offer a case in point. In the northern Rocky Mountains, clear-cutting has created habitat where white-tailed deer populations are irruptive. Cougars have increased in response and have now reached a level at which they are jeopardizing the persistence of endangered woodland caribou (Kinley and Apps 2001). All three species are native, but the fact of human-altered predator–prey dynamics raises a more general issue: Is the behavior of prey species sufficiently resilient to ensure demographic viability as conditions change?

As humans transform our world from the Holocene to the Anthropocene, the context of behaviorally mediated food webs will become increasingly distorted, given the ceaseless tension between large carnivores and humans (Berger 2005). One consequence is mesopredator release, which impairs biological diversity and is likely to lead to other unforeseen impacts (Crooks and Soulé 1999; Chapter 13, this volume). Another is the invasion of alien carnivores, although our understanding of how or whether behavior contributes to conservation in these transformed ecosystems is limited. Native prey often are poorly equipped to avoid alien carnivores, behavioral deficiencies that may arise because behavior is canalized so prey do not maintain appropriate predator avoidance responses or behavior is insufficiently responsive in the short run to adjust effectively to novel pressures (Blumstein 2006; Berger 2008).

In our world of dizzying change, understandable confusion exists about what is natural and what is not (Chapter 5, this volume). Habitat changes induced by livestock in the Great Basin Desert 130 years ago facilitated colonization by mule deer, and subsequently by cougars, which now prey on naive bighorn sheep and porcupines. Grizzly bears and wolves are currently colonizing Alaska's Seward Peninsula, although effects on food webs are not yet clear. Are these colonizers native or exotic?

What of the humans who entered the New World some 13,000 years ago or the dingoes introduced into Australia 4,000–5,000 years ago by seafaring Asians? What sense do we make of the consequent loss of thylacines and Tasmanian devils from the mainland? And to what extent did these losses leave niches to be filled by other species? Similar thorny concerns about ecological function arise from recent and historic losses of mammoths, bison, and native horses from North America (Donlan et al. 2006). The past is past, but we still have opportunities to implement conservation now and in the future.

Large terrestrial carnivores can be keystones of natural ecological function, but apart from high northern latitudes, the proportion of the earth's terrestrial realm in which they can continue to live is greatly reduced, perhaps to only a few percent of what it once was. Knowledge about the importance of carnivores and their effects on other species will be important as we argue for their survival, whether on ecological, ethical, or economic grounds. Understanding how behavioral responses to predation help to structure food webs is one additional way to appreciate the diversity and complexity of nature.

CONCLUSIONS

Fear is a powerful motivator, perhaps the most powerful of all motivators, and is felt by animals as well as humans. Animals thus modify their behavior in a variety of ways when they know or sense that they are at risk of being preyed upon. This chapter explores the behavioral responses of prey species, primarily ungulates, to three experimental conditions: large carnivores present currently and historically, carnivores at greatly reduced abundance or absent, and carnivores restored after a period of absence. Some prey responses are apparently learned, such as associating ravens with recent kills and hence predator presence. Other responses may be instinctive, such as prey aggregation under increased risk of predation and avoidance of foraging areas exposed to predator ambush. These fear-mediated behaviors can generate indirect effects at both higher and lower trophic levels. Effects at higher levels will emerge in the altered habitat use and perhaps hunting strategies of predators in response to fear-mediated behavior of prey. At lower trophic levels, prey behavior can affect spatial and temporal aspects of habitat use and foraging impacts. Complicated dynamics can arise when carnivores interact among themselves in a hierarchical fashion, sometimes through intraguild predation, so as to affect the survival of particular prey species. In the modern world humans have become a major factor in this scenario by persecuting predators and hunting prey species. Predators and prey may react in distinct ways to the presence of humans, even in parks, thereby adding a layer of complexity to normal trophic interactions. There remains much to learn about the three-way interaction between humans, predators, and prey as each seeks to optimize its advantage in a spatially and temporally complex environment.

CHAPTER 15

Trophic Cascades in African Savanna: Serengeti as a Case Study

A. R. E. Sinclair, Kristine Metzger, Justin S. Brashares,
Ally Nkwabi, Gregor Sharam, and John M. Fryxell

In this chapter, we examine the role of trophic cascades in tropical savannas, with particular attention to the Serengeti ecosystem of East Africa. Africa is perhaps the last place where the Pleistocene megafauna is still largely intact and thus where "natural" interactions can be observed among animals at the top of the trophic web. Serengeti presents a unique case study because of some 50 years of detailed information on current trophic processes supported by extensive knowledge of the region's paleoecology going back 4 million years (Sinclair et al. 2008; Peters et al. 2008). The Serengeti ecosystem also features megaherbivores, in particular elephants (*Loxodonta africana*), that have outgrown their predators and so are unlikely to be predator regulated like the many smaller herbivores in their community (Fritz et al. 2002). Here we consider how these megaherbivores of Africa create or modify trophic cascades while also discussing the contributions of apex predators, birds, migratory herbivores, small ungulates, forests, and grasses to top-down and bottom-up processes in the Serengeti ecosystem.

Because plants form the base of food chains in terrestrial ecosystems, their productivity must necessarily influence all higher trophic levels. This is the basis for bottom-up regulation of food webs. However, higher trophic levels, if self-regulating, can impose top-down regulation on lower levels, or both bottom-up and top-down regulation can occur simultaneously (Sinclair et al. 2003). Regulation occurs when there is a density-dependent loss imposed on a population (Sinclair 1989). Strong evidence for top-down regulation comes from the removal of higher trophic levels resulting in a marked increase in prey populations or a reduction in prey if the predator population is experimentally increased. Second, compelling evidence for top-down regulation derives from evidence that mortality is largely or completely due to predation. Third, further evidence for a trophic cascade comes from indirect responses at the plant trophic level (e.g., increase or decline in growth, density) resulting from changes in predator numbers.

Evidence for bottom-up regulation is most powerful where density-dependent mortality is caused ultimately by lack of food. Second, evidence comes from density-dependent reduction in the instantaneous rate of increase in the absence of predation. Third, but less powerfully, evidence comes from predator removal experiments, in which there is little or no response in the prey population because the latter is regulated from below.

We present here evidence for both top-down and bottom-up processes, based on the aforementioned criteria, from a series of natural experiments in the Serengeti ecosystem resulting from perturbations, at different times, of the predator, herbivore, and plant trophic levels. First, the appearance of a viral disease, rinderpest, reduced herbivore numbers in the 1890s. Later, when the virus disappeared in the 1960s, it released herbivore numbers. Second, a predator removal took place in the 1980s, and comparisons of the effects of this removal could be made with simultaneous controls before and after the removals. Third, the trophic cascade that resulted from the rinderpest virus could be measured at the plant trophic level, and finally the effects on tree populations of the removal of the African elephant in the 1980s, due to poaching, could be compared with simultaneous undisturbed areas before and after the megaherbivore removals. A detailed account of these disturbances is reported in Sinclair et al. (2007, 2008). We compare these results with other areas in Africa and Asia.

TOP-DOWN REGULATION

Food webs can be thought of as modules of closely interacting species, and these modules are less closely connected to others. In the Serengeti system we see the large mammal community as one such module, and this is only loosely connected to one involving birds, insects, and plants, for example. Here we examine trophic cascades in the grassland mammal community, the grassland bird community, and the forest bird community.

The Small Ungulates

The strongest evidence for top-down processes affecting mammalian herbivores comes from situations in which predators have been removed and their prey display ecological release. An unplanned predator removal, largely of lions (*Panthera leo*), took place in northern Serengeti in the period 1980–1988 because of illegal hunting. Immediately adjacent to this area, on the Kenya side of the border, is the Masai Mara National Reserve, where predators were not removed, and so this area acted as a control. Population densities of small ungulates such as oribi (*Ourebia ourebi*), Thomson's gazelle (*Gazella thomsoni*), warthog (*Phacochoerus aethiopicus*), impala (*Aepyceros melampus*), and topi (*Damaliscus korrigum*) have been documented since the 1970s in both the predator removal and control areas (Figure 15.1). In all these species there was a major increase in density during the period of low lion numbers relative to the period before removal and to the period subsequently (1989–1998), when hunting ceased and predators returned. In contrast, no such changes took place in the adjacent

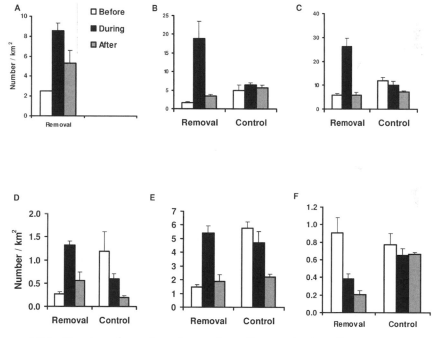

Figure 15.1. Density of ungulates in northern Serengeti where lions were removed in 1980–1988 (black bar) compared to preremoval (1967–1978) (open bar) and postremoval (1989–1998) (gray bar), and compared with the control in the Mara Reserve. **(a)** Oribi (not present in the control); **(b)** Thomson's gazelle; **(c)** impala; **(d)** warthog; **(e)** topi; **(f)** giraffe. Vertical bars = 1 *SE*. (From Sinclair et al. 2003)

control area. African buffalo had been removed by poaching some 6 years before the reduction in lion numbers. Furthermore, buffalo have remained absent both during the predator removal and after predators returned. Densities of smaller ungulates were similar in the 1970s, before predator removal and with high buffalo density, and in the 1990s after predators returned but with no buffalo present. Thus, the absence of buffalo did not explain the observed changes in other ungulates. These results provide compelling evidence for top-down effects (Sinclair et al. 2003).

More circumstantial evidence for trophic cascades in the mammal community comes from private reserves and ranches in southern Africa where predators have been removed. Wild ungulates have increased in number and are controlled by game ranching programs (Walker et al. 1987; Teer 1990). In contrast, where predators have been allowed to return on ranches in the Laikipia area of Kenya, populations of smaller ungulates have been reduced by predation (Georgiadis et al. 2007a, 2007b).

In Kruger Park, South Africa, predation was the dominant cause of mortality for small and medium-sized ungulates (Mills et al. 1995; Owen-Smith and Mills 2008). The creation of new water holes in the mid-twentieth century promoted the expansion of zebra (*Equus burchelli*) populations into new areas that, in turn, allowed lions to move in and drive down rare secondary prey, particularly roan antelope (*Hippotragus equinus*) and tsessebe (*Damaliscus lunatus*) (Harrington et al. 1999). Warthog (*Phacochoerus aethiopicus*) have also declined in numbers through intensified predation (Owen-Smith et al. 2005).

Top-down or bottom-up regulation of resident ungulates is determined by body size (Sinclair et al. 2003). Natural selection has pushed large ungulates to outgrow their predators, and predation is a minor component of adult mortality (Figure 15.2). However, below a certain threshold body size (about 150 kilograms body mass in Serengeti but likely to vary between ecosystems) there is a rapid switch from bottom-up to top-down regulation. As prey become smaller they are exposed to a greater number of predators until all adult mortality is caused by predation. The increased number of predators of small prey results from the food niches of small predators lying within those of larger predators (Sinclair et al. 2003; Radloff and du Toit 2004). Although the mode of preferred prey body size is different for each predator, the overlap in niche results in higher incidental predation of smaller prey compared with that of larger prey. This pattern of inclusive food niches for large mammal predators appears in both African and Asian savanna systems, such as at Nagarahole in southern India with tigers (*Panthera tigris*), leopards (*Panthera pardus*), and dhole (*Cuon alpinus*) (Karanth and Sunquist 1995). In essence, the size diversity of predators

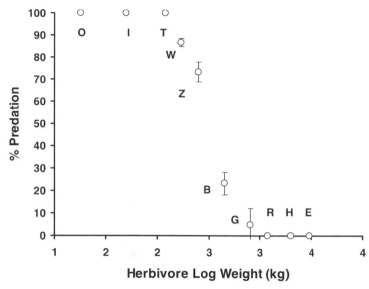

Figure 15.2. The proportion of adult mortality accounted for by predation relative to body mass in Serengeti. There is a threshold at about 150 kilograms where mortality switches from largely predation below to undernutrition above (from Sinclair et al. 2003). B = buffalo; E = African elephant; G = giraffe; H = hippo; I = impala; O = oribi; R = black rhino; T = topi; W = wildebeest; Z = zebra.

and prey results in both top-down and bottom-up processes occurring in the same system.

Overlap between dietary niches of predators is also likely to affect the response of prey to predator removal. Where the removal of one predator is compensated by an increase in another predator or predators of overlapping diet, we might expect prey release to be dampened or absent. Such is often the case in observations of mesopredator release, in which smaller predators increase in abundance and impact after the decline or extirpation of apex predators (Chapter 13, this volume). In such scenarios, the threshold prey body size that determines whether prey are bottom-up or top-down regulated should shift downward as smaller predators replace larger ones. This prediction is supported by observations of prey dynamics in altered wildlife communities or after extirpation and reintroduction of apex predators (e.g., gray wolves [*Canis lupus*] in the western United States; Chapter 9, this volume).

Impacts of Ungulates on Plants

The impact of ungulates on their plant prey has been documented for hundreds of years, most powerfully through the use of herbivore exclosures.

Recently, Pringle et al. (2007b) excluded ungulates from six *Acacia* savanna sites in Kenya for 5–9 years. Relative to control areas, there was a significant increase in tree densities but only a marginal increase in the herb layer cover. Associated with tree abundance was an increase in gecko and beetle numbers, suggesting the presence of strong indirect effects of ungulate predation on young trees.

In Serengeti, the most striking example of a trophic cascade reaching the plant level is the widespread increase in grass biomass that followed the collapse of wildebeest (*Connochaetes taurinus*) numbers in the 1890s as a result of disease outbreak (Sinclair 1979). This increase in vegetation biomass had two distinct consequences depending on habitat. On the eastern Serengeti plains, currently a shortgrass sward about 10 centimeters high, there was a change in structure that was observed at least in the 1930s–1950s (M. Leakey and A. Root, personal communication, 1977) with grass 50–70 centimeters high. A 10- by 10-meter grazing exclosure set up on these plains for 10 years generated a similar tallgrass sward in 1986 (personal observation). The change in grass biomass on the plains from shortgrass to tallgrass has several indirect cascading consequences. Among these changes were a major drop in dicot diversity from seventy species on grazed swards to fifteen species in ungrazed swards because grass, released from grazing, can outcompete small prostrate dicots for nutrients and light (Sinclair, unpublished data).

The second consequence of the lack of grazing during the 1890–1963 period occurred in the savanna habitats. During the dry season, wildebeest grazing maintains a grass biomass of about 700 (±150) kilograms per hectare and 10 centimeters in height. Without grazing this biomass is about 6,200 (*SE* ±380) kilograms per hectare (Figure 15.3) and provides fuel for widespread grass fires. Before the wildebeest increased in the 1960s, on average 80 percent of the northern Serengeti savanna burned each year (Norton-Griffiths 1979), whereas after wildebeest reached high numbers in 1977, the area burned dropped to below 20 percent (Sinclair et al. 2007). Although there was an initial increase in tree densities in 1890–1900 (because of the exodus of people because of rinderpest), burning increased after 1920 when people returned. Burning inhibits the regeneration of *Acacia* tree seedlings so that over the 50 years of widespread fires (1920–1970) there was a progressive reduction in savanna trees. When grazing impacts returned and burning was reduced after 1970, there was widespread regeneration of trees that formed, by the 2000s, a dense stand of trees (Sinclair et al. 2007). Further indirect consequences of this trophic cascade are now being documented on birds, rodents, and insects.

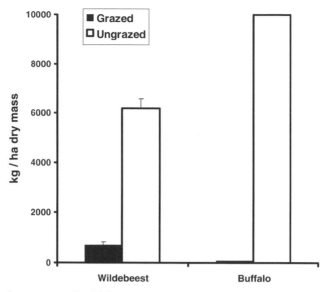

Figure 15.3. The impact of wildebeest grazing on the biomass of grass in the savanna of Serengeti and African buffalo grazing on swards of montane grassland at Mt. Meru, Tanzania, both relative to plots protected from grazing, showing top-down impacts. (A. Sinclair unpublished)

Impacts of Birds on Insects

Empirical evidence exists for top-down impacts of birds on insects. Of the 640 species of birds in the Serengeti, some 104 are vertebrate feeders, many on birds, and 343 are insectivores, compared to 115 seed or fruit feeders. Thus, the dominant food chain in terms of avian species and their prey is raptors ↔ bird insectivores ↔ insects.

Insight on the role of top-down and bottom-up regulation of insectivorous birds can be gained in areas where grazing and burning act as natural experimental removals of the grass level. A bottom-up process would predict lower insect biomass and consequently lower insectivorous bird biomass from such grass removals. In contrast, a top-down process would predict no decline in bird populations. Nkwabi (2007) compared burnt and grazed areas with matched undisturbed sites. Insect biomass decreased on disturbed relative to undisturbed sites (Figure 15.4a), but bird numbers increased (Figure 15.4b), and the bird community changed. The inference is that the change in plant structure facilitated the ability of birds to find prey (i.e., improved their functional response), and so birds at least contributed to the lower insect biomass, a top-down effect.

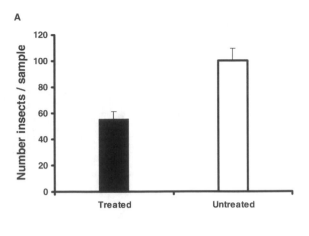

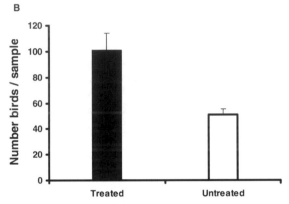

Figure 15.4. Top-down impacts of birds on insects measured from a plant biomass removal. **(a)** The mean (±1 *SE*) number of insects per sample from pitfall, tray traps, and sweepnets combined, on burnt or grazed plots (treated) compared with ungrazed or unburnt plots (untreated). **(b)** The mean number of ground- or grass-living insectivorous birds on the same sites counted over 10-minute intervals through the dry season after grazing or burning. (Data from Nkwabi 2007)

Few other studies have examined the effects of burning and grazing on bird populations in Africa (Parr and Chown 2003). In Kenya, burning had little effect on grassland bird abundance (O'Reilly et al. 2006), whereas in South Africa grazed areas resulted in an increase in some ground-living francolin (*Francolinus* sp.) species (Jansen et al. 1999), both results suggesting a role for top-down processes.

Impacts of Birds on Forest Communities

A small but growing body of research has examined the top-down effect of birds on forest dynamics, specifically seed dispersal and regeneration (e.g.,

Pejchar et al. 2008). Evergreen closed-canopy forest grows along the major rivers of Serengeti. On the Mara River, regeneration of forest occurs through seedlings growing up underneath the closed canopy. In contrast, in light gaps where the canopy has been opened through disturbance, grass invades and tree seedlings are excluded. In addition, seedlings are also absent in such open canopy forest, even under canopy where grass does not occur (Sharam 2005). Fire and browsing by elephants act as natural experimental removers of the canopy, leaving some riverine patches with open canopy and others closed. Sharam et al. (2009) compared forest patches with different degrees of openness. Obligate forest bird species were lost as the canopy opened up, with frugivores being more adversely affected than insectivores. Frugivores consume tree fruits and leave the seeds (without pericarp) on the ground, where they could germinate. If fruits are not eaten they eventually fall to the ground with pericarp still present. Bruchid beetles attack seeds with pericarp, and these fail to germinate. Seeds without pericarp are attacked less, and germination is high. Thus, in open canopy forest, without frugivores, insect attack is higher, seedling density declines, and the forest canopy continues to open until all trees have disappeared, leaving only shrubs. This positive feedback unraveling of the forest has been tracked in one patch since 1966 (Sharam et al. 2009). In essence, birds depress beetles indirectly, which promotes seedling regeneration and a stable closed-canopy forest, a top-down cascade.

BOTTOM-UP REGULATION

Several conditions allow bottom-up regulation of the food web modules. One condition depends on body size, particularly in mammal herbivores that have outgrown their predators. Another condition is that of the behavior of the herbivores through both migration and group living. A third condition depends on position in the food web; top predators are necessarily dependent on food supply, but the effects of resources are modified by whether the predators live in groups or are lone hunters.

The Megaherbivores

Between the 1890s and 1964, the African buffalo (*Syncerus caffer*) population in Serengeti suffered severe mortality in the first year age group due to rinderpest (Sinclair 1977; Dobson 1995). The disease disappeared in 1964, as an ancillary benefit of a vaccination campaign on cattle. Buffalo, thus released from disease, increased in number until 1976, when they began to level out as a result of lack of food, whereupon the instantaneous rate of increase declined (Figure 15.5a). Reduced population growth resulted largely from density-dependent mortality of adults (Figure 15.5b). Analysis of the carcasses showed that undernutrition accounted for some 75 percent of the mortality. Predation by lions accounted for the other 25 percent, but most of this involved prey that were already in a

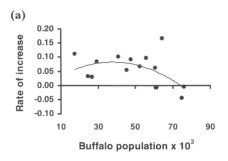

(a)

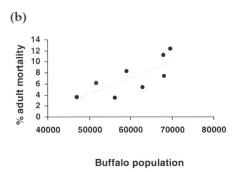

(b)

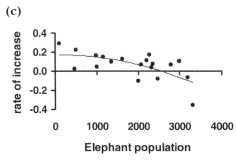

(c)

Figure 15.5. Evidence for bottom-up processes in resident megaherbivores. **(a)** The instantaneous rate of increase of the Serengeti buffalo population after reductions by rinderpest or poaching. **(b)** Density-dependent percentage adult mortality of buffalo due to undernutrition (from Sinclair 1977). **(c)** The instantaneous rate of increase of the Serengeti elephant population due to both immigration and births after reductions by poaching. (Data from Sinclair 2003 and unpublished)

state of undernutrition, as was judged objectively by inspection of the marrow in leg bones. In ungulates, fat is stored to cover periods of food limitation, the main stores being in the gut mesentery and the bone marrow. The last fat to be used lies in the leg bones, which after it is depleted leaves a translucent gelatinous marrow clearly visible in dead animals (Hanks 1981; Sinclair and Arcese 1995).

Another form of evidence for bottom-up regulation in buffalo allows a generalization that applies across eastern Africa. Buffalo eat grass only, and the best predictor for the biomass of grass is annual rainfall. Densities of buffalo close to their equilibrium number are significantly related to rainfall in twelve East African protected areas (Sinclair 1977). The interpretation is that rainfall, acting through the food supply, determines the equilibrium level of buffalo populations (Sinclair 1977). This observation can be generalized across Africa where ungulate biomass and productivity are positively related to their food supply (Coe et al. 1976; Fritz and Duncan 1993).

African elephant populations also demonstrate bottom-up regulation. Evidence comes from the subsequent population responses to the removal of elephants by humans. In Serengeti this has occurred twice. The first took place in the 1880s, when the ivory trade effectively removed all elephants from Serengeti until the 1950s. The second occurred in the 1977–1988 period, when illegal hunting was rampant in the system (Sinclair et al. 2008). After each population reduction, elephants have increased, showing a density-dependent slowing in the rate of increase (Figure 15.5c). Similarly, in Hwange National Park, Zimbabwe, elephants were released from culling in 1986, and subsequent population growth rates were negatively related to population density (Chamaille-Jammes et al. 2008; Sinclair et al. 2008). In Kruger National Park, South Africa, continuous culling of elephants from 1965 to 2000 (Whyte et al. 2003) can be regarded as experimental removals. Again, a density-dependent change in the rate of increase subsequent to each removal was exhibited by the population (Sinclair and Metzger 2009); reductions to lower density resulted in faster rates of increase.

Regulation of elephant numbers takes place through lack of food in the dry season, and predators are not involved. The mechanism was seen most clearly in Tsavo National Park, Kenya. Populations there were increasing after the imposition of protection from hunting in the 1950s. Their food in the dry season consisted of shrubs and trees within reach of water supplies, because water is an essential resource. As numbers increased they first consumed trees near water, then progressively further away until they were just able to reach their food. In 1971 a drought resulted in insufficient food within reach of water, and a large proportion (perhaps 30 percent) died of starvation (Corfield 1973; Myers 1973; Croze et al. 1981). The spatial distribution of elephants relative to food within reach of water supplies was the important feature regulating elephants in Hwange Park, Zimbabwe (Chamaille-Jammes et al. 2008). Evidence for food regulation of megaherbivores is also reported for white rhino (*Ceratotherium simum*) in South Africa (Owen-Smith 1981, 1988) and for elephants and hippopotamus in Uganda (Laws et al. 1975; Eltringham 1980).

The predator removal experiment in northern Serengeti in the 1980s (described earlier) detected no difference in the population trend of giraffe (*Giraffa camelopardalis*) in the removal area compared with that in the nonremoval area. Giraffe are regulated through food supply both in Serengeti (Pellew 1983) and in Laikipia, Kenya (Georgiadis et al. 2007a).

In summary, mammal herbivores over 400 kilograms (the size of buffalo or larger) in Serengeti suffer little or no predation as adults and are regulated ultimately through food supply. In other areas such as Kruger, the larger ungulates are also food limited (Owen-Smith 1990; Owen-Smith and Ogutu 2003). However, studies are showing that both disease (Cleaveland et al. 2008) and predation (Sinclair and Arcese 1995; Owen-Smith et al. 2005; Owen-Smith and Mills 2008) can act synergistically with food scarcity by expediting mortality. This increases the sensitivity of herbivore populations to fluctuations in bottom-up processes and provides stability to the system.

The Migrant Ungulates

Like the buffalo, the calves of migrant wildebeest of Serengeti experienced high mortality from rinderpest before 1963. As in buffalo, the disease died out, but in 1962, and the population increased sixfold until 1977. At that time it leveled out and has remained between 1.2 and 1.4 million animals since then, with the exception of 1993, when 25 percent died due to the most severe drought yet recorded (Mduma et al. 1999).

Because of the release from disease mortality, the wildebeest population showed density-dependent changes in the rate of increase (Sinclair and Krebs 2002). Adult mortality showed two trends (Figure 15.6a), initially an inverse density-dependent phase in which mortality was largely caused by predation, and a later phase in which mortality was strongly density-dependent. The later phase was caused by an increasing proportion of starving animals as density relative to food increased (Figure 15.6b) and per capita food supply declined (Figure 15.6c), resulting in lower body fat reserves (Sinclair et al. 1985; Mduma et al. 1999).

Migration is an adaptation to access ephemeral but highly nutritious food

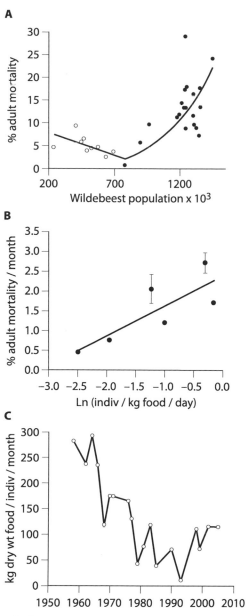

Figure 15.6. Evidence for bottom-up processes in migrant wildebeest. **(a)** The percentage adult mortality relative to population size. At low numbers mortality was largely inverse density-dependent predation, as explained in the text (open circles); at higher numbers density-dependent mortality was caused largely by undernutrition (closed circles) (data from Mduma et al. 1999). **(b)** The percentage adult mortality in the dry season is related to the density of animals with respect to food supply (from Sinclair et al. 1985). **(c)** Per capita food supply has declined over time as the population increased and leveled out. (Data from Mduma et al. 1999 and unpublished data)

supplies, a feature common to most if not all migrating ungulates, whales, and birds. In the case of migrating wildebeest, these ephemeral foods occur on the shortgrass plains where the plants are very high in nutrients but are green for only a few months each year. The wildebeest move to these shortgrass plains while the grass is still green but then must return to the tallgrass savanna when the plains dry out. The extra food on the plains provides for the growth of fetuses and lactation and results in a much higher density of animals than is observed in sedentary populations; we can compare this migratory population with a sedentary population in western Serengeti. Fryxell and Sinclair (1988a) have produced a model to show how migrant animals can escape from top-down effects of predation and so become food regulated. Similar tracking of ephemeral resources by migrant ungulates is seen in white-eared kob (*Kobus kob*) of Sudan (Fryxell and Sinclair 1988b), wildebeest, springbok (*Antidorcas marsupialis*), and gemsbok (*Oryx gazella*) in southern Africa (Viljoen 1993).

However, the migrant zebra populations of Serengeti and southern Africa are an exception. They appear to be top-down regulated (Grange et al. 2004). Although the mechanism is not entirely clear, one possibility is that that they do not all migrate; a portion of the population does migrate, but the rest are spread across the ecosystem, where they are exposed, like other residents, to high predation rates. Combined with their low reproductive rates, this predation is sufficient to hold them down.

The Major Predators

Lion and hyena (*Crocuta crocuta*) populations, the main predators on the Serengeti plains, have also increased as their prey increased, particularly the wildebeest population (Van Orsdol et al. 1985; Hanby et al. 1995; Scheel and Packer 1995; Packer et al. 2005). Lion densities are correlated with prey availability in the "poor season," when migrant populations are absent. However, migrant wildebeest leave behind them sick and aged members, which do not migrate and become part of the resident prey. As wildebeest increased in number and became resource limited, a greater number of stragglers became available, thus providing a bottom-up enhancement for the major predators.

In general, this conclusion applies elsewhere in Africa and India. East (1984) showed that the biomass of the major predators—lion, spotted hyena, cheetah, and leopard—was positively correlated with the biomass of their main ungulate prey over savanna Africa. Similarly in eleven savanna and woodland reserves of India the density of tigers is positively related to the density of their ungulate prey (Karanth and Nichols 1998; Karanth et al. 2004), which is evidence for bottom-up regulation.

RECIPROCAL TROPHIC INTERACTIONS

Although large resident ungulates such as buffalo and elephant are bottom–up regulated, they can also have reciprocal top–down impacts on plants, but such impacts depend on the season when they occur. Thus, buffalo in the Serengeti are food limited in the dry season when grasses are dormant. Even total removal of the aboveground dry material has no impact on the viability of the grass, and consequently there is no impact on either plant composition or productivity. The same argument applies with wildebeest grazing in the dry season. In short, there is no top–down effect on plants when plants protect their growing points and storage organs in the soil during the dormant season.

A different situation prevails when herbivores feed on dicots or on green monocots during the period of food scarcity. Buffalo occur on Mt. Meru, some 300 kilometers east of Serengeti, where they are the only major grazer and have no natural predators. Continuous grazing on year-round green swards at 2,000 meters altitude maintains a green shortgrass sward (less than 10 centimeters height) of 72 (SE ±10.5) kilograms per hectare of prostrate grasses and dicots with a stable species composition (Figure 15.3). Tall tussock grasses (70 centimeters) replaced the low sward at a biomass of 10,000 kilograms per hectare in exclosures within 5 years (Sinclair 1977). Similarly, on the shortgrass plains of Serengeti, the grazing impact of wildebeest takes place during the growing season and determines plant species composition, structure, and productivity. In these cases, the removal of herbivores releases the vegetation from top–down forces and permits large increases in plant biomass. Species that feed on dicots, such as giraffe, are bottom–up limited but also have structuring effects on their own food supply by inhibiting the escape of small trees into the adult stages and changing species composition (Pellew 1983; Bond and Loffell 2001; Scholes et al. 2003).

Elephant, white rhinos, and hippopotamus also have reciprocal top–down impacts on plants, as has been well documented in many areas such as Uganda, Kenya, Tanzania, Botswana, Zimbabwe, and South Africa (Laws 1970; Laws et al. 1975; Norton-Griffiths 1979; Eltringham 1980; Owen-Smith 1988; Cumming et al. 1997; Eckhardt et al. 2000; Scholes et al. 2003; Chamaille-Jammes et al. 2008). Both white rhino and hippopotamus convert tall tussock grasslands into shortgrass communities. Top–down effects occur with African elephants when they feed on trees year-round and have no alternative food sources, as in Tsavo National Park, Kenya. Asian elephants (*Elephas maximus*) can also affect vegetation in savanna or woodland areas, such as Satyamangalam, Bandipur, and

Kaziranga in India; Chitawan, Nepal; and Wilpattu and Gal Oya, Sri Lanka (Sukumar 1988).

Evidence implicating elephants in trophic cascades derives from situations in which they were completely removed during the ivory trade of the mid-1800s, as in Chobe Park, Botswana; Kruger, South Africa; and Tsavo, Kenya (Patterson 1907; Myers 1973; Whyte et al. 2003; Skarpe et al. 2004). In the absence of elephants, extensive tree and shrub communities developed, especially along riverbanks. When elephants returned after the 1950s, the plant composition changed and the vegetation became more open (often with aesthetic, ethical, and conservation repercussions). In another demonstration, woody vegetation experienced strong recruitment in Tsavo after 6,000 elephants died of starvation in 1971 (Croze et al. 1981), and in Chobe broad-leaved woodlands originated only during episodes of regeneration when browsing by elephants and impala was low (Moe et al. 2009).

Reciprocal interactions between carnivores and their prey have been documented in Kruger Park, South Africa. Predation was synergistic with the bottom-up effects of nutrition in larger ungulates, especially for the larger buffalo and giraffe (Owen-Smith and Ogutu 2003), as occurs in Serengeti.

MULTIPLE STATES: TOP-DOWN OR BOTTOM-UP

We have also recorded a situation in Serengeti where top-down and bottom-up processes involving elephants and plants alternate as two different states (Dublin et al. 1990; Sinclair and Krebs 2002). This situation was elucidated by two different disturbances to the system. First, during a period when there was extensive *Acacia* tree canopy (more than 30 percent cover), monitoring of tree mortality attributable to elephants showed that elephants were unable to reduce the tree population. However, both monitoring of burning and experiments with fire demonstrated that widespread burning over a period of some 50 years (1930s–1970s) reduced tree densities parkwide (Norton-Griffiths 1979; Dublin et al. 1990; Sinclair et al. 2007), effects that have been recorded elsewhere in savannas (Chapter 16, this volume). In the far northern part of the ecosystem, the Mara Reserve of Kenya, less than 1 percent cover remained by 1980. In this area elephants changed their feeding from large trees to seedlings, often less than 15 centimeters high (Dublin 1986), and they were so efficient at weeding them out of the grass sward that they prevented tree regeneration. This conclusion was derived from experimental elephant exclosures and marked plants (Dublin et al. 1990). Thus, elephants were able to impose top-down regulation on the tree population, but only once trees had been reduced to very low numbers. This is the classic Type III functional response of a predator (elephants) regulating prey at low densities but not at high densities (Pech et al. 1992; Sinclair et al. 1998).

The second disturbance occurred in the 1980s when illegal hunting for ivory removed 80 percent of elephants in Serengeti but left untouched the elephants in the Mara Reserve (Sinclair et al. 2007). The consequence of elephant removal was the resurgence of *Acacia* seedlings in Serengeti but not in the Mara. Thus, in summary, there are two possible states with the same density of elephants as now occurs in the ecosystem: one with trees and elephants by bottom-up processes and, in another part of the system, a grassland with elephants imposing a top-down restraint on tree regeneration.

Management interventions carried out in Kruger National Park (South Africa) also hint at multiple states. Smuts (1978) reported that when wildebeest were in high numbers, lion fed on them but did not regulate them. However, when wildebeest were experimentally reduced in number, their diminished grazing allowed tall grass to grow, which improved the functional response of lions. The combination of lower prey numbers and greater capture rates allowed lions to reduce wildebeest numbers even further.

SYNTHESIS: INTERACTIONS OF BOTTOM-UP AND TOP-DOWN FOOD CHAINS

Trophic relationships in Serengeti are complicated by the presence of both migratory and sedentary herbivores and by the presence of herbivore species regulated from the top by predators and others regulated from the bottom by forage availability. Herding greatly lowers the per capita predation rate on wildebeest and other species that associate with them because it can bring hundreds of thousands of prey into the territory of a single lion pride and saturate its resource needs. At the same time, the vast concentration of prey contained in herds alleviates predation pressure on resident herbivores, allowing them a respite in which to reproduce. Thus, different components of the system interact in ways that confer stability to the whole (Figure 15.7). Several observations support this conclusion.

First, there is a bottom-up current of regulation from grass to wildebeest and wildebeest to lions, as was demonstrated when wildebeest were recovering from rinderpest. A concurrent increase in lions demonstrated that lion populations are supported by migrant herbivores. In turn, lions have a top-down

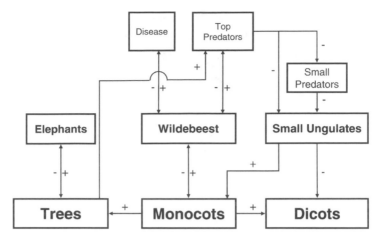

Figure 15.7. Flow diagram of trophic processes in Serengeti showing reciprocal interactions and the connection between bottom-up regulation of migrant wildebeest and top-down regulation of resident ungulates.

regulating effect on both smaller predator species, such as cheetah (*Acinonyx jubatus*) (Laurenson 1995), and smaller ungulates. Consequently, small ungulates have little impact on their food resources in Serengeti. In short, the bottom-up chain intensifies the top-down chain.

Second, the reciprocal impact of wildebeest on their own food supply has the indirect effect of promoting and maintaining both the tree population in savanna habitats and the dicot communities on the plains. In the savanna, trees then regulate elephant numbers, which in turn have a reciprocal effect on tree regeneration, as was demonstrated by elephant removal in Serengeti (Sinclair et al. 2007).

Third, the increase in tree density as a result of the recovery of wildebeest improved the ability of lions to catch their prey in savanna habitats. Since the leveling-out of wildebeest numbers in 1977, lion numbers have continued to increase, a fact attributed to greater success of lions in catching their prey (Packer et al. 2005). Thickets of regenerating trees provide cover from which lions can ambush prey (Hopcraft et al. 2005). In summary, lion numbers have increased whereas resident and migrant prey have not, which is further evidence of a top-down regulation of resident prey by predators.

Fourth, the functional responses of group-living predators depending on group-living prey provide strong stabilizing effects on the trophic cascade (Fryxell et al. 2007). Herding by prey reduces the search efficiency of predators because there are large areas of the habitat with no prey through which the predators have to move to reach a herd. Social behavior of prey therefore reduces the frequency with which predators encounter prey (Cosner et al. 1999; Nachman 2006). Grouping by predators limits search efficiency to a level similar to that of a solitary predator because they operate largely as a single entity. Although cooperation may compensate by improving the chance of capture, especially for large prey such as African buffalo, this compensation is small (Packer and Ruttan 1988). Most individual lions refrain from contributing to group hunts except when pursuing buffalo, which are inaccessible to solitary individuals or small groups (Scheel and Packer 1991). In summary, group formation and seasonal migration in ungulates are behavioral mechanisms that contribute substantially to the stability of top-down interactions in Serengeti.

Similar interactions of bottom-up processes supplementing countercurrent top-down pathways are reported for wild ungulates in the Laikipia district of Kenya: The dominant grazers (zebra) and browsers (giraffe) are regulated through food supply, but they support large predators that regulate several other smaller ungulate species (Georgiadis et al. 2007a, 2007b; Chapter 18, this volume).

The savanna systems of Africa and Asia are dominated by ungulates. The larger species are regulated by their food supplies, and predator populations are in turn determined by ungulate densities, a bottom-up process. However, there are reciprocal top-down processes through which predators affect the numbers of medium and small ungulates. The Serengeti ecosystem is an example of a migratory system embedded in a community of resident species. This has resulted in a complex interaction of a bottom-up process supplementing a top-down trophic cascade. We have elucidated these regulatory mechanisms in the large mammal community through a series of perturbations to producer, consumer, and predator trophic levels. First, bottom-up processes regulate the megaherbivores, a feature also observed in many other savanna systems in Africa. Second, bottom-up processes usually regulate migrant populations, in this case wildebeest, a feature also observed in other migration systems. Third, top predators are supported by these migrant populations, and these predators then impose a top-down regulation on both other predators and small resident ungulates. Thus, not only are both types of food chain present, but they influence each other. Fourth, megaherbivores such as elephants also have a reciprocal top-down effect on the vegetation, changing its structure and species composition, a feature observed widely in Africa. Finally, although elephants are always bottom-up regulated, their top-down impacts are determined by the density of trees; only at low tree density do they have top-down effects. This results in multiple states in which elephants, at the same density, are merely responding to their food supplies and savanna prevails, or they impose top-down regulation on trees and grassland. Therefore, the density of trees determines whether elephants can impose top-down regulation on trees.

CONCLUSIONS

Serengeti is unique and invaluable for the insights it provides for the understanding of how natural systems work. It is one of the last places on the planet where an intact system of megafauna and migratory ungulates survives intact. It illustrates the richness and complexity of trophic interactions that are the norm in the absence of anthropogenic effects. At one time the entire world supported megafauna, and much of it supported migratory ungulates before humans spread around the earth and exterminated most large animals. Animal communities throughout most of evolutionary history must have been stabilized by dynamics similar to those in Serengeti. The residual systems that are all we have nearly everywhere else are anthropogenic artifacts.

ACKNOWLEDGMENTS

We thank Tanzania Wildlife Research Institute and Tanzania National Parks for permission to work in Serengeti. We have relied on many colleagues who have provided logistic support and data. In particular we thank Simon Mduma, who runs the Serengeti Biodiversity Program; Grant Hopcraft; and Markus Borner. We also thank John Terborgh and Jim Estes for organizing the meeting and editing our chapter, and three referees for their constructive comments. The Canadian Natural Sciences and Engineering Research Council and Frankfurt Zoological Society funded the majority of this research.

CHAPTER 16

Consumer Control by Megafauna and Fire

William Bond

Biogeographers have long been fascinated by the challenge of predicting vegetation. The prevailing assumption has been that terrestrial vegetation is determined primarily by climate, with local modification by soils (Walter 1985). Proceeding on this assumption, Whittaker (1975) encountered a major discrepancy when he ordinated the world's major biomes on a temperature–precipitation plane and delineated climate envelopes for each major formation. Significant portions of the planet did not conform to the predicted climate envelopes. He referred to these areas as "ecosystems uncertain." The vegetation of these areas varies from forest to woodlands, savannas, shrublands, or grasslands. Although they occupy a small envelope on the temperature–precipitation plane, the global extent of these formations is vast (Bond 2005; Figure 16.1). The existence of "ecosystems uncertain" should be deeply unsettling to those who think the terrestrial world is strictly controlled from the bottom up (by resource limitations).

More recently a new generation of physiologically based simulation models, dynamic global vegetation models (DGVMs), has been developed for global change science (Woodward et al. 1995; Haxeltine and Prentice 1996). DGVMs attempt to predict ecosystem properties from physiological principles using only climate and soil texture and depth as inputs. Unlike Whittaker's correlative prediction of biome distribution, these models do not use mean climate variables but incorporate daily or monthly weather input for the twentieth

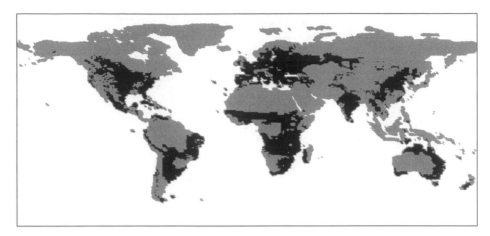

Figure 16.1. Areas of the world falling into the "ecosystems uncertain" climate envelope of Whittaker's global biome ordination on temperature–precipitation gradients. Large parts of warm temperate and tropical regions fall into the "ecosystems uncertain" climate envelope. These are areas in which either grassland or one of the types dominated by woody plants (Whittaker 1975) can occur in the same landscape. Source of climate information (mean annual precipitation [MAP], mean annual temperature [MAT], Climate Research Unit 10-minute grid aggregated to 0.5-degree pixels). Ecosystems uncertain (darker areas) were mapped for all pixels with MAP > 7.143 MAT + 286 and MAP < -1.469 MAT^2 + 81.665 MAT + 475. (These equations were derived from Figure 4.10, Whittaker 1975. Reproduced from Bond 2005, with permission from Wiley-Blackwell, the publishers)

century, thereby accounting also for climatic variability. DGVMs provide a mechanistic view of the world as it should be if controlled from the bottom up.

Bond et al. (2005) simulated world vegetation using the Sheffield DGVM (Woodward and Lomas 2004). The simulated world matched the real world in places but was vastly different in others. Areas of mismatch conformed quite well to Whittaker's "ecosystems uncertain," with forests predicted where the realized vegetation is grassland, savanna, or shrubland. The clear implication, from both correlative and physiologically based analyses, is that very large parts of the world are not as green as they could be. There should be much more forest.

For "ecosystems uncertain," biomass is a convenient measure of the mismatch between potential and actual vegetation and provides a quantifiable response variable for assessing trophic cascades, as called for by Polis et al. (2000). Biomass has been neglected by system ecologists, who have instead emphasized global patterns in net primary production. But the relationship between biomass and net primary production is complex and nonlinear (Keeling and Phillips 2007), whereas differences in biomass, such as between a grassland and a forest, are often obvious. Large shifts in biomass carry major repercussions for biodiversity and ecosystem function.

Precipitation is the major environmental variable influencing biomass (and productivity) in the biomes of the seasonal tropics, equivalent to phosphorus content of lakes as a master environmental variable for comparing different ecosystems (Walter 1985; Carpenter and Kitchell 1993). Large discrepancies between potential and realized woody biomass require explanation. These may include top-down control (e.g., herbivore-limited tree recruitment), bottom-up control (e.g., extreme soil conditions), or interactions between them. Grasslands and savannas are by far the most extensive biomes with too few trees for their climate potential, especially those dominated by C_4 grasses. These occur in a broad latitudinal band extending 30 degrees north and south of the equator. Estimates of their areal extent vary, depending on definition, with one estimate of 30 million square kilometers, or about 25 percent of the vegetated terrestrial land area (Ramankutty and Foley 1999). The world is not as green as it should be, and grassy ecosystems are the biggest exception to bottom-up control.

The anomaly of grassy ecosystems occurring in climates that can support forest has long been recognized. Explanations for the mismatch have been fiercely debated by generations of ecologists. Most debates have been parochial, restricted to a particular country or region, but the problem is global. A very common presumption in many different regions is that grassy ecosystems are anthropogenic, early successional habitats created by felling, burning, and grazing. This raises awkward questions. Is the biota of these open ecosystems a subset of woodlands and forests? If not, where do the plants and animals come from? Are open habitat species colonists from deserts and other extreme environments with no trees? Or have they evolved in similar environments that have been maintained in an "early successional state" for the eons needed for grassland specialists to evolve? It is surprising how seldom these questions have been asked. In Madagascar, for example, grasslands have long been viewed as anthropogenic artifacts, carved out of forests by cutting and burning in the 2,000 years since humans colonized the island. Yet the grasslands contain many endemic species of plants, ants, termites, birds, reptiles, and mammals that do not occur in forested habitats (Bond et al. 2008). It seems that grasslands are a natural feature of the island and have been so for a very long time.

Paleoecological studies also provide useful insights into the origin of early successional habitats. Carbon isotope analyses of paleosols and fossils show that C_4 grassy biomes evolved 7 to 8 million years ago (Cerling et al. 1997). Palynological studies indicate that these ecosystems were even more extensive in the last glacial in Africa (e.g., Dupont et al. 2000) and South America (Mayle et al. 2004), when human activities were low or nonexistent. "Ecosystems uncertain" existed long before humans began to alter terrestrial vegetation. Yet the anthropogenic hypothesis for the origin of such ecosystems remains pervasive. It has provided part of the rationale for transforming extensive areas of grassy ecosystems to crops and plantations.

TROPHIC CONTROL AND "ECOSYSTEMS UNCERTAIN"

In their green world hypothesis, Hairston et al. (HSS 1960) emphasized predation as the central process influencing plant biomass through control of herbivores. With predators removed, herbivore populations should irrupt, consuming plant biomass and creating a heavily grazed state whose greenness has been greatly reduced. Critics expressed doubt that primary consumers could ever reduce plant biomass enough to significantly affect vegetation structure and composition. Instead they argued that the world is green because most plants are inedible, full of indigestible components such as cellulose and lignin (Ehrlich and Raven 1964; Hartley and Jones 1997). Herbivores are asserted to be nitrogen limited and incapable of attaining population levels sufficient to exert significant top-down control (White 2005).

Contrast this with aquatic systems, in which phytoplankton are the base of the food chain and almost uniformly edible. Of all terrestrial growth forms, grasses are perhaps most similar to plankton. C_4 grasslands form extensive uniform feeding swards, growing very rapidly when water is available and recovering rapidly from defoliation. Although plant biomass remains low, grasslands can support very large mammal biomass, producing the closest terrestrial equivalent to the inverted biomass pyramid of aquatic ecosystems. African savannas support about ten times more mammal biomass than forests on similar soils receiving similar rainfall (Barnes and Lahm 1997, Figure 4). Intense grazing can produce grazing lawns, shortgrass swards with species that spread horizontally by stolons that enable them to cover the ground, even under intense grazing. Despite low plant biomass, grazing lawns support large numbers of mammalian grazers (McNaughton 1984) and are terrestrial analogues to phytoplankton in lakes.

TROPHIC CONTROL IN C_4 GRASSY ECOSYSTEMS

As measured by tree cover, most grasslands and savannas are not as green as they should be if controlled either from the bottom up or, as HSS (1960) suggested, by predators from the top down. Large mammals suffered mass extinctions in most parts of the world in the Late Pleistocene, but their legacy persists (Janzen and Martin 1982; Vera 2000), and remnants of the fauna survive in Africa and parts of Asia. Owen-Smith (1988) suggested that very large herbivores, megaherbivores (more than 1,000 kilograms body weight), escape predator control by being difficult and dangerous prey. If megaherbivore populations were not predator controlled, they would expand until self-regulated by resource constraints. Owen-Smith (1988) suggested that lack of predator control, coupled with large size, would have made these animals particularly influential agents of top-down control. Neither HSS nor the hypothesis of exploitation ecosystems (Oksanen et al. 1981) considers trophic control where prey size effectively eliminates predator control.

Another top-down agent, fire, is a ubiquitous "consumer" of C_4 grassy ecosystems. Fire invites analogies to herbivory, not least because, unlike other physical disturbances, fire consumes complex organic compounds and converts them to combustion byproducts. Fire has no "predator control" and in that sense escapes the constraints of the green world hypothesis. It is a very influential global consumer of vegetation. The implication is that where fire, megaherbivores, or both are important, ecosystems will not be as green as they could be because of direct consumer control of vegetation dynamics.

Although mammalian herbivores and fire are both major consumers of grass, surprisingly few studies explore their interaction (Archibald et al. 2005; Archibald 2008; Hobbs 1996; Fuhlendorf and Engle 2004). Neither can be considered in isolation. Fire is a complex beast whose activities depend on weather, ignition, the amount and type of fuel (food), and the landscape configuration of its preferred "diet." Topographic barriers to fire include physical obstructions, such as rivers and lakes, but also heavy grazing. For example, recovery of the migratory wildebeest population in the Serengeti National Park from a rinderpest epidemic was associated with a steep decline in the annual area burnt, from more than 90 percent when wildebeest numbers were small to ~20 percent when numbers approached a million (McNaughton 1992; Chapter 15, this volume). The implication is that changes in mammal populations can alter fire regimes with far-reaching (cascading) consequences.

Flannery (1994) suggested that extinction of the megafauna in Australia released fire on a continental scale, helping to convert Australia into perhaps the most flammable continent on Earth. The hypothesis has been highly controversial and, at least at the continental scale, wrong, because flammable formations were widespread millions of years before megafaunal extinctions (Lynch et al. 2007). However, Flannery's thesis has been productive in forcing ecologists to view herbivory and fire as alternative consumers exerting top-down control on ecosystems. The fire release hypothesis has received some support from paleoecological studies showing increased fire activity after extirpation of the megafauna, with cascading consequences for the vegetation (Miller et al. 2005; Burney et al. 2003; Robinson et al. 2005).

ALTERNATIVE STATES AND "ECOSYSTEMS UNCERTAIN"

The concept of alternative ecosystem states has long been applied to range-land ecosystems (Noy-Meir 1975; Dublin et al. 1990), with positive feedbacks causing rapid transitions between contrasting plant communities such as forest and flammable grasslands (Wilson and Agnew 1992). Developments of the theory and its application to trophic cascades are discussed in this book (Chapter 17, this volume; see also Scheffer et al. 2001). I have suggested that "ecosystems uncertain" can be interpreted as alternative ecosystem states broadly divided into three colors: green, brown, and black (Bond 2005). The green world represents vegetation grown to its resource-limited potential biomass, typically forests or dry thickets. The brown world occurs where grazing pressure is intense and trees are limited by herbivores. Finally, the black world is the very widespread domain where trees are suppressed by fire. Brown and black worlds are consumer-controlled in that vegetation is prevented from reaching its climate-limited potential by browsing and burning. Each state is maintained by positive feedbacks (Table 16.1) but can be tipped into alternative states under extreme conditions such as extended drought, decimation of herbivores by disease, or megafires that burn down forests. A regime shift from one state to another is rapid and accompanied by cascading changes in species composition and ecosystem structure and function (Scheffer 2009; Scheffer et al. 2001). Landscape boundaries, such as from black, frequently burnt savannas to green forests, are typically abrupt and occur with or without soil or site changes (Wilson and Agnew 1992). Regime shifts from one state to another are characteristically difficult to reverse and demonstrate hysteresis, where the tipping point in one direction of ecosystem change is not the same in the reverse direction. For example, a drought may trigger woody invasion in a savanna because of absence of grasses providing a respite from fire, browsing, and grass competition for woody saplings. But subsequent high rainfall will not cause the system to revert to savanna.

TRANSITIONS BETWEEN ALTERNATIVE ECOSYSTEM STATES

Biome switches, from black to green, have been produced by fire exclusion experiments in many parts of the world (Bond et al. 2005). Grasslands and savannas have been replaced by closed forests, with cascading changes in ecosystem structure and function (Tilman et al. 2000) and with large changes in plant species composition. Biome switches from brown to green are not as well docu-

Table 16.1. Summary of factors that help stabilize three distinct ecosystem states in C_4 grassy landscapes and mechanisms that can cause transitions from one state to another.

States	Vegetation	Stabilizing Factors	Mechanisms Causing Transitions
Green world	Tree-dominated forests, thickets	Shading excludes shade-avoiding grasses.	Death of tree layer facilitates shade-avoiding grass entry.
		No grass layer to support fire, grazers.	Death from drought, wind, fire, elephants.
Black world	Tall grass, fire-tolerant trees, shrubs	Landscape continuity of tall grass promotes fire spread.	Landscape continuity of grass fuels is disrupted by grazing, drought, spread of woody patches, etc.
	Pole saplings	Frequent large fires disperse grazer activity, preventing short-grass firebreaks from forming.	Fire frequency and intensity decrease.
Brown world	Short grass, grazing lawns, browse-tolerant trees, shrubs	Herbivores are attracted to high leafy fraction; high forage quality is promoted by dung and urine addition.	Grazing pressure decreases, including loss of short grass grazer species, allowing flammable fuel load to accumulate.
	Cage saplings	Heavy browsing by mixed feeders maintains low tree cover and good visibility, facilitating predator avoidance.	High rainfall episodes with grasses outgrowing grazers favor fires.
			Reduction in browsing leads to woody invasion and patch abandonment.

mented (but see Chapter 9, this volume). Influenced by observations of elephant impacts on savanna landscapes, Owen-Smith (1987) suggested that extinction of the Pleistocene megafauna partly accounted for end-Pleistocene changes in landscape structure, with increasing woodland and forest patches in postextinction landscapes. Elephants may be the only surviving megaherbivore capable of producing biome switches by toppling fully grown trees. When poaching reduced elephant populations in East Tsavo in Kenya, thorny savannas supporting a grazing fauna were replaced by *Commiphora* thickets (nongrassy) and the loss of grazers (Inamdar 1996). Some of the most celebrated historical examples of large-scale switches from a brown to a black world have been caused by outbreaks of diseases such as rinderpest (McNaughton 1992; Chapter 15, this volume) or, more locally, anthrax (Prins and van der Jeugd 1993). In many grassy ecosystems, browsing on seedlings and saplings suppresses woody plant recruitment and helps maintain open ecosystems. In some of these ecosystems, predation on herbivores can promote the green world by reducing herbivore pressures on plants (Ripple and Beschta 2004a).

Switches between black and brown worlds have been little studied, largely because of disciplinary boundaries (fire is studied by fire ecologists, herbivory by animal ecologists). Concentrated and repeated heavy grazing is needed to convert a bunch grassland to a shortgrass sward and then to a grazing lawn with an entirely different species composition. Frequent fires can prevent grazing lawn formation by drawing grazers off a grazed patch and onto the green post-burn flush. For example, burnt areas as distant as 2 kilometers from experimentally created grazing patches in a South African savanna drew grazers off the patch, resulting in elimination of the shortgrass sward (Archibald et al. 2005). Frequent large fires result in a positive feedback loop that prevents grazing lawn formation by maintaining tallgrass vegetation that supports frequent fires. A brown world can thus be replaced by a black world through frequent large fires.

The extent and frequency of fires in a given region depend on the presence of large continuous areas of suitable fuel uninterrupted by firebreaks, such as stream corridors. Heavily grazed patches break up the continuity of fuel and reduce the spread of fires. Fewer, smaller fires reverse the feedback loop, allowing the expansion of shortgrass grazing lawns. The white rhino is one of the very few surviving megagrazers. White rhinos are shortgrass feeders capable of creating and maintaining shortgrass swards that act as firebreaks. Waldram et al. (2008) showed that rhino removal (for game translocation) caused increased grass sward height in grazing lawns, reduction in use of the lawn by smaller shortgrass grazing antelope, and much larger fires. This is the first experimental study of extant megaherbivores supporting Flannery's (1994) hypothesis that removal of megaherbivores would have resulted in the increasing importance of fire.

White rhinos have been translocated to many savanna parks in Africa. As populations grow, unlimited by predators (they are too big and fierce), we should see changes in fire regimes as the lawns they create break up the continuity of tallgrass fuels. The brown world thus affects the black world, and vice versa. Although the effects of herbivores may be quite localized in a landscape, they influence the fire regime and therefore affect landscape dynamics at scales much larger than their individual grazing ranges.

CONSUMER CONTROL AND CASCADES

Cascading changes in species composition and ecosystem structure and function are a feature of transitions between alternative ecosystem states (Scheffer et al. 2001; Chapter 17, this volume). Species cascades are implicit in many experimental studies of fire exclusion and replacement of savannas by forests. However, the link with trophic cascade theory has seldom been made, and few studies have documented both floristic and faunal changes.

Cascading changes associated with brown world states are particularly important and controversial for ecosystem management. Rangeland scientists have traditionally seen grazing lawns and denuded patches as symbols of degradation and poor management. Yet grazing lawn grasses may be a distinctive legacy of the megafauna and a rich source of lawn grass species for the parts of the world that suffered megafaunal extinctions. In African savannas, different species of acacias occur in alternative ecosystem states. In heavily grazed patches, the saplings of brown world acacia species form cagelike structures with densely ramified shoots that protect the central stem. In contrast, species with polelike saplings characterize frequently burnt savannas. These possess large underground storage organs that facilitate rapid postburn sprouting and rapid growth above flame height (Archibald and Bond 2003). Switches between types of consumers lead to switches in tree species composition because of differences in sapling tolerance to fire or herbivory (Bond et al. 2001). There is also evidence for a specialized fauna of gazing lawns in South African savannas. There are not only shortgrass specialist mammal grazers, such as the white rhino, but also birds, grasshoppers, and spiders restricted to heavily grazed areas that do not occur in frequently burnt tall grasslands (Krook et al. 2007; Mandisa Mgobozi 2007, unpublished). The presence of a distinct biota suggests a long evolutionary history of this kind of habitat in African savannas. Far from being marks of poor management, overgrazed patches may be entirely natural components of the African savanna, contributing to the total pool of biodiversity with their unique habitat specialists. It would be interesting to examine the biota of other continents, which suffered megafaunal extinction, for habitat specialists restricted to very heavily grazed patches. Perhaps "rewilding" will be impossible on continents that lost not only their Pleistocene megafauna but also the rest of the brown world biota dependent on megafauna. It is interesting to note that the boreal region may also have switched from brown to green states in response to megafaunal extinction (Zimov et al. 1995). There are also analogous debates as to whether contemporary tundra ecosystems are overgrazed and degraded or represent alternative ecosystem states maintained by heavy grazing from reindeer/caribou (van der Wal 2006). Perhaps the domain of "ecosystems uncertain" extends also to the boreal zone, with alternative ecosystem states molded by remnant large mammal herds, fire, and the unique stress of the extreme climate.

SEMANTICS AND THE APPLICATION OF TROPHIC CASCADES TO MANAGEMENT

Trophic cascades have been defined as changes in species composition in eco-systems with two or more trophic links, meaning interactions between preda-tor, herbivore, and primary producer. In grassy ecosystems, predators may or may not exert significant direct control on herbivores. As Owen-Smith (1988) suggested and Sinclair et al. (2003) elegantly demonstrated, predator control decreases with increasing body size. Predators may also have negligible effects on populations of grazing mammals that aggregate in large herds (Mduma et al. 1999). Where megaherbivores still exist, predation is unlikely to control her-bivory and top-down control is imposed via a one-link system, from herbivore directly to plants.

The second semantic difficulty is that primary consumers can generate multiple effects via direct trophic interactions and via activities that alter habitat structure and fire regimes (McNaughton 1992). Elephants topple trees but may not feed on them. The rhino, a megagrazer, can alter landscapes by changing the fire regime, with cascading consequences for plants and animals. These effects have been described as ecosystem engineering (Jones et al. 1994) and are not necessarily propagated through the food web.

Despite these semantic difficulties, the trophic cascade literature seems highly relevant for understanding and analyzing "ecosystems uncertain" (Car-penter and Kitchell 1993; Scheffer et al. 2001). The trophic perspective con-tributes significantly to managing ecosystems far from their climate potential. It helps integrate knowledge across diverse areas of interest in these very dynamic ecosystems. Ecosystems controlled from the top down, whether by fire or her-bivory or both, cannot be managed by preservationist philosophy alone. If you add herbivores, you could be changing the fire regime, with surprising conse-quences. If you burn, fire management is bound to affect herbivores, their patch choice, and return time, with unintended consequences. If you do nothing, the grasslands may be lost altogether, replaced by forest with an entirely different biota. The changes may not be as rapid as in aquatic ecosystems, but they are of a magnitude that demands attention for effective conservation.

CONCLUSIONS

Consumer control by large mammals or fire appears to be globally widespread. Many terrestrial ecosystems are thus not as green as they should be. For example, the extent of tropical rainforests should be more than twice what it is if they were to occupy all climatically suitable areas (Bond et al. 2005), but large areas of potential forest are occupied by grassy alternative ecosystem states. Boreal ecosystems may also diverge from their potential state because of consumer activities (Zimov et al. 1995; van der Wal 2006). The extent of consumer control can be measured by ecosystem structure and composition. In tropical and subtropical regions, removal of consumer control can lead to rapid shifts to resource-controlled forests and thickets, with massive cascading losses of open habitat species. Switches in the dominant type of consumer, whether it be herbivores or fire, may have similarly cascading consequences, but transitions between the brown and black worlds are more easily reversible. That both have occurred in the historical landscape is indicated by the specialist fauna and flora associated with alternative states, at least in parts of Africa. I hesitate to call these trophic cascades, but cascades they are, and the trophic cascade literature is most aptly consulted for ideas on methods of analysis, concepts, and management implications. Far from being a small, localized phenomenon, major portions of the terrestrial realm are uncertain, with widely divergent ecosystem states depending on the interplay between resources and consumers. Uncertain ecosystems probably were even more extensive in the past, before the extinction of megafauna, and face a future only indirectly influenced by climate. They warrant much greater attention in global biogeography.

CHAPTER 17

Alternative States in Ecosystems

Marten Scheffer

Usually ecosystem changes such as the ones occurring as the result of the cascading effects of removing predators are reversible in the sense that the system recovers to the original state if the predator population is allowed to recover. However, there are exceptions to this rule. It has been shown that a range of ecosystems can have alternative stable states. This implies that there is a tipping point where runaway change propels the system to a contrasting state that is self-stabilizing. In such cases recovery to the original state is more difficult and can entail some form of shock therapy. Another tricky aspect of such critical transitions is that it is very difficult to predict them.

The study of critical transitions and alternative stable states is well rooted in mathematical theory. It had a particularly fruitful period about half a century ago. Work in the 1960s and 1970s by René Thom (1993) about catastrophe theory triggered a wide interest. With respect to ecology, Richard Lewontin (1969) pointed in a clear and systematic article to the theoretical possibility of alternative stable states. Crawford (Buzz) Holling (1973) linked these theoretical arguments to ecology in a more intuitive way, and Robert May (1977) wrote an influential review on thresholds and breakpoints in ecological systems. The recent resurgence of interest in the topic is caused by advances in linking these ideas in a solid way to the dynamics of ecosystems rather than an extension of the theory.

In this chapter I provide a summary of the topics covered in a book that I wrote recently, where much more background and examples from ecology and other branches of science may be found (Scheffer 2009). Here, I first briefly explain the theory of alternative stable states and critical transitions. Subsequently I give a few examples, referring mostly to cases described in other chapters of this book. Finally, I discuss the main implications of alternative stable states for the effect of altered trophic cascades for conservation and management.

THE THEORY

Mostly the equilibrium of a dynamic system moves smoothly in response to changes in the environment (Figure 17.1a). Also, it is quite common that the system is insensitive over certain ranges of the external conditions while responding strongly around some threshold condition (Figure 17.1b). For instance, mortality of a species usually increases sharply around some critical concentration of a toxicant. In such a situation, a strong response happens when a threshold is passed. Such thresholds are obviously important to understand. However, a very different, much more extreme kind of threshold than this occurs if the system has alternative stable states. In that case the curve that describes the response of the equilibrium to environmental conditions is typically folded (Figure 17.1c). Such a catastrophe fold implies that for a certain range of environmental conditions the system has two alternative stable states, separated by an unstable equilibrium that marks the border between the basins of attraction of the alternative stable states. The behavior of a ball on uneven surfaces of differing configurations provides a simple physical analogy (Figure 17.1, right panels).

This situation is the root of critical transitions. When the system is in a state on the upper branch of the folded curve, it cannot pass to the lower branch smoothly. Instead, when conditions change sufficiently to pass the threshold (F_2), an abrupt catastrophic transition to the lower branch occurs (Figure 17.2). Clearly this is a very special point. In the exotic jargon of dynamic systems theory it is called a bifurcation point. As we will see later, there are several different kinds of bifurcation points that all mark thresholds at which the system's qualitative behavior changes. For instance, the system may start oscillating, or a species may go extinct at a bifurcation point.

The point we have in our picture marks a catastrophic bifurcation. Such bifurcations are characterized by the fact that an infinitesimal change in a control parameter (e.g., reflecting the temperature) can invoke a large change in the state of the system if it crosses the bifurcation. Clearly, this kind of change is

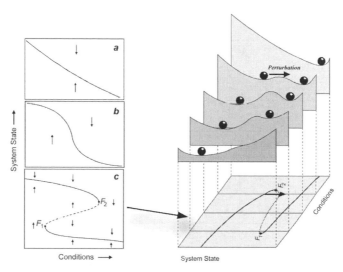

Figure 17.1. (a–c) Schematic representation of possible ways in which the equilibrium state of a system can vary with conditions such as nutrient loading, exploitation, and temperature rise. In **(a)** and **(b)** only one equilibrium exists for each condition. However, if the equilibrium curve is folded backwards **(c)**, three equilibria can exist for a given condition. The arrows in the graphs indicate the direction in which the system moves if it is not in equilibrium (i.e., not on the curve). It can be seen from these arrows that all curves represent stable equilibria, except for the dashed middle section in **(c)**. If the system is pushed away a little bit from this part of the curve it will move further away instead of returning. Thus, equilibria on this part of the curve are unstable and represent the border between the basins of attraction of the two alternative stable states on the upper and lower branches. Right: Stability landscapes depicting the equilibria and their basins of attraction at five different conditions. Stable equilibria correspond to valleys; the unstable middle section of the folded equilibrium curve corresponds to hilltops. If the size of the attraction basin is small, resilience is small, and even a moderate perturbation may bring the system into the alternative basin of attraction. (Modified from Scheffer et al. 2001)

among the most interesting ones from the perspective of this chapter. Although all kinds of bifurcations correspond in a sense to critical transitions, catastrophic bifurcations are the mathematical analogue of the dramatic transitions we are trying to understand. The bifurcation points in a catastrophe fold (F_1 and F_2) are known as fold bifurcations. (They are also called saddle–node bifurcations because in these points a stable node equilibrium meets an unstable saddle equilibrium.)

The fact that a tiny change in conditions can cause a major shift is not the only aspect that sets systems with alternative attractors apart from the normal ones. Another important feature is the fact that in order to induce a switch back

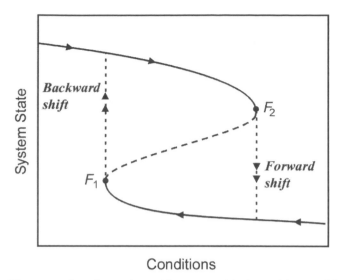

Figure 17.2. If a system has alternative stable states, critical transitions and hysteresis may occur. If the system is on the upper branch but close to the bifurcation point F_2, a slight incremental change in conditions may bring it beyond the bifurcation and induce a critical transition (or catastrophic shift) to the lower alternative stable state (the forward shift). If one tries to restore the state on the upper branch by reversing the conditions, the system shows hysteresis. A backward shift occurs only if conditions are reversed far enough to reach the other bifurcation point F_1. (Modified from Scheffer et al. 2001)

to the upper branch, it is not sufficient to restore the environmental conditions that immediately preceded the collapse (F_2). Instead, one needs to go back further, beyond the other switch point (F_1), where the system recovers by shifting back to the upper branch. This pattern in which the forward and backward switches occur at different critical conditions (Figure 17.2) is known as hysteresis. From a practical point of view hysteresis is important because it implies that this kind of catastrophic transition is not so easy to reverse.

The idea of catastrophic transitions and hysteresis can be nicely illustrated by stability landscapes. To see how stability is affected by changes in conditions, we make stability landscapes for different values of the conditioning factor (Figure 17.1, right-hand panel). For conditions in which there is only one stable state, the landscape has only one valley. However, for the range of conditions in which two alternative stable states exist, the situation becomes more interesting. The stable states occur as valleys, separated by a hilltop. This hilltop is also an equilibrium (the slope of the landscape is zero). However, this equilibrium is unstable. It is a repeller. Even the slightest change away from it will lead to a self-propagating runaway process moving the system toward an attractor.

To see the catastrophic transitions and hysteresis, imagine what happens if you start in the situation of the landscape up front in the series (Figure 17.1, right-hand panel). The system will then be in the only existing equilibrium. There is no other attractor, and therefore this state is said to be globally stable. Now, suppose that conditions change gradually, so that the stability landscape changes to the second or third one in the row. Now there is an alternative attractor, implying that the initial state has become locally (rather than globally) stable (i.e., as long as no major perturbation occurs, the system will not move to this alternative attractor). In fact, if we monitored the state of the system, we would not see much change at all. Nothing would reveal the fundamental changes in the stability landscape. If conditions change even more, the basin of attraction around the equilibrium in which the system rests becomes very small (fourth stability landscape) and eventually disappears (last landscape), implying an inevitable catastrophic transition to the alternative state. Now, if conditions are restored to previous levels, the system will not automatically shift back. Instead it shows hysteresis. If no large perturbations occur it will remain in the new state until the conditions are reversed beyond those of the second landscape.

In reality, conditions are never constant. Stochastic events such as weather extremes, fires, or pest outbreaks can cause fluctuations in the conditioning factors but may also affect the state directly, for instance by wiping out parts of populations. If there is only one basin of attraction, the system will settle back to essentially the same state after such events. However, if there are alternative stable states, a sufficiently severe perturbation may bring the system into the basin of attraction of another state. Obviously, the likelihood of this happening depends not only on the perturbation but also on the size of the attraction basin. In terms of stability landscapes, if the valley is small a small perturbation may be enough to displace the ball far enough to push it over the hilltop, resulting in a shift to the alternative stable state. Following Holling (1973) I use the term *resilience* to refer to the size of the valley (or basin of attraction) around a state that corresponds to the maximum perturbation that can be taken without causing a shift to an alternative stable state. A crucially important phenomenon in systems with multiple stable states is that gradually changing conditions may have little effect on the state of the system but nevertheless reduce the size of the attraction basin. This loss of resilience makes the system more fragile in the sense that it can easily be tipped into a contrasting state by stochastic events.

Resilience can often be managed better than the occurrence of stochastic perturbations. Although such a resilience-based management style is starting to be used in some systems, it requires a major paradigm shift in many other areas.

The last part of this chapter deals with practical questions such as how insight into the mechanisms that determine resilience may help us manage systems with less effort to reduce the risk of unwanted transitions while promoting desired ones.

AN EXAMPLE

A particularly well-studied example of an ecosystem with alternative stable states is shallow lakes (Scheffer et al. 1993; Scheffer and van Nes 2007). Submerged vegetation can greatly reduce turbidity through a suite of mechanisms such as control of excessive phytoplankton development and prevention of wave resuspension of sediments. However, the submerged vegetation also needs low turbidity in order to get sufficient light. As a consequence, there can be situations in which loss of vegetation leads to an increase of turbidity sufficient to prevent recolonization by submerged aquatic vegetation. In situations in which plants can promote their own growth conditions it seems intuitively straightforward that there may be two alternative stable states: one vegetated and another one without submerged aquatic vegetation. However, things are more complex than that. First, alternative equilibria arise only if the feedback effect is strong enough. Second, stability of one of the states can be lost if external factors such as climate or nutrient input change (cf. Figure 17.1).

To see how such loss of stability can happen, consider a simple graphic model of the response of shallow lakes to nutrient loading (Figure 17.3). An overload with nutrients such as phosphorus and nitrogen derived from wastewater or fertilizer use tends to make lakes turbid. This is because the nutrients stimulate growth of microscopic phytoplankton that make the water green and prevent light transmission. Although this eutrophication process can be gradual, shallow lakes tend to jump abruptly from the clear to the turbid state. This behavior can be explained from a simple graphic model based on only three assumptions: Turbidity increases with the nutrient level because of increased phytoplankton growth, vegetation reduces turbidity, and vegetation disappears when a critical turbidity is exceeded.

In view of the first two assumptions, equilibrium turbidity can be drawn as two different functions of the nutrient level: one for a macrophyte-dominated and one for an unvegetated situation. Above a critical turbidity, macrophytes will be absent, in which case the upper equilibrium line is the relevant one; below this turbidity the lower equilibrium curve applies. The emerging picture shows that over a range of intermediate nutrient levels two alternative equilib-

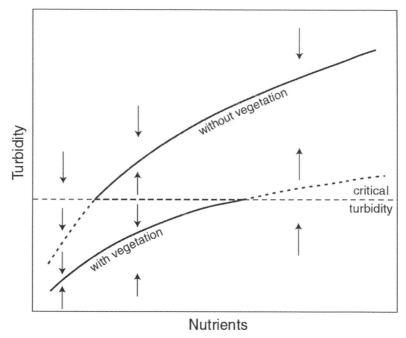

Figure 17.3. Alternative equilibrium turbidities caused by disappearance of submerged vegetation when a critical turbidity is exceeded. The arrows indicate the direction of change when the system is not in one of the two alternative stable states. (Modified from Scheffer et al. 1993)

ria exist: one with macrophytes and a more turbid one without macrophytes. At lower nutrient levels, only the macrophyte-dominated equilibrium exists, whereas at the highest nutrient levels, there is only an equilibrium without macrophytes.

The zigzag line formed by the stable and unstable equilibria in this graphic model corresponds to the folded line in Figure 17.1c and the gray panel below the stability landscapes (Figure 17.1). However, this simple example may illustrate the way in which a facilitation mechanism may cause the system to respond to environmental change showing hysteresis and catastrophic transitions. Gradual enrichment starting from low nutrient levels will cause the lake to proceed along the lower equilibrium curve until the critical turbidity is reached at which macrophytes disappear. Now, a jump to a more turbid equilibrium at the upper part of the curve occurs. In order to restore the macrophyte-dominated state by means of nutrient management, the nutrient level must be lowered to a value where phytoplankton growth is limited enough by nutrients alone to reach the critical turbidity for macrophytes again. At the extremes of the range of nutrient levels over which alternative stable states exist, either of the equilibrium lines approaches the critical turbidity that represents the breakpoint of the system. This corresponds to a decrease of resilience. Near the edges, a small perturbation is enough to bring the system over the critical line and to cause a switch to the other equilibrium.

MECHANISMS

The key ingredient to obtain alternative stable states is a positive feedback that drives the system toward either of the states. There are many different positive feedback mechanisms that can cause this effect. The shallow lakes example illustrates the effect of a positive feedback of plants on their own growth through a change in a key environmental factor. If the environment is harsh, self-facilitation through amelioration of conditions may lead to alternative attractors provided that the feedback effect is strong enough. For instance, in dry environments the microclimate may be ameliorated in the shade of large plants where temperatures are lower and humidity of soil and air is somewhat higher. If conditions are sufficiently arid, it is possible that seedlings may survive only under the canopies of such nurse plants. Obviously this implies that in a completely barren situation it is difficult to get vegetation started, even if the vegetated state can be stable because of the nursing effect of adult canopy (Holmgren and Scheffer 2001). The feedback of vegetation on moisture can also happen on much larger scales. In some regions such as the Sahel and part of the Amazon area, vegetation may promote precipitation. Loss of vegetation from such regions can lead to a climate that is too dry to support vegetation needed to keep up the precipitation (Scheffer et al. 2005). Erosion prevention may be another way in which plants keep their environment suitable. Fertile soils can form with time under vegetation. Loss of vegetation in some situations may lead to soil erosion that makes vegetation recovery difficult. Similarly, on smaller scales erosion of intertidal mudflats and lake sediments may be prevented by a layer of attached algae and other microorganisms. Initial consolidation can happen only in quiet periods, as normal resuspension and erosion prevent the organisms from establishing (Scheffer et al. 2003a; Van de Koppel et al. 2001).

Another way in which plants may create a positive feedback on their own growth is by affecting food web interactions. Again, this is well documented in shallow lakes where submerged vegetation induces important structural changes in the food web, promoting water clarity (Jeppesen et al. 1998). For example, the submerged plants provide a hiding place for water fleas against fish predation, and those water fleas swim out at night to filter the water clear of phytoplankton. In this way the plants protect the enemy (water fleas) of their enemy (phytoplankton). In other situations plants may protect predators that control herbivores that would otherwise eat the plants. For example, kelp forests may protect crabs, which subsequently control sea urchins that could otherwise devastate the kelp (Steneck et al. 2004).

So far I have highlighted examples of positive feedback in vegetation development. However, alternative stable states in nature can arise from a range of other mechanisms. For instance, classic Lotka–Volterra models show that it may arise if competition within a species is less strong than competition between species (Begon et al. 1996), and there is evidence that this mechanism may explain persistent cyanobacterial dominance in lakes (Scheffer et al. 1997). Also, models with large sets of species suggest that alternative stable community states should commonly arise in complex competitive networks (Van Nes and Scheffer 2004). Other studies indicate that alternative stable states can occur if a predator shares a resource with its prey (Diehl and Feissel 2000; Mylius et al. 2001) or controls the natural enemies of its offspring (de Roos et al. 2003; Walters and Kitchell 2001) and if populations live in scattered habitat fragments (Hanski et al. 1995).

Importantly, whenever alternative stable states are present any change in the system can invoke a difficult-to-reverse shift to an alternative state. This book provides examples of how affected predator populations can lead to profound alterations of the ecosystems of lakes (Chapter 4), temperate reefs (Chapter 3), tropical reefs (Chapter 5), tropical forests (Chapter 8), temperate and boreal forests (Chapter 9), and arctic and subarctic tundra and taiga systems (Chapter 10). The potential for alternative stable states implies that reestablishment of predator populations may not always lead to a straightforward return to the original system state. For example, reintroduction of northern pike in shallow lakes has been suggested as a way to invoke a trophic cascade leading to a clear state with submerged vegetation. However, the lack of such vegetation in the turbid state to begin with implies a lack of habitat that can serve as a refuge for small pike, essential for building up a sufficient predator population to control planktivorous and benthivorous fish (Scheffer 1998).

EVIDENCE

Although mechanisms can be complicated, usually a positive feedback in the development toward one of the states can be identified as the basic driver of alternative stable states. However, whether alternative stable states will arise from a positive feedback depends on the strength of the feedback. Thus, identification of a positive feedback does not prove that alternative stable states will occur.

There are different ways to infer from field data, experiments, and models whether alternative stable states are present in a system (Scheffer and Carpenter 2003). The observation of sudden radical changes in state, known as regime shifts, suggest catastrophic transitions between alternative states but do not constitute definitive proof (Carpenter 2003). Particularly strong cases such as the shallow lakes example are typically built by combining these approaches, including repeated experimental manipulation of entire ecosystems. However, just as proof of deterministic chaos or the effects of density-dependent regulation or competition in field situations has turned out to be elusive, irrefutable proof that alternative stable states occur on meaningful spatial and temporal scales in field situations is usually difficult to obtain. For instance, the repeated whole ecosystem manipulations that make the shallow lakes case so convincing cannot be used to study ocean ecosystems (Chapter 6, this volume) or regional vegetation systems such as the Amazon where alternative stable states are predicted to result from a feedback between vegetation and the regional climate (Oyama and Nobre 2003). An important question in this respect is what our null model should be. Because it would be remarkable in light of what we know now if ecosystems lacked alternative stable states, one could argue that instead of focusing on proving that a system has alternative stable states, it might be better to attempt to falsify this hypothesis. Thus we should assume that all systems have alternative stable states separated by critical thresholds, unless it can be demonstrated that from any initial condition the system eventually settles to the same state (Scheffer and Carpenter 2003). Obviously, this is very difficult in practice. However, one may question where the burden of proof should be. Erroneously assuming that a system has no alternative stable states may lead to dangerous false assumptions, such as the idea that effects of suppressing predator populations can be easily reversed.

IMPLICATIONS FOR MANAGEMENT

Broadly speaking, there are two reasons why knowing whether a particular ecosystem has alternative stable states and understanding how they are regulated can be useful in management. First, it may help to design strategies to prevent a catastrophic shift from a good to a bad state; second, it may help to find smart ways of invoking a shift back to a good state. Clearly, it is not always straightforward to determine whether a particular state is good or bad. Some states are widely considered bad. For instance, most do not like the collapse of cod in the North Atlantic Ocean. However, "good" is often not good in all aspects and for all stakeholders. Similarly, "bad" is often not bad for all either. For instance, the cod collapse has led to better prawn and lobster harvests (Worm and Myers 2003; Steneck and Wilson 2001). Nonetheless, the good–bad caricature is a useful simplification to see the two sides of the coin of alternative stable states from a management perspective: preventing bad transitions and promoting good transitions.

A crucially important phenomenon in systems with multiple stable states is that gradually changing conditions may have little effect on the state of the system but nevertheless reduce the size of the attraction basin (Figure 17.1). As noted earlier in the chapter, this loss of resilience makes the system more fragile in the sense that it can be easily tipped into a contrasting state by stochastic events. Although recent studies suggest smart ways to infer from the dynamics of a system whether it is losing resilience (Carpenter and Brock 2006; Van Nes and Scheffer 2007), it remains difficult to predict tipping points in practice. Loss of resilience is also one of the most counterintuitive aspects when it comes to explaining change. Whenever a large transition occurs, the cause is usually sought in events that might have caused it. The idea that systems can become fragile in an invisible way through gradual trends in climate, pollution, land cover, or exploitation pressure may seem counterintuitive. However, intuition can be a bad guide, and this is precisely where good and transparent systems theory can become useful. Resilience can often be managed better than the occurrence of stochastic perturbations. For instance, a lake that is not loaded with nutrients is less likely to shift to a turbid state in a climatically extreme year than a lake that has a near-critical concentration of nutrients. We cannot prevent heatwaves or storms, but we can manage the long-term trends in pollution and nutrient load.

Designing ways to promote a self-propagating runaway shift from a deteriorated state to a good state is perhaps the most rewarding part of the work on alternative stable states in ecosystems. The nice thing is that it can be really easy once you find the Achilles' heel of the system. In its most beautiful form it goes like this: Find out how to reduce the resilience of the bad state first, and then flip it out with little effort. Fish removal as a shock therapy to make turbid shallow lakes clear again is a classic example. First the resilience of the turbid state is reduced (and that of the clear state enhanced) by decreasing the nutrient load to the lake. Subsequently, a brief intensive fishing effort flips the system into the clear state (Hosper 1998; Scheffer et al. 1993).

An innovative idea related to managing ecosystems with alternative stable states is that we can often make smart use of natural variation in resilience. Recognizing this is important in strategies for promoting wanted transitions and preventing unwanted transitions. Natural swings may open windows of opportunity to induce a transition out of an unwanted state. For instance, a rainy El Niño year may be a window of opportunity for forest restoration (Holmgren and Scheffer 2001).

CONCLUSIONS

In conclusion, the mathematical theory of critical transitions between alternative stable states has a long history. Although it appeared difficult at first to tie the theory well to practice, this situation has changed radically over the past decade. The applications highlighted in this chapter are just a small sample, and readers interested in the issue can find much more background in another recent overview (Scheffer 2009). As elaborated in other chapters, removal of top predators may have profound cascading effects on ecosystems. The existence of alternative stable states in some ecosystems implies that reversal of such changes is not always easy. Nonetheless, insight into the mechanisms governing the stability of alternative states may sometimes open possibilities to trigger a shift from a deteriorated situation back to a more desirable state with little effort.

PART IV

Synthesis

Critics and skeptics of the larger ideas in science commonly think of supporting case studies as just that: idiosyncratic accounts that do not provide enough evidence for an inductive logic that is sufficiently general in application. But if an idea is indeed true, at some point the weight of supporting evidence becomes so forceful that all but the harshest critics and deepest skeptics are forced to pay homage. Along the way, as part of such paradigm shifts, there must be efforts to synthesize the evidence.

The four concluding chapters in this volume provide such a modern synthesis, but in quite different ways. Chapter 18 (Holt et al.) explores the consequences of predation and top-down forcing processes from a theoretical perspective, through the deductive process of developing a formal logic and then conducting analyses based on that logic. This is accomplished in two general ways: by reviewing the dynamic properties of three-trophic-level food chains under top-down control and by expanding these models to include various real-world complexities such as production, food chain length, diversity, and spatial and temporal scale. The results point to a rich potential (if not the inevitability) of top-down forcing and trophic cascades as key drivers in the organization of nature.

Chapter 19 (Shurin et al.) takes a different and more inductive approach of assembling the empirical evidence from studies of various ecosystems and then analyzing and interpreting that evidence through meta-analyses. Although this effort confirms that trophic cascades occur in all major types of ecosystems, it also indicates substantial variation in their strength across these systems. However, as the authors point out, what remains to be seen is the degree to which this variation is caused by such methodological factors as scaling effects or the degree to which people working in different systems have looked for trophic cascades rather than true differences in their strength and breadth of occurrence on the other.

Chapter 20 (Soulé) expands the science of trophic cascades into the applied arenas of conservation and natural resource management. At first glance, and after the preceding chapters, this may seem so obvious as to be hardly worth mentioning. But conservation and management distinguish themselves from basic science through human conditions and constructs, and Chapter 20 considers how these conditions and constructs have served as barriers not only to changes in policy but to progress in science itself. The unfortunate consequence is that although the conservation of large predators and the restoration of top-down forcing processes are clearly an essential cog in reversing the biodiversity crisis, that realization is still largely absent from mainstream conservation visions. Chapter 20 ends on a more hopeful note by identifying simple solutions to repatriating large predators and providing guidance for how they might be implemented.

Chapter 21 (Terborgh and Estes) is not a summary and overview of the preceding chapters but instead a synthesis based on the materials covered earlier in the book and the authors' own perspectives on ecology and conservation. This synthesis led to an echoing of Soulé's conclusion in Chapter 20 that the top-down forcing processes initiated by large vertebrates (and especially apex predators) are essential elements in the fabric of nature, and therefore their conservation and restoration are essential if we are to have any hope of curtailing or even reversing the current global trend in biodiversity loss.

CHAPTER 18

Theoretical Perspectives on Trophic Cascades: Current Trends and Future Directions

Robert D. Holt, Ricardo M. Holdo, and F. J. Frank van Veen

What is a trophic cascade, and why do we care? Pace et al. (1999, p. 483) define a trophic cascade as "reciprocal predator–prey effects that alter the abundance, biomass, or productivity of a population, community, or trophic level across more than one link in a food web." A more specific definition is provided by Persson (1999, p. 385): A trophic cascade is a "propagation of indirect mutualisms between nonadjacent levels in a food chain." Estes et al. (2001, p. 859) state that "a trophic cascade is the progression of indirect effects by predators across successively lower trophic levels." A folk definition might be "the enemy of my enemy is my friend."

The introductory chapter of this volume included a brief history of what we might call classic theoretical models of trophic cascades, emphasizing in particular qualitative messages and implications that have yet to be addressed in detail (or at all) in empirical studies, and it touched on the importance of alternative stable states related to cascades. The other chapters of this volume are replete with dramatic case histories illustrating why we should care about trophic cascades and aim toward a deep conceptual understanding of the factors that determine variation in the strength of cascades between biomes. Ultimately, this understanding has to be grounded in an empirical basis of well-crafted observational and experimental studies, but theoretical explorations can help clarify how known processes might be expected to govern cascade strength and provide pointers for fresh directions of empirical inquiry. Here, we provide an overview of a few likely extensions and modifications of the basic theory of trophic cascades that in our view warrant sustained attention by both theoreticians and empiricists over the next few years.

Classic food chain theory ignored many real-world complications. This is not by any means a critical observation. By recognizing what predictions change when we relax or modify assumptions in these simple models, we use these "perfect crystal" models (May 1973a) to identify axes of variation (in environmental variables or organismal traits) that can then be used to interpret differences in food web dynamics between different empirical systems. There are many directions in which one can imagine theory developing, and it is impossible to really do justice to all the possibilities in the space of this chapter. Instead, we present a few vignettes to briefly explore several important features left out of classic trophic cascade theory.

VIGNETTE 1: DIRECT PLANT–PREDATOR INTERACTIONS

The classic Lotka–Volterra food chain model (e.g., Equation 1.1 in Chapter 1 and related models such as those in Oksanen et al. 1981 and others cited in Chapter 1) assumes that to understand food chains, one can separately analyze the rate of attack of the predator on the herbivore and the rate of consumption of the plant by the herbivore and then splice these two interactions together to describe the dynamics of the full system. One current exciting area of study emphasizes the interdependence of different trophic linkages under the rubric of "trait-mediated interactions" and "interaction modification" (Bolker et al. 2003).

For instance, the rate of consumption of the plant by the herbivore may depend on traits of the herbivore that also influence its risk of predation, and these traits may respond plastically to the balance between predation risk and foraging reward. Foraging activity might itself expose an herbivore to predation, for instance because visual predators more easily cue on moving prey. In this case, herbivores should shift their behavior, feeding less when predators are perceived to be abundant and then foraging more intensely when predators are rare. There is a substantial and growing literature on the ecology of fear (Brown et al. 1999; Brown and Kotler 2004), showing both that it is a significant element in many natural predator–prey interactions and that it can provide a mechanism that mediates strong trophic cascades (Beckerman et al. 1997). Many studies suggest that trait-mediated effects are at least equal in strength to density-mediated effects (Werner and Peacor 2006; Schmitz et al. 2004). The magnitude of trophic cascades may thus reflect the scope for plastic responses by prey to predators and by plants to herbivores as well (e.g., induced chemical or structural defenses; Van der Stap et al. 2007). Plastic responses of prey to predators may straddle generations, with far-reaching effects on density and on spatial distributions at large scales. This occurs in aphids (sap-feeding insects), which respond to the presence of natural enemies by secreting alarm pheromones, which in turn induce wingless adults to give birth to winged offspring (Kunert et al. 2005). Wings are costly, so these offspring have lower fecundity, but they more than make up for this cost because they can leave the host plant, disperse over large distances, and thus escape intensifying predation or parasitism at their birthplace. This in turn reduces aphid density on the original host plant, potentially weakening the impact of the aphid on the host in future generations. *Daphnia* (water fleas) can show a similar delayed response to chemicals released by fish; in this case the females are more likely to produce diapausing eggs when fish are sensed, allowing the offspring to escape high predator densities, not in space but in time (Slusarczyk 2005). Again, this reduces the densities of zooplankters floating in the water column, which should lower the rates of herbivory on phytoplankters in subsequent generations. Moreover, in both cases herbivores can appear at places and times at rates not strongly coupled to local production, leading to mismatches in predictions relating primary productivity to the intensity of trophic cascades (Holt 2008b). Induced defenses therefore can have complicated effects on the spatial patterning and long-term temporal dynamics of trophic cascades.

By contrast, much less attention has been given to interactions between the plant level and rates of predation. A priori, one could easily imagine that the foraging efficiency of the top predator should sensitively depend on vegetation structure and biomass. For example, Ripple et al. (2001) report that in Yellowstone predation risk for elk is elevated in aspen copses (because of the substantial risk of a fleeing elk tripping over fallen logs, dramatically hindering escape from a wolf or cougar), and so elk feed less there. In the Serengeti, lions more readily capture wildebeest by ambush in tall grass, where a lion can easily hide, than in short grass, where a large predator is more conspicuous (Packer et al. 2005; Hopcraft et al. 2005). This in turn has consequences for lion population dynamics: During the 1994 drought in the Serengeti, the wildebeest population declined severely as a result of lower food abundance. This resulted in an increase in tall grass that persisted for several years and allowed the lion population to hunt more effectively and thus to increase in its numbers (Packer et al. 2005), potentially intensifying predation on ungulates. In general, in some systems predators may more easily capture prey in thick cover than in thin.

One consequence of this is that when vegetation grows in stature, there might be an intensification in predation, leading to a reduction of herbivory, which in turn promotes further vegetation growth. This positive feedback potentially has important consequences for community dynamics. To explore this effect more formally, we can mimic an effect of vegetation on predator efficacy by making the predator attack rate in the Lotka–Volterra model (Equation 1.1 of Chapter 1) a function of vegetation biomass, as follows:

$$\frac{dP}{dt} = P[b'a'(R)N - m']$$

$$\frac{dN}{dt} = N[abR - a'(R)P - m] \qquad\qquad (18.1)$$

$$\frac{dR}{dt} = R[r - dR - aN]$$

This algebraic rendition of the plant–predator interaction seems to describe the Serengeti lion–wildebeest–grass example, where grass cover has a direct effect on the predator attack rate. In the Yellowstone example, by contrast, changes in vegetation biomass alter herbivore perceptions of predation risk rather than attack rates, so the role of predation is to mediate the spatial pattern of habitat use by the herbivore rather than to increase herbivore deaths by direct consumption as in Equation 18.1. To account for this behavioral response, the "mean field" assumption of these equations would have to be expanded to account for spatial variation in perceived risk and movement between habitats. We revisit the Yellowstone case later (albeit without new algebra). Equation 18.1 is the same as the Lotka–Volterra model discussed in Chapter 1, except that we have replaced a constant attack rate by the predator on its prey with a variable attack rate that depends directly on plant biomass, R. The attack rate may either increase or decrease with R, depending on the detailed behavioral tactics used by the predator in finding and subduing its prey. An analysis of this model (Holt et al. in preparation) shows that a wide range of dynamic outcomes are possible, including unstable dynamics and alternative stable states, which are not observed in the simple Lotka–Volterra model. The introduction of a relationship between the predator attack rate and the biomass of the lowest trophic level in effect leads to a long feedback loop, and this feedback can be destabilizing. Here, we present some illustrative numerical examples rather than analytic details.

For simplicity, let us assume that there is a linear relationship between the predator attack rate and the thickness of the vegetation (truncated at 0 if necessary, because the attack rate cannot be negative):

$$a'(R) = \max(\alpha + \gamma R, 0) \tag{18.2}$$

Figure 18.1a shows an example of alternative states in the trophic organization of a community, when there is a positive relationship between predator attack rates and vegetation cover. At one equilibrium (on the left), the predator exists stably with the herbivore. A disturbance occurs (at time 0), which knocks predator numbers down. This leads to a surge in herbivore density and a temporary increase in predator numbers, but because herbivore numbers are high, the

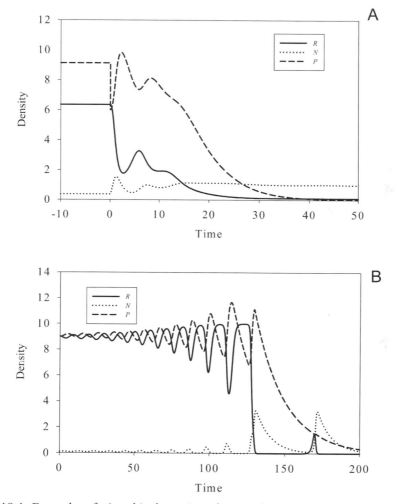

Figure 18.1. Examples of tritrophic dynamics, when predation depends on the biomass of vegetation. **(a)** An example of alternative stable states in the food chain model described by Equation 18.2, with a linear increase in predator attack rates with increasing vegetation biomass. **(b)** An example of strongly unstable dynamics in a food chain. Technically, the dynamics seem to approach a heteroclinic cycle, which in practice would amount to inexorable extinction.

vegetation continues to be reduced, which leads (with a lag) to a reduction in the attack rate on the herbivore. This reduction in consumption eventually begins to drag down predator numbers. The final state of the system is one in which herbivores crop vegetation down to a low level, which increases the herbivores' ability to escape predators, and so the predators cannot persist. Figure 18.1b shows another example, where the aforementioned functional dependence of predator attacks on vegetation density leads to wildly unstable dynamics. Adding small quantities of refuges for the herbivore can moderate these oscillations without making them disappear. If the functional relationship between attack rate and vegetation is nonlinear (e.g., sigmoid, increasing with vegetation biomass but saturating) (details not shown), alternative stable states may arise, each with a full food chain. But in one state, the predator is rare (but persistent), and the plants are overexploited by herbivory, whereas in the other the predator is common, the herbivore is kept in check, and the vegetation is thickly abundant.

In other systems, prey may find cover when vegetation is thicker, so attack rates would decline with increasing vegetation biomass (Arthur et al. 2005). For instance, Ayal (2007) argues that the sparse cover of deserts makes prey more vulnerable to visually hunting, roaming predators. An effective herbivore that can sharply reduce vegetation stature or biomass can thereby increase the likelihood that it will itself be controlled in a top-down fashion by predation. It is less likely that the food chain exhibits alternative states in this case (details not shown).

There are many other ways in which a basal resource can directly influence predator dynamics. For instance, plants may provide top predators with resources directly (e.g., fruit, nectar), with benign microclimates (e.g., patches of shade for a thermoregulating lizard), or with domiciles (e.g., ants on acacias with swollen thorns). Plants that feed predators in effect exert apparent competition on herbivores (van Rijn and Sabelis 2005). Plants experiencing herbivory may exude chemicals that help lure mobile predators (Dicke and Vet 1999); if the effectiveness of this tactic varies with plant density, then a model such as Equation 18.1 can be used to describe the resulting dynamics. Depending on the detailed way in which these effects are woven into a food chain model, one may observe enhanced stability, or the reverse.

VIGNETTE 2: TROPHIC CASCADES IN SPACE

A largely unexplored issue is how trophic cascades in local communities influence dynamics at larger spatial scales, how spatial processes in turn modulate trophic cascades at local scales, and how these processes are reflected in the emergent spatial structure of landscapes.

In the theoretical examples presented earlier, changes in vegetation biomass resulted in changes in predation rates in a homogeneous landscape. The wolf–elk–aspen example provides a case study for a spatial response by herbivores to predation risk in a heterogeneous landscape, where the heterogeneity itself emerges from the trophic interactions. In this case, a spatial patchwork mosaic of aspen groves and open habitat areas is reinforced over time by the presence of the predator. In the absence of wolves, elk could in theory produce a spatially homogeneous and open landscape with low aspen cover. With the introduction of wolves, small initial differences in aspen cover across the landscape become reinforced by elk perceptions of predation risk, as they avoid aspen patches and thus permit the growth of saplings and seedlings. In addition to leading to localized trophic cascades within aspen groves, wolf predation can thus theoretically magnify habitat heterogeneity at landscape scales, with important implications for the maintenance of biodiversity of other taxa such as bird assemblages at these scales.

If unstable dynamics readily emerge in local food web dynamics (a topic that has been one of the principal themes of the predator–prey literature over the last century), then for the food web to persist, stabilizing mechanisms not built into local interactions must exist. The most generic stabilizing factors in food webs that do not involve mechanisms of direct density dependence such as territoriality or interference all involve space, in various guises. At a very local scale, introducing absolute spatial refuges (where a limited number of prey are protected in the refuge) into any of the models discussed in Chapter 1, for victim or prey species at any trophic level, can be strongly stabilizing. This is also true for Equation 18.1. When refuges are sufficiently common, predation is expected to be weak, and trophic cascades should then be quantitatively trivial.

In the absence of refuges, space may still facilitate persistence if interactions are localized and dispersal is limited. For instance, dispersal permits colonization–extinction dynamics, so that an interaction persists at a regional scale, despite local extinctions caused by unstable food web dynamics. Even if consumers do not drive their prey locally to extinction, prey may go extinct because of disturbances. In an unlinked food chain, by assumption the top predator does not have alternative prey, and so it too should go extinct. Elsewhere (reviewed in Holt 2009), island biogeographic and metapopulation models for food chains have been developed, incorporating sequential colonizations and coupled extinctions. Here we simply recount some key conclusions from these models that are pertinent to trophic cascades.

First, food chains tend to be shorter on smaller or more isolated islands or habitat patches. In addition to colonization and extinction dynamics, there can be other reasons to expect the length of food chains to be shorter on smaller islands or patches (Holt 2009). A growing body of empirical studies shows that area indeed influences the trophic organization of communities.

The upshot of this work on area effects on food chain length is that there should be predictable spatial variation in the strength of trophic cascades. On very small islands, it may be difficult for even the herbivore to persist. On somewhat larger islands, the herbivore may be present, but the predator may be absent. These are sites where one might expect to observe intense herbivory (as seen in the transient islands of Lago Guri; Terborgh et al. 2006). On yet larger islands, predators should be predictably present and moderate the impact of herbivores on their food base. So there may be a hump-shaped relationship between the average intensity of herbivory and island size, arising because of spatial variation in the likelihood of top-down trophic cascades imposed by predation.

When there are strong top-down effects of predators on herbivores, the dynamic consequences of these local cascades can feed back to larger spatial scales. Within a continent, no area may be a defined source, but instead all local sites may experience extinction over a sufficiently long time span. Population persistence then requires that occupied sites supply emigrants who can colonize empty sites fast enough to replenish these losses. In such metacommunities (Holyoak et al. 2005), alternative stable states can arise in food chain dynamics (Holt 2002). Even in simple Lotka–Volterra models, top predators with strong direct density dependence (e.g., territoriality) can stabilize otherwise strongly unstable plant–herbivore interactions (Rosenzweig 1973). What this implies is that sites with the full chain may enjoy lower herbivore extinction rates than sites without a top predator. Moreover, these sites can provide a steady supply of colonists of both the top predator and the herbivore, to colonize sites where extinctions have occurred (e.g., because of disturbances such as hurricanes). So if a landscape starts out with most sites occupied by the full food chain, overall extinction rates will be low and colonization rates high. But if instead it starts out with few sites containing the predator, and the herbivore goes extinct rapidly in the sites it occupies alone, few patches will contain the herbivore, and the predator will not be able to increase when rare. The net effect of this is that the landscape as a whole can exist in alternative states (Holt 2002, 2009).

There are many important spatial dimensions of trophic cascades other than those captured in island biogeographic and metapopulation theory. For instance, in a heterogeneous landscape, productive habitats may be juxtaposed to unproductive habitats. If predators are mobile, the intensity of predation in the unproductive habitat may greatly exceed that expected from just local productivity because of spillover predation from the adjacent productive habitat (Holt 1984; Oksanen 1990; DeBruyn et al. 2004). This pattern of predation can lead to source–sink dynamics, which can help stabilize otherwise unstable trophic dynamics (Holt 1984). Organisms with complex life histories necessarily couple distinct habitats, leading to trophic cascades transcending ecosystem boundaries (McCoy et al. 2009). For instance, Knight et al. (2005) reported that fish in Florida ponds greatly depressed larval dragonfly abundances compared with fishless ponds. This in turn was reflected in the abundance of adult dragonflies buzzing around those ponds, which in turn influenced the abundance of insects such as bees and butterflies (due to both direct mortality and behavioral avoidance). Also, allochthonous resources can provide a spatial subsidy for consumers, sustaining resident predators at numbers well above that expected from local production, which can strengthen trophic cascades inflicted on resident herbivores in the recipient habitats (Polis et al. 1997). An important objective of research on trophic cascades should be to firmly integrate food web interactions and spatial ecological dynamics so as to better gauge the causes and consequences of spatial variation in the strength of cascades.

VIGNETTE 3: PATHOGENS AND FIRE AS MODULATORS OF TROPHIC CASCADES AND LANDSCAPE PATTERNS

Elsewhere in this volume, Sinclair et al. and Bond explore both the role of rinderpest as a trigger for a trophic cascade in the Serengeti and the role of fire as a "consumer" and determinant of vegetation biomass in the savanna biome. The Serengeti presents a situation in which a consumer (the wildebeest) is able to circumvent the regulatory pressure of predation by being highly mobile in large herds, diluting predation (Fryxell et al. 1988). This allows wildebeest to increase to high numbers, where it is regulated by its food resources (Mduma et al. 1999). The top-down impact of wildebeest herbivory on plant biomass is one of the principal drivers of temporal dynamics and spatial patterning in the Serengeti ecosystem (Holdo et al. 2009a, 2009b, in press). The consumer–resource equilibrium between this dominant herbivore and its resource base was disrupted by the emergence of rinderpest more than a century ago and then reestablished only after rinderpest eradication by deliberate human intervention (Sinclair 1979; Dobson 1995; Holdo et al. 2009b). Fire has played a key role in this cascade, which features not only the vertical propagation of trophic perturbations (the classic model) but also lateral effects between two resource guilds, grasses and trees (Holdo et al. 2009b). Trees and grasses compete in savannas (Scholes and Archer 1997), and fire often acts as a consumer that can mediate this competitive interaction by affecting the slow-growing but competitively dominant trees to a greater extent than the fast-growing grasses, which in turn feed the fire—a striking if nonstandard example of apparent competition (Holt 1977). Fire in effect is a transient megaherbivore, which can inflict massive changes on vegetation, independent of any higher-up top-down controls. The wildebeest population explosion that followed from rinderpest eradication in the 1960s led to a reduction in grass cover, a decline in fire, and an expansion of the tree population (Packer et al. 2005; Sinclair et al. 2007; Holdo et al. 2009b).

This transient trophic cascade from rinderpest to wildebeest down to vegetation, modulated by fire, exemplifies several broadly applicable themes. Trophic cascades have traditionally been viewed through the conceptual lenses of classic food web ecology, which focuses on the feeding relationships, escape tactics, and so on of the macroorganisms that to our eyes dominate natural ecosystems. Yet there is increasing evidence that hidden players—microorganisms—left out of the usual food web diagram can have enormous impacts on how communities are structured and ecosystems function (Thompson et al. 2001). Pathogens themselves may generate trophic cascades, as did rinderpest in the Serengeti.

Pathogens can also modify the strength of trophic interactions in more subtle ways, for instance by making prey easier to catch, which could magnify the ability of a predator to generate a trophic cascade, or by keeping in check the numerical responses of predators and thereby weakening cascades. Another very significant but nonstandard cascading effect of top predators may be via the modulation of disease dynamics in hosts that are also prey. An epidemiological consequence of top predator removal may be the unleashing of host–pathogen interactions at lower trophic levels and disease emergence both within and across host species (Hudson et al. 1992; Packer et al. 2003; Ostfeld and Holt 2004). Consider a prey species harboring a specialist pathogen, and assume that this prey species is also attacked by a generalist predator. Abstractly, this is an example of intraguild predation. Predators attack infected hosts, and so they can be predators directly on the pathogen. But predators can also attack healthy hosts, and so in a sense they compete with the infection for access to the resources in those host bodies. What is the effect of predator removal on the incidence of infected prey? In a wide range of epidemiological models, predator removal increases disease incidence (the fraction of hosts infected) and sometimes paradoxically can lead to a decrease in total host numbers. The latter can occur if predators preferentially pick off infected prey (e.g., because they are easier to catch) and if the pathogen can strongly regulate its host by both a reduction in fecundity and elevation in mortality for infected individuals (Packer et al. 2003). By shifting mortality regimes in a host species and increasing the prevalence of disease, predator removal can also lead to unstable host–pathogen dynamics and recurrent epidemics (Hochberg et al. 1990; Dwyer et al. 2004). All these mechanisms can then have knock-on effects on the resource populations used by the host species. If the pathogen is not completely specialized to a given host species, top predator removal can increase spillover infection onto alternative hosts, paving the way for cascading extinctions as resistant host species suppress vulnerable host species.

Various real-world complications can change these predictions in interesting ways. For instance, if a host has a strong immune response and strong negative density dependence in recruitment, somewhat paradoxically, modest amounts of predation can actually *boost* disease prevalence. The reason this occurs is that density-dependent compensation in reproduction leads to an increase in the supply rate of fresh, young, susceptible hosts (Holt and Roy 2007; Roy and Holt 2008), which can feed the infection. Moreover, in some systems parasites exploit both prey and predators to complete their life cycles. Models of such systems suggest that the emergent interactions can be very unstable (Fenton and Rands 2006).

VIGNETTE 4: EFFECTS OF DIVERSITY, IN SPECIES AND IN INTERACTIONS

One obvious limitation of models such as Equation 18.1 is that they assume that there is just a single species at each trophic level. This is a far cry from reality. Adding diversity to these models makes them more realistic and also much more complicated. This is a huge topic, and we cannot pretend to do justice to it here.

With multiple species on each level, one has to consider the interplay of trophic cascades and the conditions for species coexistence within trophic levels. Figure 18.2 shows a schematic example. At the top we see the typical effect of removing a top predator in an unlinked food chain: The herbivore increases (at least as a transient), and the basal level decreases. This is the classic view of a cascade. At the bottom, we have assumed that two plant species are present. If these species differ in their sensitivity to herbivory, the less sensitive can sustain

Top Predator Removal

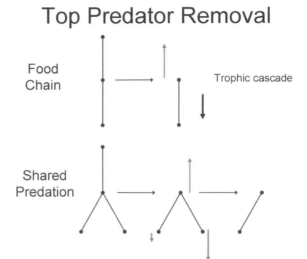

Food
Chain

Trophic cascade

Shared
Predation

Figure 18.2. Schematic depiction of a trophic cascade, when there is just one basal plant species present (top) or two (bottom). In the latter case, one consequence of the cascade may be extinction of a plant species due to intensified apparent competition.

the herbivore after release from predation, so that as the herbivore surges in numbers it greatly depresses the more vulnerable plant species. Because the herbivore is sustained at some level by the more resistant plant species, it can overexploit the vulnerable plant species to the point of extinction. This apparent competition effect is likely to be significant whenever multiple species are present at lower trophic levels. Considering interactions between consumer species, removal of a top predator can lead to an increase in intermediate predators (mesocarnivore release), with devastating consequences for prey species of the released predators (Crooks and Soulé 1999). Increasing the diversity of predators or prey also permits a rich array of additional interactions to occur, because of behavioral shifts or interference among species (Prasad and Snyder 2006).

Even when species within a trophic level coexist, heterogeneity in their properties can lead to shifts in how trophic level abundances change along productivity gradients (Chase et al. 2000). For instance, some experimental studies show that an increase in herbivore diversity can lead to a reduction in the indirect impact of top predators on plant biomass (Duffy et al. 2005). In arthropod systems, the outcome of trophic cascades may also strongly depend on the specificity of the signal provided by a target herbivore and the diversity of herbivores facing the predator or parasitoid. Vos et al. (2001) showed that a specialist parasitoid wasp that specifically targets one herbivore species can also be attracted to plants damaged by other herbivores. The presence of the other herbivores therefore can have a negative effect on the interaction strength between the parasitoid and its host; this tends to stabilize their dynamics by lowering average attack rates on the host this specialist parasitoid needs. However, Vos et al.'s model also showed that there may be a threshold diversity at which the interaction strength is reduced to such an extent that the parasitoid goes extinct. If trophic cascades in arthropod communities emerge from the concatenation of many specialist predator–prey or host–parasitoid interactions, then an increase in herbivore diversity may lower the overall strength of trophic cascades. The patterning of specialization and generalization in trophic relationships is of course central to the quantitative description of food webs. An important task for future work is to understand how the details of this trophic patterning in food webs influences the likely strength of trophic cascades.

There are many examples of responses to chemical cues from species at nonadjacent trophic levels, leading to surprising impacts on trophic cascades. For example, hyperparasitoid wasps at the fourth trophic level exude chemicals to which their host's hosts (aphids) respond by having greater fecundity, probably because they spend less time on defensive behavior and more on feeding (van Veen et al. 2001). In this case, the population growth rate of the herbivore is dependent on the density of its enemy's enemy, and having a top predator present intensifies herbivory on the plant.

Another complication is that top predators can be omnivores and intraguild predators, feeding on the same resources as do the intermediate consumers. Empirical studies show that adding diversity to the predator trophic level can both weaken (Finke and Denno 2004) and strengthen (Byrnes et al. 2006) trophic cascades, and theory can help illuminate why these disparate effects are observed. For instance, some predators might be able to use other predators as prey (intraguild predation). Models of intraguild predation (Holt and Polis 1997) predict that if the intermediate consumer does not have its own exclusive resources, to persist it must be superior at exploiting the shared resource. This implies that top predator removal will lead to a decline in the basal resource species. Moreover, intraguild predation can alter patterns of abundance along gradients in productivity. Figure 18.3 shows an example from a model where both the top predator (IG predator) and intermediate predator (IG prey) share a resource (shared prey), but the IG prey also has an exclusive resource (exclusive prey). Along a gradient in productivity, the IG prey is first absent, then increases, but then decreases once the IG predator is present. At intermediate levels of productivity, total predator abundance (adding intraguild predator and intraguild prey) declines with productivity, and the basal resource increases in

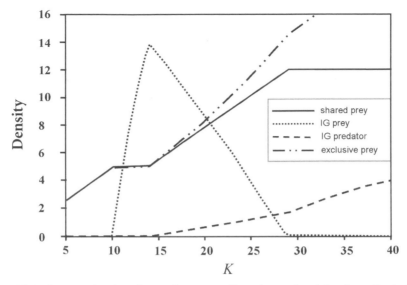

Figure 18.3. Patterns in abundance along a gradient in productivity, for a food chain in which there is intraguild (IG) predation, and alternative prey for both the top and intermediate predator. At high productivity, the intermediate consumer is excluded. In the zone where the two consumers coexist, the patterns in abundance by trophic level differ from the classic step pattern predicted in a simple, unlinked food chain (as in Oksanen et al. 1981). (Modified from Holt and Huxel 2007)

biomass because of the negative effect of the top predator on the intermediate species (see Holt and Huxel 2007 for more details). These effects differ from those predicted by classic food chain models (Oksanen et al. 1981) because there are shifts in species composition and relative abundance within trophic levels along the productivity gradient.

In complex communities, predators can influence the strength of many interactions beyond just plant–herbivore interactions, and sometimes these alternative pathways may be more significant than the classic trophic cascade. In the fish–dragonfly example sketched earlier (Knight et al. 2005), the bees and butterflies were all important pollinators for several species of flowering plants growing near the pond margins, and these plants proved to be more pollen limited (i.e., had lower seed set) near ponds without fish (and thus with abundant dragonflies) than near ponds with fish. Follow-up work is needed to determine whether these effects on plant reproduction lead to changes in local plant abundance and community structure. But even if there is no change in local abundance, it is clear that plant populations adjacent to fishless ponds with lower seed production will be poor sources of propagules in a metapopulation context.

Mutualisms and facilitation are ubiquitous features of communities, and changes in trophic cascades, via shifts in top predators, can have large effects on mutualism interaction webs (Knight et al. 2006). This could have ramifying influences both on local dynamics (e.g., on plant reproduction) and on landscape connectivity via dispersal. This topic has barely been touched in the empirical or theoretical literature.

VIGNETTE 5: THE INTERPLAY OF TEMPORAL VARIATION AND TROPHIC CASCADES

Most theory to date on trophic cascades has assumed that species interactions play out in constant environments. Yet temporal variability and disturbances can have large impacts on the strength of interspecific interactions, including trophic cascades. For instance, Spiller and Schoener (2007) report that at their Bahama study site a hurricane greatly amplified the impact of herbivores on plants because of an overall reduction in predation by lizards, which were hammered by the storm and lagged in their numerical recovery. Seasonal variation in food supply can constrain consumer dynamics, thus freeing the resource base from consumption in certain phases of the year and potentially destabilizing food chain dynamics (Oksanen 1990). A localized pulse at the base of a resource chain could lead to increases in herbivores, which in turn attract predators from a surrounding landscape. After depleting the local patch of its resources, the predators could then spread out over neighboring habitats, which would then experience transient spikes in predation and thus the strength of trophic cascades (Holt 2008b). Understanding how temporal variation and disturbance translate into temporal and spatial variation in the impact of trophic cascades is an important challenge for both theoretical and empirical studies.

VIGNETTE 6: *HOMO SAPIENS*, THE "UBER"-TOP PREDATOR?

Across the globe, humans have decimated top vertebrate predators, with ramifying consequences for ecosystem structure and human interactions (Stolzenburg 2008). Sometimes, humans themselves are significant predators, acting in many ways as the dominant determinant of top-down cascades. The Serengeti again provides a potential example of this role. Humans can hunt wildebeest and other ungulates for local consumption and the bushmeat market. Using a spatially explicit simulation model, Holdo et al. (in press) showed that increased wildebeest hunting in the Serengeti as a result of the expanding human population is predicted to have strong knock-on effects on the amount of fire and ultimately tree cover in the Serengeti. Moreover, these effects are likely to propagate across space because of the spatial coupling effect of wildebeest movement across the system. This spatial propagation can be demonstrated through simulations of the system with and without wildebeest present (Holdo et al. 2009a). Figure 18.4 shows the predicted long-term (based on a 100-year time horizon) state of the Serengeti ecosystem with and without wildebeest (i.e., with human hunting kept to a low level and allowed to increase in an unbounded manner). When wildebeest are absent from the system (top panels), standing grass biomass is high, fires are frequent, and tree cover is uniformly low. When wildebeest are present (bottom panels), the intensity and timing of grazing in relation to rainfall create a distinctive pattern of fire occurrence across the landscape, and this in turn has a dominant effect on the amount and distribution of tree cover (Figure 18.4). Intense predation on wildebeest by humans is predicted ultimately to affect not only the total amount of tree cover present but also the pattern of its spatial distribution.

CONCLUSIONS

Theory in ecology helps to clarify our understanding of the often surprising effects that emerge because of the nonlinear feedbacks inherent in ecological systems. Much of what we have presented here can be viewed as statements of hypotheses that warrant much closer examination by empiricists. There are a number of important issues we have not discussed, such as how theory might help us understand the seeming contrast between terrestrial and aquatic ecosystems in the strength and ubiquity of trophic cascades, and how evolutionary dynamics can modify trophic cascades. To end this chapter, let us reprise some of the main themes we have presented here and in Chapter 1 as a kind of road map that may be helpful in guiding future empirical studies of trophic cascades.

- We need to characterize the effects of trophic cascades not just on average abundances of populations, guilds, and entire communities but on dynamic stability, which influences extinction risks and thus the stability and durability of entire food webs.
- Linking trophic cascades to spatial ecology—ranging from metacommunity dynamics, to spatial subsidy and spillover effects, to behavioral redistributions, to ontogenetic habitat shifts—is a large and important

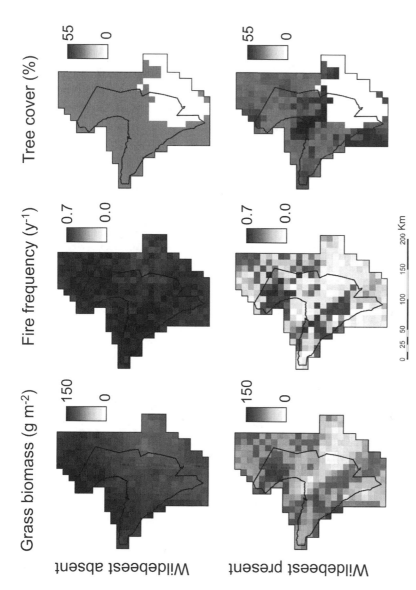

Figure 18.4. Spatial distribution of standing grass biomass, fire frequency, and tree cover with (bottom panels) and without (top panels) wildebeest in the Serengeti ecosystem after a 100-year simulation. The Serengeti ecosystem is defined here as the range of the wildebeest migration and comprises about 30,000 square kilometers, simulated in a lattice at 10- by 10-kilometer resolution (Holdo et al. 2009a). The simulation results shown here assume for simplicity that no elephants are present.

challenge. The real impact of predators may not be so much that they increase local plant biomass but that they indirectly enhance plant reproductive output, which in turn sustains metapopulations via colonization at broad spatial scales.

- Trophic cascades have many "nonstandard" dimensions, including the ecology of fear, trait modifications, and influences of vegetation on predation rates, that have yet to be fully explored, either theoretically or empirically.

- Trophic cascades can severely affect mutualism and facilitation webs and shift host–pathogen interactions in many ways. All of these effects may loom larger than the traditional emphasis on changes in levels of herbivory.

- Theoretical analyses of multispecies interactions—intraguild predation, apparent competition, keystone consumption, competitive guild interactions—can be used to elucidate many of the complexities of trophic cascades in both natural and disturbed or managed ecosystems. A particularly important complexity is the ways in which temporal variation and disturbance regimes influence the pattern of spatiotemporal variation in the strength of trophic cascades.

The other chapters in this volume recount dramatic examples of how top predators affect the structure and functioning of natural ecosystems, particularly through their influence on the intensity of herbivory. The theoretical ideas sketched in this chapter suggest that top predators could have far more pervasive impacts on community structure and dynamics than currently believed. To us, this makes a compelling case that conservation, harvesting, and land management strategies should have as a primary focus the retention and protection of the full panoply of top predators so at risk today.

ACKNOWLEDGMENTS

We thank Mike Barfield and David Hall for their assistance, the University of Florida Foundation for its support, and John Terborgh and Jim Estes for having convened a very fine meeting at White Oak, and then inviting R.D.H. to participate.

CHAPTER 19

Comparing Trophic Cascades across Ecosystems

Jonathan B. Shurin, Russell W. Markel, and Blake Matthews

The idea that predators indirectly regulate ecosystems by controlling herbivores, the foundation of the trophic cascade concept, has achieved wildly uneven acceptance among ecologists working in different ecosystems. The trophic cascade in lakes is one of the best understood examples of community interactions influencing primary production (Lawton 1999). Limnologists have demonstrated effects of piscivorous fishes on phytoplankton through a combination of field mesocosm and whole ecosystem experimentation, theoretical studies, and examination of macroscopic patterns (McQueen et al. 1986; Carpenter and Kitchell 1993; Mazumder 1994). Pervasive trophic cascades have been shown in a wide range of lake systems, but exceptions are often informative. Along with nutrient abatement, trophic cascades have proven sufficiently reliable to serve as a tool for managing lake eutrophication (biomanipulation; Shapiro and Wright 1984). The importance of trophic cascades in lakes relative to other factors such as nutrient loading, and the factors governing their expression, are still debated after four decades of research (DeMelo et al. 1992; Currie et al. 1999; Drenner and Hambright 2002). However, their role in many lakes is undeniable (Brett and Goldman 1996).

In contrast with the extensive literature from lakes, the generality and importance of trophic cascades in terrestrial and marine systems remain contentious topics despite some dramatic examples. The weight of the evidence for trophic cascades in lakes relative to terrestrial or marine habitats may reflect the amenability of lakes to experimental manipulations, the attention the subject has been given by limnologists, or real differences in the relative strength of trophic interactions as drivers of ecosystem processes. This distinction is critical to the question of whether our knowledge of lakes can inform our understanding of other systems.

Measuring the extent of predator effects in relation to other assaults on the integrity of aquatic and terrestrial ecosystems is one of the most important challenges facing ecologists. The tractability of lakes as experimental systems is inversely related to their importance for global ecosystems. Freshwaters cover around 3 percent of the world's land surface (Downing et al. 2006), and oceanic processes dominate climatic forcing and global geochemical cycles (Falkowski et al. 2000). Schindler et al. (1997) showed that the introduction of piscivorous fishes shifts lakes from being net sinks to net sources of carbon to the atmosphere. However, because lakes represent such a small contribution to the global environment, the potential implications of this effect are minor (Cole et al. 2007). By contrast, small changes in the balance between oceanic production and respiration have profound implications at the global scale because of the vastness of the pelagic ocean (del Giorgio and Duarte 2002). In addition, trophic structure among high-order marine predatory fishes is undergoing rapid and widespread change in nearly every corner of the ocean (Pauly et al. 1998; Myers and Worm 2003; Halpern et al. 2008). The removal of biomass by industrialized fishing required 8 percent of global marine primary productivity to support it in the early 1990s (Pauly and Christensen 1995), a figure that has undoubtedly risen since. If the effect of predator removal is similar in lakes and oceans, then current rates of fishing could have dramatic ecosystem effects that extend well beyond the fish themselves. Predator extirpations on land can also transform terrestrial ecosystems, shifting the balance between herbaceous plants and trees and even affecting stream geomorphology (Ripple and Beschta 2007a). The indirect consequences of altered trophic structure are potentially great but poorly understood.

Here we review the evidence for cross-system variation in the strength of trophic cascades. We focus our attention on two main contrasts: between aquatic and terrestrial systems and between lakes and marine systems. Our goal is to ask whether the roles of predators vary between ecosystems and how different environments exert selective pressures on organisms that may generate these contrasts.

EVIDENCE FOR TROPHIC CASCADES IN LAKES

Trophic cascades have been a contentious topic in ecology since Hairston, Smith, and Slobodkin (1960) first proposed that predator control of herbivores is a general explanation for the accumulation of ungrazed plant biomass (Murdoch 1966; Ehrlich and Birch 1967; Strong 1992). Lake ecosystems presented some of the first and most compelling evidence for this hypothesis, beginning with Hrbáček et al. (1961) and Brooks and Dodson (1965). Carpenter and Kitchell (1993) provided experimental evidence that increasing the abundance of piscivorous fish can have cascading effects on lake ecosystems that persist for years, are robust in the face of nutrient fertilization (Carpenter et al. 2001), and occur over a wide range of lake trophy (Vanni et al. 1990; Mittelbach et al. 1995). Lawton (1999, p. 182) called experimental studies of the role of fish predators in lake ecosystems "one of the triumphs of ecological science."

Whole-lake experiments with contrasting fish communities provide some of the most compelling evidence for the importance of trophic cascades in lakes, but the generality of trophic cascades to a larger body of lakes is still uncertain because aspects of the environment and lake communities can play a role in their expression. Correlative studies have shown that algal biomass for a given level of phosphorus is lower in lakes with abundant large *Daphnia* (Mazumder 1994) and in lakes with both planktivores and piscivores (Drenner and Hambright 2002). Although these patterns are consistent with top-down control of algal biomass via piscivores, experimental studies in a wide range of lakes show equivocal support for the generality of trophic cascades. In a review of seventeen experimental studies of piscivore effects on algal biomass, only seven (four piscivore additions, one piscivore removal, and two piscivore enhancements) supported the trophic cascade hypothesis (Drenner and Hambright 2002). Thus, despite classic examples, the evidence for piscivore-mediated cascades in lakes is mixed. Exceptions to the generality of cascades in lakes can help refine our understanding of top-down control in other ecosystems.

Comparing the environmental factors that govern the strength of cascades in lakes may facilitate contrasts between systems. First, strong imbalances between the elemental composition of grazers and algae can diminish trophic efficiency and reduce the potential for grazer control over autotroph biomass. In lakes, trophic cascades may be more prominent at high phosphorus levels (Benndorf et al. 2002; Drenner and Hambright 2002), where phosphorus-rich *Daphnia* grazers have the greatest influence on phytoplankton. Elser et al. (2000b) found that introduction of piscivorous pike into an experimentally

eutrophied lake (Lake 227 of the Experimental Lakes Area) reduced planktivorous minnow abundance, increased zooplankton grazing, and decreased phytoplankton biomass. However, the same treatment in a nearby oligotrophic lake
(Lake 110) had little effect on zooplankton and phytoplankton trophic levels
(Elser et al. 1998). Trophic cascades mediated by *Daphnia* can also occur in oligotrophic lakes with high carbon:phosphorus phytoplankton; however, the effects of predator manipulations may take years to detect (Parker and Schindler
2006). Severe stoichiometric imbalances between producers and consumers
may weaken the potential for cascades by preventing the establishment of the
most effective grazers.

The second feature of lakes that alters the intensity of cascades is habitat
structure. In lakes, structurally complex habitat can dampen top–down control
by allowing spatial refuges for prey from their predators (Crowder and Cooper
1982). Extensive littoral vegetation provides refuges for both benthic invertebrates and pelagic zooplankton that migrate horizontally throughout the day
(Jeppesen et al. 1997). Similarly, deep lakes show weaker cascading effects of
fish as a consequence of hypolimnetic refuges for vertically migrating zooplankton (Tessier and Woodruff 2002). Structurally complex habitats weaken
coupling between predators and prey, but the way in which structure affects
predator foraging or prey avoidance can be subtle and specific to particular taxa
(Warfe and Barmuta 2006). Quantifying the degree of complexity across systems demands a common measure of structure that may prove challenging with
diverse habitat types.

Finally, subsidies of prey from littoral food webs and organic matter from
terrestrial ecosystems may increase the importance of trophic cascades in lake
ecosystems as they often do in terrestrial systems (Chapter 11, this volume).
The decomposing food chain connects to the classic phytoplankton-based food
chain through bacteria that recycle terrestrial detritus and are preyed on by
protozoans and larger zooplankton (Chapter 4, this volume; Prairie et al. 2002)
and through detritivorous invertebrates that are prey for higher trophic levels.
Large terrestrial carbon subsidies may support high densities of herbivorous
zooplankton that exert increased control over phytoplankton (Vander Zanden
et al. 2005). Similarly, benthic prey contribute heavily to the diets of many
nominally piscivorous fishes (Schindler and Scheuerell 2002). Such prey subsidies from littoral habitats can enhance densities of fish predators in lakes,
leading to stronger top–down control over the pelagic food chain. Alternative
routes to the classic lake food chain through terrestrial detritus and benthic
prey may subsidize pelagic consumers and increase consumption pressure on
plankton.

EVIDENCE FOR TROPHIC CASCADES IN OCEANS

Trophic cascades have been well documented in coastal zones for marine intertidal and benthic communities but less so in the pelagic. The easy manipulation
and observation of rocky intertidal and shallow subtidal communities facilitated
early experimental and comparative approaches that provided evidence of the
strong effects of predators and grazers in these systems (Paine 1966; Paine and
Vadas 1969; Mann and Breen 1972; Estes and Palmisano 1974). However,
whole-system experiments analogous to the work in lakes are impossible in
most of the open ocean. Instead, marine ecologists have had to rely heavily on
the role of humans as predators to reveal the cascading effects of predators on
marine ecosystems. Time series of changes in biomass across trophic levels,
comparison between reserve and nonreserve areas, and spatial patterns of predator and prey abundance provide evidence for trophic cascades in marine ecosystems. Here, we review insights into trophic cascades in the ocean resulting
from these approaches.

Marine Benthic

Most examples of marine trophic cascades come from experimentally tractable
hard- and soft-bottom benthic communities (reviewed by Pinnegar et al. 2000).
The best-elaborated example involves sea otters, sea urchins, and kelp forests
(Estes and Palmisano 1974). Declining otter populations in the Aleutian Islands
released sea urchins from predation pressure, leading to widespread kelp defor-
estation and extensive urchin barrens. In contrast, kelp forest communities in
southern California responded differently to the removal of sea otters. Predator
diversity and functional redundancy in California kelp communities maintained low urchin populations until these predators also fell victim to overexploitation (Estes et al. 1989; Tegner and Dayton 2000; Steneck et al. 2002). Cascades in kelp forests may depend on diversity and functional redundancy at the
predator trophic level.

Comparison of community structure between reserve and nonreserve
areas, or along gradients of exploitation, has also revealed evidence for trophic
cascades in benthic marine systems (Chapters 3 and 5, this volume). The best
examples of trophic cascades in temperate systems come from contrasting community structure inside and outside marine reserves in New Zealand. Unprotected areas are dominated by urchin barrens that coincide with low densities
and small sizes of predatory fish and spiny lobsters. Fish and lobsters are larger
and more abundant inside reserves, urchin densities are reduced, and large
brown macroalgae are common (Shears and Babcock 2003). Trophic cascades

have been reported in coral reef ecosystems and marine reserves (McClanahan and Shafir 1990; Dulvy et al. 2004a). However, Newman et al. (2006) surveyed coral reef communities along a gradient of fishing intensity at thirty-four reserve and nonreserve areas across the northwestern Caribbean. Herbivorous and predatory fish biomass increased within marine reserves, and herbivorous fish biomass was negatively correlated with fleshy algal biomass. However, because fishing pressure falls on both predatory and herbivorous fishes, the loss of predators did not cascade to the producer trophic level. Thus, trophic cascades in coral reefs may be obscured by trophic diversity among fishes of similar size.

Marine Pelagic

Evidence for trophic cascades is sparser in marine pelagic ecosystems than benthic. The vast size of pelagic systems, high trophic complexity, and complex oceanography may dampen cascades in these systems (Chapter 6, this volume). Absence of evidence for cascades in the open waters of the ocean may also relate to the forbidding environment for experiments and observations. A growing number of long-term time series and spatial analyses provide convincing evidence that trophic cascades in marine pelagic systems are likely, at least under some conditions.

Time series of correlations in abundance between predators and prey have been used to infer the direction of trophic control. Negative correlations may indicate predator suppression of prey by predators, whereas positive correlations suggest bottom-up control of predators by their resources (Worm and Myers 2003). Shiomoto et al. (1997) provide compelling evidence of top-down control of macrozooplankton and phytoplankton biomass by zooplanktivorous salmon in a 10-year data set from the North Pacific. They show repeated oscillations between years of high pink salmon (*Oncorhynchus gorbuscha*) biomass and years of correspondingly low macrozooplankton and high chlorophyll concentrations. Similar examples of time series during declines in top predatory fishes in the Black Sea (Daskalov 2002), the northwest Atlantic (Frank et al. 2005), and the Baltic (Casini et al. 2008) also demonstrate diagnostic negative correlations between the biomass of pelagic top predators, planktivorous fish, zooplankton, phytoplankton, and nutrient availability. As top pelagic predators were removed via industrial fishing, biomass of small pelagic fishes and phytoplankton increased, while the biomass of large zooplankton and nutrient availability decreased. Examples of positive correlations between adjacent trophic levels have also been shown. Frank et al. (2006) found that piscivorous and planktivorous fishes fluctuated synchronously in lower latitudes of the northwest Atlantic and inversely at high latitude. They suggested that top-down control dominates in cold waters with few species but is weakened as more species are added to the south. This intriguing observation suggests that trophic control varies geographically as a function of temperature and species diversity as a consequence of the potential for compensation among species within trophic levels.

Another approach to assessing the importance of trophic cascades in marine pelagic ecosystems is to examine spatial correlations between primary productivity and consumer biomass. Ware and Thomson (2005) analyzed satellite observations of surface chlorophyll concentrations and mean annual yields of resident and migratory fishes of the northeast Pacific continental margin. Linear correlations ($r^2 = .87$) between these factors account for 87 percent of spatial variance in resident fish yields. Ware and Thomson interpret these patterns to indicate that bottom-up forces play a dominant role and that top-down control is negligible. However, similar spatial correlations between fish yield and algal productivity have also been demonstrated in lakes where cascades are often important (Downing et al. 1990). Measures of predation intensity such as zooplankton size structure and the presence of piscivorous fishes often explain additional variation in the relationship between phytoplankton and zooplankton densities (Mazumder 1994). Positive spatial correlations between fish predators and phytoplankton therefore are not incompatible with strong top-down effects.

TERRESTRIAL SYSTEMS

Hairston, Smith, and Slobodkin's (1960) original hypothesis was based on the idea that predators maintain terrestrial plant biomass, and indeed subsequent work has shown numerous examples of indirect trophic control. However, their argument attracted criticism from the beginning (Murdoch 1966; Ehrlich and Birch 1967; Strong 1992). Natural and manipulative experiments have identified important roles for spider predators in grasslands (Schmitz 2006), lizards on Bahamian islands (Schoener and Spiller 1999a), wolves and ungulates in western North America (Ripple and Beschta 2007a), weasels and voles in Norwegian tundra (Aunapuu et al. 2008), and tropical forest fragments on Venezuelan islands (Terborgh et al. 2001). A number of other studies have shown weak or negligible indirect effects of predators on plant biomass or performance (Finke and Denno 2004; Gruner 2004; Van Bael and Brawn 2005; Schmitz 2006). These case studies are well reviewed in other sections of this volume; however, they illustrate two main contrasts with the aquatic literature. First, few studies have been replicated in multiple terrestrial habitats of the same type. The paucity of examples offers much less information for making informative contrasts between studies. Second, the variety of habitats studied is also quite limited, with several important biomes represented by only one or two experiments. The need for more broadly replicated experiments and observations in a wider range of environments remains strong in terrestrial systems.

CASCADES OCCUR IN ALL ECOSYSTEMS, BUT WHAT IS THEIR RELATIVE STRENGTH?

The study of trophic cascades reveals a rich complexity of interactions between producers, grazers, predators, and decomposers. Some of these complexities confound Hairston, Smith, and Slobodkin's (1960) classic view of a world organized like a food chain with a few discrete trophic levels. Nevertheless, the trophic cascade concept has proved remarkably robust as a useful metaphor, a management tool, and an organizing principle of food web ecology. In this section we review quantitative evidence for variation in cascade strength and potential contrasts between ecosystems that may generate such broad differences.

Meta-Analysis

Whole-system experiments with large, wide-ranging predators are feasible in lakes but not in many marine or terrestrial systems (but see Chapter 11, this volume); however, extensive smaller-scale predator exclosure experiments have been carried out in lentic, terrestrial, and marine environments. Such experiments lack the realism of the whole-system studies (Chapter 4, this volume) and are therefore less useful for establishing the utility of predator manipulations as a management tool. However, these experiments are useful for elucidating mechanisms of effects, and in fact they have given a remarkably consistent picture of the role of predators in aquatic ecosystems (Brett and Goldman 1996; Micheli 1999).

Syntheses of experimental studies of trophic cascades point to potential differences in top-down control between aquatic and terrestrial ecosystems, benthic and pelagic habitats, and freshwaters and marine systems (Shurin et al. 2002). A recent meta-analysis considered cascade strength in terms of the response by plant community biomass to predator removal and therefore may neglect important shifts in plant community composition, decomposers, nutrient cycling, diversity, or other aspects of ecosystem structure and functioning. The main contrasts identified indicate that the effect sizes of predator removal manipulations on plant community biomass were stronger in aquatic than terrestrial ecosystems, stronger in freshwater than marine plankton, and stronger in marine benthos than marine plankton.

A recent meta-analysis of a completely independent data set of herbivore removal experiments also supports the aquatic–terrestrial contrast by showing weaker grazing control of autotroph biomass in terrestrial systems compared with aquatic systems (Gruner et al. 2008). In addition, analysis of the biological and methodological factors associated with cascade strength in experimental studies showed that invertebrate herbivores tended to exert stronger control over plant biomass than vertebrates (Borer et al. 2005). A meta-analysis of herbivore removal experiments arrived at the same conclusion (Bigger and Marvier 1998). The lower mass-specific metabolic rate of invertebrates may allow them to reduce plant biomass to a greater degree than vertebrates (Shurin and Seabloom 2005).

Although the contrasts identified here were strongly supported by the meta-analysis of 114 published experiments, the literature on cascades reveals a number of important holes and biases in the kinds of studies that have been performed to date. First, only a narrow range of terrestrial habitats have been examined where plant community biomass is monitored. Some of the most comprehensive examples of terrestrial cascades (Schoener and Spiller 1999a; Terborgh et al. 2001; Ripple and Beschta 2007a) were not included in the meta-analyses because they did not report plant community biomass. The experiments included in the analyses represented a wide range of grassland and agricultural environments and a great diversity of predators and herbivores. However, it is entirely possible that terrestrial predators regulate ecosystem processes in ways that are not apparent from measures of plant standing stock or that larger effects would be observed in systems that are not amenable to experimentation. Synthetic studies of a greater range of ecosystem attributes and processes are necessary for robust comparisons across ecosystems.

In addition, the comparison of cascade strength via meta-analysis focused on effects on plant community biomass as a comparable unit for measuring effect size across systems. However, predators in all systems may influence other aspects of ecosystem structure and performance, and such variation has not been quantitatively compared between systems. For instance, Schmitz (2006) found that spider predators had no effect on the biomass of an old-field plant community, but they affected plant community composition in ways that accelerated nitrogen cycling and increased light penetration. Lensing and Wise (2006) showed that predators affected forest soil decomposer communities in subtle and variable ways that depend on rainfall conditions. Turnover in community composition at lower trophic levels or effects on the decomposition food web may obscure responses in terms of plant biomass in terrestrial systems, leading to the apparently weaker cascades compared with those in aquatic systems. However, such effects are not confined to the terrestrial realm. For instance, Tessier and Woodruff (2002) showed that planktivorous fishes in Michigan lakes had no effect on the biomass of the phytoplankton community but instead shifted its composition toward more edible forms, resulting in greater consumption efficiency for zooplankton. The potential for complexity within the producer community to dampen effects of predators at the level of overall plant biomass is therefore common across systems. Whether species turnover in response to changes in predator density is more pronounced in some systems than others remains to be tested.

What Factors Differentiate Ecosystems, and How Might These Affect Cascade Strength?

DIVERSITY

Diversity of plants, herbivores, and higher trophic levels varies greatly between ecosystems and may be a major factor regulating the strength of trophic interactions in ecosystems. One of the most apparent contrasts between marine and freshwater systems is the much greater diversity of marine plankton and fish communities. The classic examples of lentic trophic cascades come from temperate lakes with one planktivorous and one piscivorous fish species (Carpenter and Kitchell 1993) or where planktivores are the juvenile stages of piscivores (Persson et al. 2003). Reviews of similar studies in a wider range of lake ecosystems revealed much more equivocal effects (Drenner and Hambright 2002). Diversity among species that share similar resources may weaken the potential for transmission of strong top-down effects (Hillebrand and Shurin 2005). If the degree of generality (i.e., the number of prey species consumed) among consumers does not depend on diversity, then adding more species to a food web should diminish the strength of top-down regulation by any particular predator.

Frank et al. (2006) provide intriguing evidence that greater diversity among fishes weakens the intensity of trophic cascades in the northwest Atlantic. Cold, high-latitude areas contain few species of demersal fish predators or their forage fish prey compared with warmer areas. The sign of the temporal correlation in biomass between predatory and forage fishes shifts from negative to positive with increasing temperature and diversity. Frank et al. propose that the higher potential for compensation among species within a trophic level weakens top-down control at high diversity. Compensatory dynamics among planktivorous prey may be accelerated in warmer waters by a higher growth rate or a larger pool of available species (Shackell and Frank 2007). Whether the decoupling of trophic level dynamics by high diversity due to warmer temperatures extends to the plankton remains an important outstanding question. Lake plankton and fish communities are geologically young, geographically small, and isolated, all of which may contribute to lower species diversity relative to marine systems. Common groups of marine plankton including foraminiferans, coccolithophores, tunicates, euphausiids, and cnidarians are either absent or much less abundant and diverse in lakes. If Frank et al.'s (2006) interpretation of the role of diversity in marine trophic cascades is correct, it may provide a general mechanism whereby top-down control is expected to be stronger in freshwater systems than in the ocean.

FOOD WEB COMPLEXITY: OMNIVORY

The strength of evidence for trophic cascades in lake ecosystems is interpreted by nonlimnologists as an indication of the fundamental simplicity of lake ecosystems or of their tractability as experimental systems (as argued by Strong 1992). This distinction has important implications for whether lakes represent a special case, distinct from other ecosystems, or whether the prevalence of limnetic trophic cascades is a sign of their general importance in all ecosystems. Two arguments have been advanced for the simplicity of lake food webs: Phytoplankton are palatable and undifferentiated, so zooplankton consumers are highly generalized and function as a unified guild; and lentic consumers can be neatly classified into discrete trophic levels, whereas those in other systems are highly omnivorous. The concept of trophic levels as a descriptor of natural systems has been vigorously challenged by the high incidence of omnivory (Paine 1980; Cousins 1987; Polis and Strong 1996; Persson 1999). Whether omnivory dominates terrestrial ecosystems while simple, discrete trophic levels occur in lakes (as argued by Strong 1992) remains murky, but synthesis of published food web descriptors provides some intriguing clues.

Thompson et al. (2007) analyzed the distribution of trophic positions among species in topological food webs from different environments. The food webs characterize the dominant species in an ecosystem and thorough sampling of their diets, produce a binary matrix of feeding interactions among co-occurring species. Webs were collected from the literature and represented lake, marine, stream, and terrestrial environments. They asked whether species' trophic positions are discretely or continuously distributed, indicating trophic levels or trophic tangles, respectively, and whether the degree of omnivory varied between the ecosystems studied. The distribution of trophic position was aggregated only at the level of plants and herbivores; among carnivores, trophic position was nearly continuous. Interestingly, this pattern characterized food webs in all types of terrestrial and aquatic environments, and the four systems showed very similar levels of omnivory. The main habitat contrast that emerged was that food chain length (the number of species at high trophic positions) was greater in marine systems than lakes or terrestrial systems and shortest in streams. Schoener (1989) also concluded that marine food chains are longest. Any apparent differences in cascade strength between systems are therefore un-likely to result from systematic variation in the degree of trophic complexity. Although topological food webs incorporate detailed information on species and their diets, they ignore variations in abundance or energy flow through different trophic linkages and may lead to false impressions of trophic structure.

Omnivory may account for the apparently stronger cascades observed in marine pelagic studies than in freshwater. Stibor et al. (2004) proposed that omnivory among copepods dampens trophic cascades relative to lake plankton, which are often dominated by herbivorous cladocerans. They argued that trophic complexity within the zooplankton is mediated by system productivity such that copepods (the dominant mesozooplankters in the ocean and many lakes) function mainly as predators of protozoans in oligotrophic conditions but are largely herbivorous in more productive places where larger algae dominate. Increased productivity therefore leads to a loss of intermediate trophic steps among the zooplankton and a shortening of food chains. Cladocerans may provide a stronger, more direct link between planktivorous fishes and algae in lakes. This contrast in the trophic role of dominant zooplankton taxa may explain the freshwater–marine contrast in the meta-analysis (Shurin et al. 2002).

Another interesting ecosystem contrast in trophic complexity lies in the prevalence of mutualistic interactions mediated via pollinators. Knight et al. (2006) show that terrestrial predators often have strong indirect negative effects on plant reproductive success by suppressing pollinator populations. Because most phytoplankton reproduction is primarily asexual and higher aquatic plants are pollinated largely by movement of currents, this interaction is confined largely to terrestrial systems. The wide range of negative and positive indirect effects on land may dampen cascades relative to aquatic systems, where predator effects on producers are mediated primarily by herbivory.

Allometry

One of the most striking contrasts between pelagic, benthic, and terrestrial environments lies in the size structure of producers and consumers. Unicellular

CONCLUSIONS

Evidence for systematic differences in the importance of trophic cascades is illustrative and intriguing but incomplete. The weight of the evidence suggests that cascades occur and have the potential to be strong in all ecosystems. The published studies performed to date indicate stronger top-down control in aquatic systems, but the coverage of the terrestrial sphere is sparse. The question of how the roles of different processes vary between ecosystems continues to stimulate debate; however, without quantitative comparisons this discussion is anecdotal and inadequate for understanding similarities or differences. Syntheses and meta-analyses are fraught with difficulties and uncertainties but can lead to tremendous insight and shape the direction of future research efforts. Greater synthesis of studies across systems demands advances in empirical and theoretical directions.

CHAPTER 20

Conservation Relevance of Ecological Cascades

Michael E. Soulé

Why are ecological cascades relevant to protecting the diversity, resilience, grandeur, and beauty of the nonhuman world? This chapter is an attempt to synthesize and explain why. We have concluded that species interactions initiated by a relative handful of potent ecological actors (Soulé et al. 2005), including keystone species and suites of co-occurring predators (Paine 1966), are responsible for many of the patterns and processes that maintain diverse and resilient ecosystems.

Because strongly interactive species exert powerful influences and because their disappearance leads to the simplification of ecosystems via direct and indirect follow-on events such as changes in herbivory and the release of meso-predators, we are puzzled that top-down regulation in biotic communities is not universally accepted by professional scientists and conservationists. Why is it taking so long for ecologists to understand that the stripping of top predators from the earth's marine, freshwater, and terrestrial communities is probably a greater threat to biodiversity than other forcing agencies such as climate change, at least in the short term?

There are many reasons. First, managers are human, and when it is pointed out to them that critical processes and species have disappeared from their park or protected area, they naturally become defensive. For example, an article in the *San Francisco Chronicle* (Perlman 2008) discusses the ecological impacts of human interference with cougar predation on mule deer in Yosemite National

Park in California (Ripple and Beschta 2008). The study refers to the decreasing ecological effectiveness of cougars in the park caused by large numbers of noisy humans and the resulting lack of tree recruitment due to increased deer herbivory. Referring to the research, Nikki Nicholas, the park's chief of resource management, said, "They've brought up a really complicated issue and an interesting theory, but it needs much more rigorous study." "A lot of it is up to speculation," added Steve Thompson, senior wildlife biologist at the park (Perlman 2008: A-1). Such comments seem to be common among managers in some agencies.

Scientists are trained to be critical and to look for flaws in methods, design, analysis, and logic, but the reasons for resistance go deeper than this (see Chapters 3, 8, and 13, this volume). Terborgh et al. (1999) describe many of the reasons why scientists avoid this area of research, including the difficulty of studying rare or cryptic species and the fear of embracing findings that do not come from well-controlled and replicated experiments. Other reasons for resistance include the following:

- We are unfamiliar with past faunal truncations. Most large animals on most continents were extirpated near the end of the Pleistocene, and the process has continued during the last two centuries (Laliberte and Ripple 2004), so current ecosystems lack key, coevolved species. Less than 1 percent of protected areas retain the largest carnivores that survived the Pleistocene (Chapter 14, this volume), which means that research sites typically lack many strongly interacting species. Thus, many ecologists fail to appreciate the prior existence of strong, stabilizing interactions (Chapter 19, this volume).
- We are unaware that the behavior of many persisting animals is atypical in the absence of apex predators (Chapter 14, this volume).
- Academics are under pressure to obtain grants, to conform to disciplinary conventions, to avoid implicit or explicit criticisms of reviewers, and to publish quickly and often, all of which instill conservatism in researchers and bias their selection of systems that produce rapid and unequivocal results, that is, systems that contain abundant, small, short-lived, rapidly reproducing, sedentary species. Also, academic scientists face institutional pressures to pursue nonapplied, curiosity-driven research.
- Moreover, scientists are trained to be suspicious of new ideas and hypotheses. Because the bottom-up energy flow model of ecological determination has been the dogma in ecology for half a century, it may take another generation before the role of top-down forcing is widely

accepted. Or as Niels Bohr put it, science progresses one death at a time.

- Humans are averse to complex causation and often oversimplify phenomena by dichotomizing and polarizing. One contemporary manifestation is the academic conflict between the proponents of reductionism in science and the proponents of a radically constructionist worldview that grants equal validity to all views and narratives. The question "Which is the best of way of explaining how nature works: bottom–up or top–down?" is a typical example of a dualistic oversimplification in ecology. A similar point was made in a recent editorial (Homer-Dixon and Keith 2008) about the virtues of an atmospheric screening solution to slow the warming of the earth: "The important thing is to get scientists, environmentalists and global-warming skeptics alike out of the nonsensical all-or-nothing dichotomy that characterizes much current thinking about geo-engineering—that we either do it full scale, or we don't do it at all."

The silliness of polarized views, such as bottom–up and top–down forcing in ecology, is revealed when we consider the interaction of proximal (ecological) and ultimate (evolutionary) causation. While acknowledging the key role that apex predators play in savanna ecosystems, Sinclair et al. (Chapter 15) also describe how "the savanna systems of Africa and Asia are dominated by ungulates. The larger species are regulated by their food supplies, and predator populations are in turn determined by ungulate densities, a bottom–up process." Among other bottom–up phenomena they list are prey that are large enough to discourage predation and prey species that are able to minimize the regulatory impacts of predation by large-scale migrations and by aggregating in herds. Such a bottom–up interpretation is narrowly correct but temporally misleading: Natural selection by predators on prey for large body size, for migratory behavior, and for aggregation—top–down evolutionary forcing—appear to be the ultimate source of these community "bottom–up" processes. Therefore, an evolutionary perspective can obviate a dualism such as "top–down versus bottom–up."

ECOLOGICAL ISSUES AND PATTERNS

Ecology advances more by identifying complex interactions and emergent properties than by reductionism (Kauffman 2008). However, emergence is contextual, and context permeates all of ecology. Sadly, the context in which natural systems function is now pervaded directly and indirectly by the human enterprise. In this section I examine how factors such as human-generated disturbance interact with top-down influences in various ecosystems. This is far from a comprehensive review of civilization's impacts globally. Rather, my purpose is to highlight some of the factors that contribute to the heterogeneity of such processes.

The effects of adding or deleting strongly interactive species are sometimes nonlinear, producing phase shifts or alternative community states, some of which may not respond to management interventions (Chapter 17, this volume). Sometimes there are tipping points or thresholds generated by positive feedbacks (runaway, reciprocal, reinforcing dynamics) that, once crossed, are analogous to the solidification of water at 0 degrees Celsius. Chapters 4 and 17 include examples. In some cases alternative states may be unyielding to management interventions, as discussed later in this chapter.

On the other hand, changes in management policies and practice can return some alternative states to their former, more diverse states. Along the Pacific Coast of North America, kelp beds reduced to stony underwater barrens by voracious sea urchins have been restored by the return of sea otters (Chapter 3, this volume). In sections of Yellowstone National Park it appears that most if not all of the habitat degradation and extinction processes precipitated by the elimination of wolves more than 75 years ago are reversible, as evidenced by rapid reappearance of aspen recruitment, willow growth, and beaver wetlands upon return of the wolf (Chapter 9, this volume). Overgrowth of algae in lakes has been resolved by reintroducing large predatory fish such as largemouth bass that prey on smaller, plankton-eating fish, thus permitting the large zooplankton to flourish and filter out the offending phytoplankton. Other systems are less amenable to reversal. The fishery management community is plagued by relatively stable changes in marine systems caused by overfishing of top predators (Chapter 6, this volume).

On land, alternative stable states can occur when predators switch to a newly arrived prey. Berger (Chapter 14, this volume) describes how intensive logging of forests in the northern Rocky Mountains facilitated an invasion of white-tailed deer. In turn, well-fed cougars increased in numbers and started preying on a small, endangered population of woodland caribou, pushing them toward the irreversible state of extinction. Similarly, where plants escape herbivory, less diverse alternative community states may develop that are largely immune to change (Chapters 18 and 19, this volume), as when East African grasslands changed into scrub and woodland after a rinderpest epidemic, and when coral-dominated reefs are overgrown by unpalatable benthic algae after overfishing of herbivorous fish (Chapter 5, this volume).

Terrestrial Realms

Disturbances, natural and anthropogenic, interact with interspecies processes such as ecological cascades in complex ways. For example, fires can change the abundance and distribution of species, thus increasing or reducing the impact of those species. In turn, grazing can affect the frequency, intensity, and spatial extent of fires (Chapter 16, this volume). Fire is also released by management actions such as the introduction of exotic, fire-conducting grasses for livestock forage. Bond likens fire to herbivory, referring to fire as a ubiquitous, abiotic, global consumer that may be released where herbivory declines. Conservationists should also be aware that most megaherbivores were removed on most continents only 11,000 years ago (Chapters 1 and 16, this volume) and therefore ask whether the current fire regimes in many grassland ecosystems are natural.

Coevolved, megaherbivore-dominated grazing systems often sustain much diversity and high productivity. Bond (Chapter 16, this volume) recommends that conservationists keep in mind that "African savannas support about ten times more mammal biomass than forests on similar soils receiving similar rainfall." Heavily grazed patches—grazing lawns—are dominated by grasses that spread by stolons and can thus protect soils even under intense grazing. To the uneducated such lawns appear denuded or overgrazed.

Grazing lawns are a terrestrial analogue of phytoplankton-based ecosystems in lakes and seas, supporting large numbers and a high biomass of grazers on a low biomass of photosynthesizers. Once more widespread, such lawns are diminishing where livestock occur and where there are frequent fires that create fresh flushes of tender grass that lure herbivores off the lawns (Chapter 16, this volume). Where white rhinos still persist in Africa, they maintain grazing lawns, and these lawns attract smaller herbivores and act as natural firebreaks.

Noting the large proportion of specialist birds, grasshoppers, and spiders in a rarely burnt South African savanna, Bond (Chapter 16, this volume) recommends comparing the heavily grazed African savannas with grasslands on other continents that suffered the extinction of their megafauna and the decimation of most surviving species of native ungulates and where the presence of a distinct grassland biota suggests a long, complex evolutionary history. Mammoths, mastodons, glyptodonts, and giant ground sloths disappeared many millennia ago in the New World, and bison and other grazers have been pressed down to one tenth of their former abundances and ranges in the nineteenth and twentieth centuries. Thus we have only hints of the contributions to the diversity of species and ecosystems that large herbivores afforded. Many grassland birds in North America are now endangered because of the disappearance of native grazers, habitat destruction, and the abundance of mesopredators. Such systems beg for audacious megafaunal restoration, as described later in this chapter.

The eradication of native apex predators in most of the world has undoubtedly contributed to the ecological and extinction crisis, in part because it has facilitated the release of livestock. The population explosion of people and the increasing per capita consumption of meat worldwide are also major drivers of the cow explosion. Livestock consume huge amounts of water and produce one third to one half of human-caused greenhouse gases (Steinfeld et al. 2006).

The majority of the land surface throughout much of the world is now a livestock café. The removal of apex predators has been driven largely by pastoralism, although gratuitous, recreation-driven extirpation of wolves and other carnivores still occurs in places such as Alaska. Seventy percent of former forests in the Amazon have been converted to livestock pasture. Globally, 26 percent of the ice-free terrestrial surface is dedicated to livestock grazing. Another 33 percent is used for feedcrop production on arable lands. Thus a total of 70 percent of agricultural land on the planet is dedicated to livestock, and the percentage is growing. In the contiguous (lower) forty-eight U.S. states, roughly 61 percent of the land area is affected by livestock production; this includes rangeland, pastureland, and forestland that is grazed, along with croplands devoted to livestock feed (G. Wuerthner, personal communication, 2009; Steinfeld et al. 2006). The convenient assertion that livestock are ecological surrogates for native herbivores is almost always wrong, given the behavioral traits of domesticated animals (Wuerthner 1997) and the fact that natural systems contain both browsers and grazers. Cows and other livestock are a global ecological catastrophe.

The wet tropics have much more species diversity, more ancient and tighter mutualisms, and much more complex food webs, on average, than other terrestrial realms (Chapter 8, this volume). In part, this is related to the greater dimensionality (i.e., verticality) of tropical rainforests and to the abundance and diversity of aquatic habitats. As Terborgh and Feeley stress, diffuse webs imply weak, pairwise interactions between species. Based on these observations, the naive might cling to the hope that the addition or the removal of any given species in the tropics is less likely to cause a noticeable cascading change. However, supporters of "sustainable development" schemes must accept that by far the more typical ecological intervention in the tropics is wholesale commercial clearing of entire tracts of forest or the eradication of almost all large animals by hunters and loggers. Thus, the aforementioned tropical–temperate distinction based on food web complexity is academic.

Marine Realms

Trophic relations in the wet tropics roughly parallel those in tropical seas, but marine trophic relations are often more complex. For example, trophic triangles—in which prey species feed on the young of apex predators (e.g., fish such as cod)—distinguish marine food webs from terrestrial webs (e.g., hares and moose don't prey on wolf pups or cougar kittens) and significantly complicate the recovery of some marine apex predators. Estes (Chapter 3, this volume) gives other examples of trophic triangles that maintain alternative stable states.

In addition, many large marine animals are omnivores, further ramifying the web of trophic interactions and obscuring distinctions such as top-down and bottom-up forcing that are more obvious in less complex systems. The same probably applies to phenomena such as mesopredator and herbivore release, as described later in this chapter. Nevertheless, ocean-wide outbreaks of mesopredators, herbivorous sea urchins, fleshy algae, and jellyfish are occurring with increasing frequency. The events appear to be spreading over large areas of chronically overfished parts of the Atlantic, Pacific, Mediterranean, and Black Sea, shunting energy from economically valuable and diverse, higher trophic levels to lower ones.

The oceans today are so highly perturbed that the "normal," prehuman state is as much a thing of the past as the Pleistocene megafauna is on land (Jackson et al. 2001). It is a reasonable hypothesis that all historical ecosystem-level regime changes in the oceans have been caused by human activities. The direct and indirect effects of fishing (both commercial and recreational), collecting, and other forms of disturbance such as sewage pollution and agricultural runoff severely alter top-down interactions relevant to conservation (Norse and Crowder 2005). Large-bodied keystone predators are now rare or absent from most coastal and open ocean marine systems, just as they are on land.

According to Essington (Chapter 6, this volume), globalized commercial fishing almost always targets apex predators. Essington notes that commercial fisheries are nimble, shifting their effort from high to low trophic level species as the larger fish are depleted. Restraint of commercial fishers is nearly unknown, in part because theirs is a highly mortgaged enterprise. Ships and modern fishing gear are costly, and indebtedness leads the industry to overexploit one stock after another (Chapter 5, this volume).

In the process, as Essington notes, apex predators are often squeezed from above and below, as fishing directly clobbers the adults while starving them by depleting stocks of food fish. In marine and freshwater systems (Chapters 3, 4, and 5, this volume) every ecological variable from light penetration to the emergence and frequency of disease can be affected by the presence or absence of strongly interactive species such as large predatory or herbivorous fishes. This means that resource extraction and disturbance by human beings almost always produces major changes, including ecological cascades.

On top of this, Essington (Chapter 6, this volume) points out that models once used to manage fish stocks are now anachronistic because of the unpredictability of climate change effects and technology. Apex fish predators are becoming extremely rare in reef systems and in open, blue-water ocean habitats because commercial and sport fishers use sophisticated technologies to find them and benefit from perverse incentives from governments that subsidize unsustainable commercial fisheries.

In lakes and oceans, these nonholistic fisheries often focus on one species at a time (Chapter 6, this volume), inadvertently causing episodes of ecological meltdown. Carpenter et al. (Chapter 4, this volume) note that such fisheries often have disproportionately large effects on top predators because these species have long maturation times, large ranges, and complex life histories that may require multiple kinds of habitats or prey. Such nonecological fishing practices ignore trophic cascades (and other ecosystem phenomena) and therefore are likely to precipitate ecological meltdowns.

The good news is that marine protected areas work where fishing restrictions are enforced. As discussed by Sandin et al. (Chapter 5, this volume), "The strongest trophic consequences of the establishment of no-take marine reserves on Caribbean reefs are an increase in herbivorous fish biomass and a related decrease in fleshy algal biomass." Under intensive fishing, large carnivorous fish are the first to go, followed by herbivorous species occupying the next trophic level down. Thus, the impact of restoring top-down regulation in marine and terrestrial systems can be different, because on land (in countries with effective game legislation) mainly carnivores are persecuted. Herbivores may then become superabundant, as they did in northern Yellowstone National Park. The recent restoration of wolves initiated system recovery by reducing herbivory on aspen and willows by elk (Ripple and Beschta 2006b, 2007b). Thus, in the marine case, protection was beneficial because it led to the restoration of natural herbivory, whereas in the terrestrial case herbivory decreased when natural predation on herbivores was restored.

The unpredictability of functional and structural changes brought about by anthropogenic changes in freshwater and marine systems is a common thread in many chapters of this volume, and coral reef die-offs are a case in point. Overfishing leads to population explosions of fleshy algae that leak labile sugars into the water (Chapter 5, this volume); the sugars increase the numbers of microbes that are pathogens on reef-building corals. Thus fishing indirectly destroys coral reefs by flipping these communities from states dominated by herbivores to ones dominated by microbes. I think that no one in 1975 would have predicted such a sequence of events triggered by overfishing in shallow, tropical marine systems. Of course, some of the effects of land-based human activities in the coastal zone are predictable, including the degradation of coral reefs, mangrove forests, and other systems caused by the increasing loads of nutrients, toxic chemicals, sediments, and pharmaceuticals in runoff from the land and by direct collection of edible and marketable organisms.

Aquatic, Riparian, and Coastal Realms

In freshwater realms, as everywhere else, almost all cascades are human caused and unpredictable, at least in specifics. Carpenter et al. (Chapter 4, this volume) write that the massive changes recorded in whole-lake experiments and by quotidian management "were usually not forecast from models and small container experiments." Even in small lakes where human activities are tightly controlled, it turns out that managing food webs is extraordinarily difficult. So the notion of managing freshwater food webs is beyond the capability of current science, even in the most organized societies.

Undoubtedly, the same complexity and unpredictability apply to coastal and riparian areas where realms blend. These regions are also subject to powerful physical processes, constant input of exotic species, high rates of harvesting, and overwhelming economic pressures such as industrial, urban, and recreational development. For these and other reasons, these hybrid ecosystems are probably even more sensitive than the more homogeneous realms they border (Baxter et al. 2004). For example, cougars in Zion and Yosemite national parks avoid riparian areas of high human visitation, leaving such areas open to overbrowsing by deer (Ripple and Beschta 2006a, 2008). Analogous studies in coastal waters document some classic cases of mesopredator and herbivore release (Chapters 2 and 3, this volume).

IMPLICATIONS OF HERBIVORE AND MESOPREDATOR RELEASE

It appears that herbivore release and mesopredator release are the most salient effects of trophic cascades for conservation on land. Herbivore release is often perceived as a three-level process (e.g., apex predator reduction–increase in herbivory–deleterious changes in plant community), but the number of links in the chain is usually irrelevant to conservation. For example, random events can trigger a two-step cascade, such as when a decrease in large herbivore abundance leads to more frequent and hotter fires. The goal of conservation should be to minimize extinctions, not to obsess about step number.

Many descriptions of herbivore release on land refer to the increase in herbivory by native terrestrial herbivores, such as the spatially extensive eruptions of deer species in North America, Europe, and Japan (Chapter 9, this volume). The ecological release of deer species (Cervidae) in North America has caused overbrowsing, overgrazing, crop damage, tens of thousands of costly road accidents, a nuisance for homeowners, a threat to wild lands, and an increase in the incidence of Lyme disease (Telford 2002). The ecological release of deer has multiple causes, including the elimination of native predators such as wolves and pumas (Ripple and Beschta 2006a; Beschta and Ripple 2008), habitat fragmentation, and habitat conversion that benefits deer species. All of these may be exacerbated by the inability of hunters in many places to be effective surrogates for native apex predators (Chapter 9, this volume).

Mesopredator release is the population growth and loosening of behavioral and ecological restraints on middle-sized predators in response to reduced abundance and ecological effectiveness of larger, top predators (Chapter 13, this volume). Overabundant mesopredators are likely to be characterized by high rates of recruitment and high rates of dispersal, both of which immunize them to control efforts. The usual follow-on effect of mesopredator release is heavier predation on their prey, although the specific consequences are often unforeseen, as Berger (Chapter 14, this volume) elaborates.

A species that is an apex predator in one situation can be subordinate in another. For example, Soulé et al. (1988) suggested that the disappearance of coyotes—subordinate and prey to wolves but now the apex predator in wolf-free ecosystems—from suburban patches of coastal sage scrub in southern California leads to a rapid eradication of native bird species in these patches by house cats, a hypothesis later confirmed (Crooks and Soulé 1999). On the other hand, when reintroduced wolves achieved ecologically effective densities in Yellowstone National Park, coyote numbers plummeted. One result was a deep decline in pronghorn antelope fawn mortality attributable to coyotes, presumably because fawns are more vulnerable to coyotes than to wolves (Berger et al. 2008).

Marine conservationists face daunting problems in overcoming mesopredator release. One is the scale problem. For example, many large, marine apex predators are highly mobile, which implies that the spatial scales of management and conservation must also be large. Thus, local marine conservationists on the Atlantic coast of North America can do little to protect nearshore, shallow-water populations of scallops from released mesopredators such as cownose rays because the cause is overfishing of the great sharks in deeper water along the Atlantic coast.

The reestablishment of terrestrial apex predators in human-dominated landscapes is fraught with difficulties. Once apex predators are extirpated and mesopredators reign, the cat is out of the bag. Even in special situations in which cats or other mesopredators could be controlled cheaply and efficiently, some sophistication is needed. Brashares et al. (Chapter 13, this volume) refer to the case of oceanic islands where introduced cats actually offer a degree of security to nesting seabird colonies by preying on introduced rats. Once the cats are removed, the rats do much more harm than the cats, so the order of extirpation can be critical (Courchamp et al. 2003b).

Most conservationists are unaware that the ecological release of mesopredators is often unintentionally exacerbated by direct or indirect energy subsidies that multiply the damaging impacts of mesopredators. Subsidies can be lumped into three overlapping categories: food, elevated structures that enhance habitat quality and hunting efficiency for corvids and raptors, and the proliferation of roads and tracks that create edge habitat, increase access into interior habitats, and indirectly provide food from road kills.

Food subsidies lead to population explosions in mesopredators. Among such subsidies are garbage dumps or tips, compost piles, birdseed spilled from feeders or accessible to acrobatic rodents and carnivores, pet food left on patios and porches, agricultural spillage and residues, rejected fruits and vegetables on farms, gut piles from hunters, exposed livestock carcasses, and road kills.

All these food sources can promote population growth of mesopredators such as rats, raccoons, foxes, opossums, gulls, crows and related species, feral dogs, and house cats (Crooks and Soulé 1999), whether or not the numbers are also elevated because of reduction in apex predator populations. Animals killed on roads by motor vehicles cause population explosions in the number of crows, ravens, and other scavengers, many of which are effective predators on nestling and juvenile birds, amphibians, reptiles, beneficial insects, waterfowl, small mammals, and domesticated animals. Crow, raven, and gull populations often explode when provided with direct or indirect nutritional subsidies, even where large carnivores remain abundant.

There isn't space here to elaborate on the other categories of subsidies. A few hints must suffice.

- Structures such as transmission poles, fences, and billboards facilitate energy-conserving hunting by raptors, crows, magpies, and other

predatory birds. Culverts and bridges provide shelter and facilitate dispersal for mesopredators.

- The edges of roads provide many mesopredators, including predatory
  birds, with access to the breeding sites of forest-nesting songbirds.
- Rural and suburban developments with feeders, compost, and pet food
  are often ecological sinks that invite birds, amphibians, and other animals
  but also provide backyard and porch hunting opportunities for mesopredators such as cats.

Brashares et al. (Chapter 13, this volume) illustrate the social and ecological
complexities of reversing mesopredator release once it occurs, citing research
on olive baboons in Ghana. Large predators including lions, leopards, hyenas,
and wild dogs disappeared in several Ghanaian savanna parks by 1986 because
of hunting for pelts and persecution of predators in general. In parallel with the
disappearance of the large carnivore guild in these parks, particularly lion and
leopard, olive baboons have become superabundant. Baboons then became
consummate hunters. A result has been precipitous declines in the prey of baboons, such as other primates, ungulates, and birds. Unhindered by fear of lions
and leopards, baboons may form long lines that march with military precision
across the savanna, scaring up and killing prey as they advance. In addition, baboons have become an economic and security menace to villagers in areas adjoining reserves as they mount daring raids on crops, livestock, and villages, occasionally even preying on human infants.

Can the clock be turned back? In principle yes, but in practice no.
Brashares et al. note, "Taken together, the economic, social, and health costs of
this mesopredator release dwarf the investment that would be necessary to restore, protect, and manage Africa's apex predators." But with parks becoming
habitat islands increasingly isolated by a growing human population using ever
more land, there is little likelihood that dangerous large carnivores will be tolerated. In summary, mesopredator release induced by predator removal is now
nearly universal (see Roberts 2007).

CONCLUSIONS

> This reconciliation [with past abominations] reveals the profound
> moral perversity of a world that rests essentially on the nonexistence
> of return, *for in this world everything is pardoned in advance and therefore
> everything cynically permitted.*
>
> —Milan Kundera, *The Unbearable Lightness of Being* (emphasis added)

Humanity is failing to arrest the extinction crisis. Instead, we excuse, permit, and adapt. The reasons are complex, but three of the main problems are that people in general are not rational; decision makers and managers at local, national, and international levels are often ignorant of ecological principles such as edge effects, source–sink dynamics, and ecological cascades; and many ecologists and conservationists do not understand that herbivore and mesopredator release are among the primary drivers of the extinction and ecological crises, and most are frantically jumping on the climate bandwagon.

First, people are not rational. The evidence is overwhelming that emotions, including desire for status, control, contentment, and physiological gratification such as overconsumption of meat, dominate mental processing and that our vaunted cognitive circuitry is subordinated in ordinary social life to Pleistocene impulses. Rather than the executive functions of the frontal cortex controlling our limbic emotions, it is the other way around: Primitive drives dictate the content of our logical and analytical processing (Haidt 2006; Pinker 2006; Burton 2008).

In this postmodern world we are subject to a perfect storm of alienation from wild places and wild creatures, compounded by Olympian levels of "Disneying" distractions. Distractions and denial augment our ability to reconcile to gradual dissolution, such as the disappearance of the places we loved as children and most of earthly creation. Denial and distraction also help explain why spiritual and aesthetic arguments on behalf of nature (its beauty, grandeur, and solitude, inspiring emotions such as awe and transcendence), not to mention the popular economic arguments such as the value of ecosystem services, are generally too weak to challenge the appetites for immediate sensual gratification and the campaign contribution avarice of politicians.

Sadly, economic arguments, unless they promise immediate and culturally appropriate benefits, fail to inspire entrepreneurs and governments to invest in large-scale conservation projects. Among such failed proposals are adaptive management, sustainable development, monitoring, the prudential valuing of ecosystem services, and cost-based decision processes for designing "portfolios" of protected areas. These rational but idealistic approaches will continue to be ignored by the public, in part because they lack emotional salience; they may compute, but they don't convert. In fact, given the scale of social and economic turmoil, saving nature will not be a priority until civilization is confronted with immanent collapse that can be credibly linked to the destruction of biodiversity.

Second, scientific concepts such as ecological cascades, edge effects, and ecological sinks, though familiar to ecologists, are largely unknown to lay persons. For example, even educated people are shocked when informed that roads and exurban development (such as cottages and recreational facilities plunked down in remote areas) have deadly margins extending up to several kilometers into unfragmented habitat (Hansen et al. 2005). People are equally surprised when hearing of the silent (or empty) forest syndrome in remote tropical rainforests, the dead zones in oceans, and the general lack of enforcement of antipoaching laws in and around most national parks globally (Terborgh 1999; Terborgh et al. 2008). Even though the public is generally aware of the dangers of climate change, because it poses an imminent threat to their well-being, few people realize that it will shift the distributions of many plants and animals that must either move or perish. Besides, a recent Pew poll shows that the environment has dropped to the bottom of twenty issues of concern to voters in the United States (Revkin 2009).

Third, this volume presents copious evidence that the unpredictable, devastating downstream effects of apex predator removal, particularly herbivore and mesopredator release, are major drivers of global ecological collapse. Sadly, the siren song of funding for climate change research has spellbound funders and researchers alike, diverting attention away from the immediate challenges such as ecological cascades, habitat loss, and exotic species.

The good news is that there exists one simple, documented, and inexpensive treatment for rescuing many terrestrial and some aquatic ecosystems. Applying this treatment lowers three of the major hurdles to biodiversity protection in stable societies: habitat fragmentation, climate disruption, and ecological cascades.

This simple solution is the most effective tool for mitigating habitat fragmentation and for overcoming island effects, edge effects, genetic drift, inbreeding, and demographic accidents affecting the viability of isolated populations. It facilitates the dispersal and migration of strongly interacting species such as apex predators. Therefore, this solution facilitates top-down control of mesopredators and herbivores. This solution also moderates the near-term challenge presented by climate disruption because it facilitates geographic shifts in distribution (ranges) of flora and fauna.

The solution is connectivity. Restoration advocates and ecologists should be encouraged because the restoration of landscape and waterscape permeability promotes the recovery of strongly interactive species such as apex predators—the cheapest form of ecological restoration. Connectivity is simple, cost-effective, and even culturally popular, particularly if fish and game species benefit where dispersal and migration are facilitated by the mitigation or removal of obstacles such as roads, sprawl, coastal zone development, and reservoirs and other water diversions.

Given the recent disappearance of large native herbivores on several continents at or near the end of the Pleistocene, many coevolved interactions are now rare, and today's truncated ecosystems suffer their absence. Donlan et al. (2006) claim that it is likely that higher levels of productivity and diversity could be restored by the cautious introduction of surrogates for recently extinct species, assuming that predators are permitted to exercise effective control of the large ungulates.

Therefore, three huge challenges to nature, namely climate disruption, the global decapitation (predator extirpation) of ecosystems, and habitat degradation caused by fragmentation, are all responsive to the same treatment: landscape connectivity. "Nature's aspirin" is the freedom to disperse, particularly for strongly interacting species such as wolves, whose disappearances are likely to trigger ecological cascades but whose repatriation can restore critical interactions and processes.

If there were a Kantian imperative—"Act only according to that maxim whereby you can at the same time will that it should become a universal law"—or prime directive in conservation, then it would be "Never remove or discourage native predators or other strongly interactive, native species." At the top of the to-do list for conservationists, therefore, is the requirement to restore apex predators wherever and whenever possible. A corollary task for restoration ecologists is to control mesopredators and herbivores by restoring top-down interactions; purging artificial habitat elements, such as elevated perches used by predatory birds; and minimizing food subsidies such as road kill and pet food that contribute to mesopredator increase.

Finally, the traditional goals of conservation, including maximal production and use of natural resources including game and harvestable species, representation of nineteenth- and twentieth-century vegetation communities, and control of apex predators, are increasingly anachronistic in the face of rapidly changing conditions, social needs, and technology. Conservation planning and action nowadays must emphasize regional- and continental-scale connectivity, strongly interactive species, and ecological processes such as dispersal of apex predators.

ACKNOWLEDGMENTS

I gratefully acknowledge valuable comments, suggestions, and material assistance from Jim Estes, John Terborgh, George Wuerthner, Brian Miller, William Ripple, Kevin Crooks, Veronica Christie, and Dawn Reeder.

CHAPTER 21

Conclusion: Our Trophically Degraded Planet

John Terborgh and James A. Estes

A few far-sighted thinkers, notably Elton (1927) and Hairston, Smith, and Slobodkin (HSS 1960), were decades ahead of their times in recognizing that ecosystems must be stabilized by regulation from the top down. HSS argued that the world is green because predators limit the number of herbivores and thus prevent the destruction of vegetation. Ecologists have been arguing about this simple line of reasoning ever since, without coming to closure. Our purpose in writing this book was to bring long-delayed closure to this centrally important topic in ecology.

The Elton–HSS thesis and conceptual developments that have flowed from it in the ensuing 50 years define a body of theory that has recently come of age. The theory of trophic cascades encompasses interactions of countervalent bottom-up and top-down forces to achieve stable balances in the form of so-called alternative states. In theoretical jargon, alternative states are strong attractors. They are dynamically stable unless jolted away from equilibrium by major perturbations. Perturbations capable of triggering transitions between alternative states can be either abiotic or biotic. Increases or decreases in nutrient inputs, salinity, precipitation, temperature, or fire are examples of abiotic triggers, whereas additions or deletions of keystone species or entire trophic levels are the most common biotic triggers. The theory of trophic cascades is important because it defines the way nature works, and there is much to follow from that.

And trophic cascades are important because we cannot conserve nature unless we have a confident understanding of how ecosystems function.

The authors of this book have offered evidence of the existence and mode of operation of trophic cascades in every chapter. Strong interactions between bottom-up and top-down forces are ubiquitous in ecosystems around the world in both terrestrial and aquatic environments. We regard trophic cascades as a universal property of ecosystem functioning, a law of nature as essential and fundamental to ecology as natural selection is to evolution.

Our conclusion may seem rash to some, unwarranted to others. Every scientific revolution leaves die-hard advocates of the old thinking in its wake. Naysayers to the fundamental and universal importance of trophic cascades remain skeptical of the importance of top-down regulation and, instead, insist that bottom-up forces, in themselves, provide an adequate basis for understanding the structure of ecosystems and the functioning of nature (Strong 1992; Polis 1999; Ware and Thomson 2005). Objectively coming to grips with the relative importance of bottom-up and top-down forcing is complicated by the fact that the logical arguments for each take different forms and are supported by different kinds of evidence. The visceral appeal of the bottom-up view to many stems from the fact that eating is an essential life process, whereas being eaten is not. But there is another more important side to this logic, which is that the fitness consequences of being eaten (except for plants) are immediate and absolute, whereas the fitness consequences of what one eats typically are manifested in much more diffuse and indirect ways through survival and reproduction.

From this it follows that the interaction strengths (as defined by Paine 1980 and others) of bottom-up forcing processes are commonly weak, whereas those of top-down forcing processes are often strong. Aside from large-scale catastrophic events, such as drought or fire on land and El Niños or regime shifts at sea, most of the supporting evidence for the importance of bottom-up forcing comes in the form of geographic patterns in food chain lengths or the distribution and abundance of species. In stark contrast, geographic patterns in the distribution and abundance of species have rarely been considered from the perspective of top-down forcing and trophic cascades. Even if it is granted that top-down regulation has not been investigated via rigorously controlled experiments as extensively as one would wish, direct experimental evidence for ecosystem structuring via bottom-up forcing is even more limited.

The observations and arguments presented in the preceding chapters combine to suggest a qualitatively new perspective on the workings of nature. Physical constraints and bottom-up forcing are instrumental in defining species' ranges and ecosystem productivity; top-down forcing and trophic cascades define the dynamic relationships that determine the distribution and abundance of species within these broad bottom-up constraints.

Evidence in support of these conclusions has been presented from, quite literally, the four corners of the globe. The authors have traveled to the ends of the earth to conduct experiments and field observations in the most natural environments remaining on the planet: Serengeti, the high arctic, the Aleutian archipelago, and the Line Islands of the central Pacific. Unfortunately, travel to such remote places is necessary if one is to study nature as it evolved. The ecosystems found over most of the peopled world, more than 90 percent of it, are anthropogenic artifacts, many of them currently and often imperceptibly undergoing slow transitions between alternative states in a nearly global process of trophic degrading of the earth's ecosystems.

GHOSTS FROM THE PAST

Beginning about 50,000 years ago, our species began to peel away the layers of the earth's trophic architecture (Roberts et al. 2001). The first layer to go was the largest and most conspicuous, the megafauna. Elimination of megafauna spread in a wave around the world, occurring first in Australia, then passing to Eurasia, North America, South America, the Antilles, and Madagascar, until finally, about 1,000 years ago, reaching New Zealand and the farthest corners of Oceania (Burney et al. 2004; Burney and Flannery 2005; Steadman 2006). The megafaunal layer was removed by premodern humans who left no written record. Though hardly "mega" by continental standards, some of the last victims of this mostly prehistoric wave of extinctions were endemic faunal elements of the remote Mascarene islands (Mauritius, Réunion, and Rodriguez), unknown to humans until European sailors stumbled upon them in the sixteenth century. Eyewitness accounts document the slaughter of the dodo, solitaire, giant tortoises, and other less notable vertebrates of these islands by meat-hungry sailors. In a marine counterpart, Steller's sea cows went the way of the dodo a mere 18 years after the species was discovered in the Commander Islands by the Bering expedition in 1741. The perpetrators of these last extinctions are thus unequivocally known. Otherwise, evidence has had to be painstakingly unearthed and reconstructed by paleontologists (Koch and Barnosky 2006; Steadman 2006). The accumulated evidence associating megafaunal extinction with the arrival of humans at widely scattered times over the last 50,000 years is so consistent and voluminous as to be unarguable (Martin and Steadman 1999; Burney and Flannery 2005).

Extermination of megafauna constituted only the first stage of the assault on nature. As modern industrial humans spread around the globe in a second wave of colonization and exploitation, the use of guns, snares, hunting dogs, poisoned baits, and other sophisticated means enabled the successful persecution of remaining top carnivores (lions, tigers, bears, wolves, jaguars) from all but a minor and shrinking portion of the terrestrial realm. Thus, a second layer of the earth's original trophic organization has been widely removed. And currently, a third layer—smaller herbivorous mammals, including ungulates, primates, and large rodents—is being depleted or eliminated over major portions of Africa, Asia, and tropical America by the bushmeat trade (Fa et al. 2002). When the larger animals are gone, all that is left is an "empty forest" (Redford 1992; Terborgh et al. 2008), and even that is under the pressure of logging and agricultural development. Thus, the assault on nature began at the top but is ending at the bottom.

A similar story can be told of the world's oceans. The equivalent in the marine realm is "fishing down marine food webs" (Pauly et al. 1998: 860). The marine equivalent of megafaunal overkill has almost run its course (Myers and Worm 2003; Roberts 2007) but with a shifted time frame. The technology needed to overexploit the oceans has existed only for the past several centuries. The slower-swimming whales went first, then pinnipeds, sea turtles, manatees, and now sharks, tuna, and billfish (Jackson 2001; Springer et al. 2003).

It would be naive in the extreme to imagine that such a wholesale dismantling of the earth's terrestrial and marine food webs would be without consequence. In fact, the consequences are all around us, but we are largely blind to them, being parties to the "shifting baseline" phenomenon (Pauly 1995). Thus, to pursue this discussion further, we must begin by defining a "state of nature," the condition of being "natural" as that which evolved before the advent of humans and their highly effective technology of extermination.

The natural state of the earth's biota does not exist today, apart from a few special places such as Serengeti, the Okavango Delta, and the Line Islands of the central Pacific. Elsewhere it has to be reconstructed from the fossil record and other accounts of Earth history. Throughout the record of life on land, the trophic organization of animal assemblages has been consistent, except for periods of recovery from major extinction events (Flannery 2002). Producers, consumers, and carnivores have been a constant since the early Paleozoic in the sea and since at least the late Paleozoic on land. At any time, herbivores and carnivores have ranged in size from small to large. *Tyrannosaurus rex* was much larger than any living terrestrial carnivore, but it was dwarfed by contemporaneous herbivores. The largest herbivores extant today—elephants, rhinos, buffalo, hippos, giraffe—are so large as to be immune or nearly immune as adults to the largest carnivores in their environments, be they lions, tigers, or wolves. We shall use the term *megaherbivore* to refer to predation-immune herbivores, although Owen-Smith (1988) advocated a more restricted use of the term.

Megaherbivores have been a constant feature of terrestrial animal assemblages since the late Paleozoic. This can be deduced from relative sizes of herbivore and carnivore. A 200-kilogram lion easily brings down a 150-kilogram zebra or wildebeest but has difficulty killing an 800-kilogram buffalo or giraffe (Sinclair et al. 2003). Therefore, extinct herbivores, with estimated weights several to many times greater than that of the largest contemporaneous carnivore, can be presumed to have been megaherbivores. Using this criterion, it is clear from the fossil record that megaherbivores evolved again and again throughout the history of life, with new families of reptilian megaherbivores arising in the Permian, Triassic, Jurassic, and Cretaceous. Mammal evolution in the Cenozoic was particularly rapid, such that new radiations of megaherbivores appeared in the Eocene, Oligocene, Miocene, and Pleio/Pleistocene. Megaherbivores, considered by those of us in the New World to be exotic and far away, are entirely natural components of terrestrial ecosystems and have been so throughout evolutionary history. To the contrary, a world without megaherbivores, such as more than 90 percent of the earth's terrestrial realm today, is decidedly unnatural.

In Chapter 16, Bond raises the intriguing possibility that much of the world's vegetation was in an alternative state before the extinction of megafauna. This revolutionary proposal emerged from comparison of actual vegetation maps with potential vegetation maps based on soils and climate, that is, bottom-up considerations. Expectations based on climate predict a much greater fraction of the world to be in closed-canopy forest than is actually the case (Bond et al. 2005). The conventional explanation for the occurrence of "natural" grasslands in regions that can support forest (in the state of Illinois, for example) is fire. But Bond shows that fire and megafauna interact in complex ways and that the close-cropped grazing swards of megafauna can serve as firebreaks (McNaughton 1984). When megafauna are absent, taller species of grass replace grazing-dependent species and generate fuel that can sustain fire. In short, there is a regime shift from a grazing-dependent ecosystem to a fire-dependent one.

After humans occupied a landscape, fire frequency would invariably have increased because aboriginal peoples the world over set fires to drive game

(Flannery 1994). A shift in fire frequency might have done more to alter vegetation in some regions than the loss of megaherbivores per se. Yet the presence of various species of serotinous pines and numerous associated fire-dependent plant species in large sectors of North America and elsewhere bespeaks a long history of selection by fire. Much more remains to be learned about the three-way interaction between vegetation, megaherbivores, and fire in parts of the world not now occupied by megaherbivores.

An arctic version of Bond's argument has been put forward by Sergei Zimov and colleagues (1995). Pollen profiles show that before extinction of the wooly mammoth and other high arctic megafauna, northern portions of Asia and Alaska were occupied by semiarid steppe, not by the mossy tundra and taiga found in these places today (reviewed in Guthrie 1990). In fact, only traces of what Zimov called the "mammoth steppe" survive today (on arid south-facing bluffs), although Zimov is conducting experiments with grazing animals in an effort to recreate it (Zimov 2005). Here is another example of an alternative state that may have occupied a major portion of the terrestrial realm. If the areas of vegetation identified by Bond (Chapter 16, this volume) as inconsistent with climate-based predictions are combined with the mammoth steppe of Zimov, the total approximates 50 percent of the earth's land surface, suggesting that, indeed, we live in a transformed world that would have been unfamiliar to the first human migrants.

Although the presence or absence of megafauna can transform landscapes, as research in contemporary Africa has amply shown, there is more to megafauna than the immediate impacts of their herbivory. Alternative states are typically stabilized by positive feedback loops. Although large herbivores impose a powerful top-down force on vegetation, capable of changing both the structure and composition of plant communities, they also provide bottom-up forcing in the form of urine and feces. This effect is known as grazing optimization (McNaughton 1985). Grazing optimization involves positive feedbacks because rapid return of nutrients accelerates plant growth and productivity, and these in turn increase the biomass of herbivores. Grazing optimization was recently demonstrated in a forest ecosystem, so herbivore enhancement through bottom-up processes may prove to be a general property of terrestrial ecosystems supporting high herbivore biomasses (Feeley and Terborgh 2005). If so, loss of megaherbivores around the world would have had multiple effects on vegetation: altered structure and species composition via reduced herbivory and increased fire frequency, and reduced productivity and increased recycling of primary productivity via the detritivore pathway (Wardle and Bardgett 2004). It is an unfamiliar and rather disorienting thought to imagine the tundra and taiga as anthropogenic habitats, but current evidence is pointing to this startling reassessment. Needless to say, the structure and composition of plant communities have myriad secondary effects on the distribution, abundance, and behavior of other species and ecosystem processes (Berger et al. 2001b).

If Bond and Zimov et al. are right in their deductions, their insights open the door to a much larger set of issues: the extent to which contemporary landscapes are anthropogenic artifacts. Analysis of pollen and macrofossils provides robust support for the existence of plant and animal communities in the Pleistocene that were quite distinct from those found anywhere today. Apart from extinct megafauna, most of the same species were present, but they were often combined in ways that today seem incomprehensible in terms of climate. For example, small mammal assemblages composed of species whose ranges now lie far apart co-occurred in midcontinent North America during the late glacial maximum (Graham et al. 1996). Conversely, some species whose current ranges overlap broadly occupied Pleistocene ranges that were far apart (Davis and Shaw 2001). How is one to interpret such results?

The conventional wisdom on these riddles has been to attribute them to differences in the Pleistocene climate and novel vegetation formations associated with them (reviewed in Martin 2005). But, as science, this is hardly better than arm waving. Of course, Pleistocene climates were different, and some of them certainly have no counterparts in today's world. Moreover, it is to be expected that climate differences influenced vegetation and the biogeography of many species. But was climate the whole story? Thanks to Bond, Zimov, Flannery, and others, we are beginning to appreciate that factors in addition to climate may have been involved: the prevalence of fire as a landscape engineer and the top-down impacts of megaherbivores on vegetation and their bottom-up contributions to plant productivity. If any of these forces was an important determinant of Pleistocene vegetation, the use of Pleistocene data to extrapolate future effects of climate change could be quite misleading.

ECOLOGICAL BIFURCATIONS AND THE ANTHROPOGENIC WORLD

Loss of megafauna in continental and insular sectors of the terrestrial world took place millennia ago. Pollen cores and other preserved evidence can provide some clues to the preextinction vegetation, but aside from such fragmentary scraps of information, we may never know many of the details because confounding variables would cloud the process of interpreting any evidence.

Given that plant succession converges in a millennium or so, the world's terrestrial ecosystems have long since adjusted to the loss of megafauna. The post-megafaunal alternative state is what at least Western science has long regarded as "pristine" nature (Jackson 2001). Yet it is nothing of the sort: It is "nature" in the first stage of anthropogenic degradation.

With the loss of the megafauna, the top-down force of herbivory on vegetation must have diminished dramatically over more than 90 percent of the earth's continental landmass. What impacts did this have on the biota, apart from redrawing the boundaries of major ecosystems (e.g., between savanna and forest, taiga and steppe, desert and grassland)? The answer is not known with any precision. What we do know is that a large majority (more than 90 percent?) of the plants and animals recorded from the late Pleistocene (apart from megafauna) persisted into modern times (Coope 1987). Ancillary extinctions seem to have been few.

Can we celebrate this as good news, an indication that losses of further trophic levels from the earth's biota will also have minimal consequences for biodiversity? Almost certainly not; the news next time is likely to be dismal. Why?

Loss of megaherbivores and other megafauna over most of the terrestrial realm ushered in an entirely unprecedented trophic organization consisting of predators, herbivores too small to resist them, and producers. This is the world of Hairston, Smith, and Slobodkin (1960), but it is not the world that evolution created. It is a world of moderate and patchy herbivore pressure in which forests and other woody vegetation are preeminent and in which fire plays a major ecosystem role (Chapters 9, 14, and 16, this volume). The HSS world is the world most ecologists and the public at large accept as their frame of reference for "pristine" nature. It is the world we know, and it is collapsing because additional layers are being stripped away from the residual trophic structure of the earth's terrestrial and aquatic ecosystems.

Concerted persecution by humans has rid much of the terrestrial realm of top predators. There is a global geographic pattern to this dichotomy: Remaining wilderness regions generally retain top predators; settled regions generally do not, independent of habitat, latitude, or national income statistics. This dichotomy has existed for a long time, decades to centuries in various parts of the world. However, a second bifurcation is more recent in origin and is spreading around the world as this is written. The second dichotomy results from the fact that some countries have implemented effective game management programs and others have not. The former tend to be among the developed nations of middle and higher latitudes, whereas the latter are concentrated around the equator.

In portions of the world now lacking top predators, the matter of whether game laws are effectively implemented or not leads to large and contrasting departures from the HSS world. Game laws provide a cover of protection under which animals are released from most top-down control except regulated hunting. In most industrialized countries, the popularity of hunting has been declining steadily. Consequently, many industrialized countries with effective game laws are suffering from ungulate irruptions and mesopredator release (Chapter 9, this volume), whereas those without such laws are suffering from the "empty forest" syndrome (Redford 1992; Fa et al. 2002). Where game laws are strong, herbivores have become rampant (Chapter 9, this volume). Where game laws are weak or unenforced, human hunters tend to eliminate all fauna larger than rats and squirrels, including birds, mammals, large reptiles, and even some amphibians (Peres and Palacios 2007). Unfenced livestock often substitute for native fauna, and systematic overgrazing and overbrowsing are widespread. Thus, whether game laws are implemented or not, normal top-down trophic control is lacking. What are the likely consequences of these now globally widespread conditions?

Let us begin with the case of well-implemented game management. Before the institution of game laws in the United States in the early twentieth century, the white-tailed deer (*Odocoileus virginiana*) was the only surviving native ungulate in much of eastern North America. It survived in large expanses of forest, and in many agricultural landscapes it had been extirpated by overhunting. A near absence of deer made it easy to impose hunting regulations, because rural people felt they had little to lose. Deer recovery under conservative regulations (bucks only, one per hunter per season) took many decades. Meanwhile, hunting pressure failed to increase in proportion to an exponentially growing deer herd. Nowadays, even when enticed with expanded seasons and take limits, hunters no longer regulate many deer populations. Deer (and other mammals, such as beaver and several mesopredators) have consequently exploded in abundance (McShea et al. 1997; Chapters 9 and 13, this volume). Overbrowsing by deer is having devastating impacts on plant diversity and patterns of forest regeneration (Rooney and Dress 1997; Rooney et al. 2004), and overabundant beaver are denuding riparian zones, leading to bank erosion and stream degradation. Abundant mesopredators such as raccoons, opossums, and feral house-cats are severely reducing bird populations, especially of ground-nesting species (Crooks and Soulé 1999). Similar scenarios are unfolding in other parts of the world.

The HSS world is a world of moderation. For thousands of years since the demise of the megafauna, HSS has conserved biodiversity via the dynamic feedbacks that operate between predator and prey. Under intact predation regimes, herbivores and mesopredators are held to population densities at which they exert only mild pressures on the next level down. But when carnivores are severely persecuted or eliminated, populations of their prey surge into overshoot as they consume resource capital (Beschta and Ripple 2008). The result is severe top-down pressure on ground vegetation and lesser prey, such as lizards, frogs, and ground-nesting birds (Chapter 13, this volume). Cascading local extinctions are one predictable outcome (Chapter 20, this volume). Other, less predictable outcomes such as the spread of disease can be anticipated (Ostfeld and Holt 2004). Is this the fate that awaits "nature" in much of the industrialized world? It is a disquieting thought, but large declines in biodiversity and its associated ecosystem services seem inevitable wherever top-down regulation is lacking.

We must now ask what happens when both predators and consumers are removed from the system, as was the case over much of North America before game laws were instituted. This history is now repeating itself in much of the developing world, where market forces are motivating the systematic elimination of animals from elephants and gorillas to cane rats over large portions of tropical America, Africa, and Asia (Fa et al. 2002). The resulting "empty forest" is only beginning to be studied. The preliminary results are not encouraging (Wright 2003; Wright et al. 2007). The consumer and producer levels of mainland tropical forests are linked together by countless mutualisms between pollinators and seed dispersers and plants because nearly all tropical plants need the services of animals to complete their reproduction. The animals may be vertebrates or invertebrates, depending on the service and plant species, but very few tropical plants can reproduce successfully if these mutualisms are disrupted (Chapter 8, this volume).

In tropical forests depleted of large vertebrates, critical functions are either diminished or exaggerated. These functions affect plants in both positive and negative ways. Pollination and seed dispersal confer positive benefits on plants, whereas herbivory and seed predation impose negative costs. In the HSS world to which we are accustomed, plant species composition is maintained from generation to generation by a balance between these positive and negative animal-mediated functions (Terborgh et al. 2006). If the animal community is perturbed by human interventions, large shifts in plant composition will ensue (Wright et al. 2007; Terborgh et al. 2008). A negative spiral can be expected to play out in overhunted forests. In the first stage, hunters deplete birds and mammals that provide critical services to plants, such as seed dispersal. Then the reproduction of the trees they disperse is curtailed. The next generation of trees will then be deficient in the species that large avian and mammalian seed dispersers depend on for food, and so forth. At this point, even if hunting is curbed, recovery becomes a long, slow process, requiring tree generations to complete.

So far we have omitted any mention of the role of arthropods as predators and consumers. There are few comparative measurements, but in a megaherbivore-free HSS world, vertebrate and invertebrate herbivores may play roughly equivalent roles because both are under strong top-down control (Chapters 7 and 11, this volume). Arthropod herbivory has not been shown to transform landscapes as can megafaunal herbivory, because arthropods tend to be more selective feeders and to be under top-down control except during occasional outbreaks. But if the top-down regulation that stabilizes the HSS world could be experimentally suspended, the resulting explosion in herbivory might indeed have the power to transform landscapes (Chapter 8, this volume).

The fundamental experiment of removing *all* predators from an ecosystem, both vertebrate and invertebrate, has never been executed because the technical challenges have simply been too great. We have only the experiment of removing the predators of vertebrate herbivores, and the consequences are dramatic, as detailed in this volume. If all invertebrate predators (including parasitoids) could be removed, the consequences might be equally dramatic. We shall not know with confidence until the experiment is performed.

TROPHIC CASCADES IN AQUATIC SYSTEMS

So far this discussion has focused on terrestrial ecosystems. Aquatic ecosystems differ in many important respects. They tend to be open and strongly influenced by transport phenomena. The principal producers are microscopic algae. Consequently, macroherbivores have never been as diverse and abundant in aquatic systems as they are on land. Marine vertebrate herbivores of all sizes are essentially restricted to the coastal margins. Aquatic food chains tend to be longer than their terrestrial counterparts. At least two important attributes of aquatic systems contribute to this fact. First, the primary producers are tiny algae that can be eaten only by slightly larger zooplankton that, in turn, are eaten by small fish, and so on. Food chain length is also enhanced by the fact that all fully aquatic forms (those that respire with gills) are ectothermic, a trait that greatly increases the efficiency of energy transfer between trophic levels.

Aquatic food chains are less linear than terrestrial ones and may contain complicating features. For example, the larvae of many species, including fish, often feed at a different trophic level from adults, and the adults of one species may feed heavily on the juveniles of another. Different species use distinct modes of defense and aggression, leading to intransitive competition rings (Buss and Jackson 1979). Organic matter from primary producers enters the food web not only through the classic grazing food chain but also via the microbial loop (Azam 1998). Despite all these differences between terrestrial and aquatic systems, evidence reviewed in the preceding chapters affirms a universal role of top-down regulation in structuring and stabilizing aquatic ecosystems, in both freshwater and marine environments.

Ocean ecosystems are or once were liberally festooned with large predatory vertebrates, including the great sharks, billfishes, great tunas, large toothed whales, baleen whales, dolphins and porpoises, pinnipeds, and seabirds (Myers and Worm 2003; Roberts 2007). Many of these have been reduced to scattered remnant populations while large, nonpredatory vertebrates, including marine turtles, dugongs, and manatees, have been reduced an order of magnitude or more from levels of abundance described by early sailors (Jackson et al. 2001). In contrast with the history on land, these high trophic status consumers were not stripped away until recently, and so far there have been few human-caused extinctions of large marine vertebrates. The few extinctions that have occurred have been species with an obligate association with land (seabirds and pinnipeds). Great whale populations probably were never regulated by killer whale predation, but killer whales influence their behavior (Corkeron and Connor 1999), and the great whales are or once were an important prey resource for killer whales (Springer et al. 2003).

The decimation of large marine vertebrates has had wide-ranging effects on the structure and function of ocean ecosystems, even leading to changes in waves and currents, primary productivity, and fish abundance (Duggins et al. 1989; Reisewitz et al. 2005). The systematic trophic degradation of coral reefs has led to spectacular losses in coral abundance and species diversity through algal overgrowth and the spread of disease (Hughes 1994; Chapter 5, this volume). Even linkages between aquatic and terrestrial realms and between the coastal oceans and the high seas are commonly controlled by apex consumers (Croll et al. 2005; Estes et al. 1998). Although dissenters may continue to protest, we have reached a point in the acquisition of knowledge where there can be no doubt: Top-down forcing initiated by species at the highest trophic levels influences the workings of our world in profound ways. This much is clear, even though there is a great deal about trophic cascades yet to be discovered and understood.

TROPHIC SHUNTS

When system-wide regime changes lead to alternative states, species that were previously maintained at low to moderate abundance by top-down regulation can explode into excessive abundance, such as white-tailed deer in the eastern United States or sea urchins in higher-latitude coastal oceans. Exploded populations of species low on the trophic ladder, or even plants, can assume the role of trophic shunts. Trophic shunts can be defined as organisms that intrinsically or potentially short-circuit the hierarchical food chain by usurping primary productivity at low trophic levels.

There are both natural and human-induced trophic shunts. Natural trophic shunts include whales and elephants (i.e., megafauna). These are creatures too large as adults to be vulnerable to predators, and consequently they tend to increase to high biomasses at which they are bottom-up limited. For example, it has been estimated that at least 60 percent of the primary productivity of the prehuman ocean was harvested by whales (Croll et al. 2006). We know very little about the direct and indirect effects of reducing whale populations by more than 90 percent, but they must be substantial (Estes et al. 2006). Similarly, Owen-Smith (1988) has estimated that megaherbivores (e.g., elephants, giraffes, hippos, rhinos, buffalo) typically account for at least 50 percent of the consuming biomass of intact terrestrial systems that still retain megafauna. Megaherbivores effectively shortcut the food chain and deprive predators of potential prey while recycling huge amounts of harvested biomass directly back to the soil in the form of dung and urine. Even plants can become a trophic shunt, as when floating macrophytes take over a pond otherwise dominated by phytoplankton. Macrophytes absorb most of the light striking the pond and release oxygen, not into the water mass but directly into the atmosphere, causing anoxia and effectively shutting down the normal fish-dominated aquatic food chain (Scheffer et al. 2003b).

When human interventions alter the natural regulation of ecosystems, unnatural trophic shunts can appear unexpectedly, often with consequences that are considered highly undesirable by humans. The sea urchin barrens that result when sea otters are extirpated along North Pacific shores are a prime example. Exploded populations of sea urchins consume all macroalgae, reducing on-site productivity to lower levels, depriving fish and other organisms of a food supply and refuges from predators, and many other adverse impacts. Overabundant deer consume seedlings, reducing tree recruitment and altering the composition of forests on large scales (Rooney et al. 2004). And they can drastically reduce the biodiversity of herbaceous plant communities, even (or especially) in protected areas (Rooney and Dress 1997). Moreover, deer are a crucial resource for black-legged ticks, thus increasing the occurrence of Lyme disease (Ostfeld and Holt 2004). In another example, overfishing of great sharks along the East Coast of the United States has allowed a common shark prey, the cownose ray, to increase perhaps a hundred-fold. Cownose rays are now destroying commercial shellfish beds and wreaking economic havoc (Myers et al. 2007). These are well-understood cases of consumers that have become trophic shunts upon reduction of top-down control.

But for each example of a trophic shunt operating via a well-understood chain of cause and effect, there are others that are not understood at all, and these can be fraught with surprises (Doak et al. 2008). In some regions of the ocean we are beginning to see evidence for the wholesale reorganization of marine ecosystems. For example, bacterial blooms observed in the western Line Islands by Sandin et al. (Chapter 5, this volume) appear to be an indirect effect of overfishing at the top of the food chain, though exactly how the effect of overfishing propagates across two or more lower trophic levels remains unresolved. Reports of large-scale outbreaks of shrimp, squid, and, especially, jellyfish have appeared in both the scientific and popular literature in recent years (Worm and Myers 2003; Essington 2006; Jackson 2008). These outbreaks are completely novel and possess the hallmarks of trophic shunts. Are such blooms mere aberrations, or are they the harbingers of an ocean in transition to alternative states? If swarms of squid and jellyfish can constitute alternative states to the normal fish-dominated food web, are these states stabilized by hysteresis? If so, the world's fishing industry is in for some big surprises and big adjustments. Catastrophic regime changes in the world's oceans are a matter of grave concern, for not only are the oceans an important source of protein for a large fraction of humanity but they regulate climate and the composition of our atmosphere.

CODA

As we embark on the twenty-first century, the anthropogenic world most of us occupy is trophically degraded to a degree that foretells widespread biological changes, including losses of biodiversity. Terrestrial megafauna disappeared over most of the globe thousands of years ago. More recently we have systematically exploited or otherwise persecuted top predators in both the terrestrial and aquatic realms. And concurrently in the tropics, we are rapidly eliminating herbivores, seed dispersers, and other mutualists to feed an insatiable market for bushmeat. Intact ecosystems have become the rarest of the rare. These last few remnants of the world evolution created are absolutely precious. Lose them, and we lose all reference to hundreds of millions of years of evolutionary history.

Of all the trophic layers we have lost or are losing, the top carnivore layer is the most crucial to the survival of contemporary nature, because the top-down regulation it provides stabilizes the interactions between consumers and producers. Strip away the top predators, and ecosystems convulse through harsh transitions to simpler alternative states, exemplified by the sea urchin barrens.

There is little public awareness of impending biotic impoverishment because the drivers of collapse are the *absence* of essentially invisible processes (such as predation and seed dispersal) and because the ensuing transformations are slow and often subtle, involving gradual compositional changes that are beyond the powers of observation of most lay observers, who are beguiled by the shifting baseline.

We view the ongoing impoverishment of the earth's biological systems with grave concern. The well-being of humanity depends intimately on healthy, productive, and predictably functioning ecosystems of all kinds, from forests to grasslands and from headwater streams to the deep ocean. In some way or another, humans use every ecosystem on Earth; none is expendable. We thus view the continuing trophic degradation of the planet as a crisis every bit as serious, universal, and urgent as climate change. And, as with climate change, ignoring the problem won't solve it. It will only grow worse until appropriate and sufficient measures are taken to correct it.

Throughout history, humans have had to learn their lessons the hard way, through bitter experience and through repeating the mistakes of others (Diamond 2005). Will ignoring or refusing to accept the ideas presented in this book become another case in point? Only time will tell.

References

Aanen, D. K., P. Eggleton, C. Rouland-Lefevre, T. Guldberg-Froslev, S. Rosendahl, and J. J. Boomsma. 2002. The evolution of fungus-growing termites and their mutualistic fungal symbionts. *Proceedings of the National Academy of Sciences* 99:14887–14892.

Abrams, P. A. 1995. Implications of dynamically variable traits for identifying, classifying and measuring direct and indirect effects in ecological communities. *American Naturalist* 146:112–134.

Abrams, P. A. 2002. Will small population sizes warn us of impending extinctions? *American Naturalist* 160:293–305.

Abrams, P. A., B. A. Menge, G. G. Mittelbach, D. A. Spiller, and P. Yodzis. 1996. The role of indirect effects in food webs. Pp. 371–395 in G. A. Polis and K. O. Winemiller, eds. *Food webs: Integration of patterns and dynamics*. Chapman and Hall, New York.

Agrawal, A. A. 1998. Algal defense, grazers, and their interactions in aquatic trophic cascades. *Acta Oecologica, International Journal of Ecology* 19:331–337.

Aide, T. M. 1988. Herbivory as a selective agent on the timing of leaf production in a tropical understory community. *Nature* 336:574–575.

Aizen, M. A., and P. Feinsinger. 1994. Forest fragmentation, pollination, and plant reproduction in a chaco dry forest, Argentina. *Ecology* 75:330–351.

Ale, S. B., and C. J. Whelan. 2008. Reappraisal of the role of big, fierce predators! *Biodiversity and Conservation* 17:685–690.

Aleksandrova, V. D., V. N. Andreev, T. V. Vahtina, R. A. Dydina, G. I. Kareva, V. V. Petrovsky, and V. F. Šamarin. 1964. *Kormovaja Haraktristika Rostennij Krajnego Severa USSR*. Nauka, Moscow.

Alexander, R. D. 1974. The evolution of social behaviour. *Annual Review of Ecology and Systematics* 5:325–383.

Allen-Morley, C. R., and D. C. Coleman. 1989. Resilience of soil biota in various food webs to freezing perturbations. *Ecology* 70:1127–1141.

Allombert, S., A. J. Gaston, and J.-L. Martin. 2005a. A natural experiment on the impact of overabundant deer on songbird populations. *Biological Conservation* 126:1–13.

Allombert, S., S. Stockton, and J.-L. Martin. 2005b. A natural experiment on the impact of overabundant deer on forest invertebrates. *Conservation Biology* 19:1917–1929.

Alverson, W. S., D. M. Waller, and S. L. Solheim. 1988. Forests too deer: Edge effects in northern Wisconsin. *Conservation Biology* 2:348–358.

Ames, E. P. 2003. Atlantic cod stock structure in the Gulf of Maine. *Fisheries Research* 291:10–19.

Anderson, P. J., and J. F. Piatt. 1999. Community reorganization in the Gulf of Alaska following ocean climate regime shift. *Marine Ecology Progress Series* 189:117–123.

Anderson, R. C., and A. J. Katz. 1993. Recovery of browse-sensitive tree species following release from white-tailed deer (*Odocoileus virginianus*) Zimmerman browsing pressure. *Biological Conservation* 63:203–208.

Anderson, R. C., and O. L. Loucks. 1979. White-tailed deer (*Odocoileus virginianus*) influence on structure and composition of *Tsuga canadensis* forests. *Journal of Applied Ecology* 16:855–861.

Anderson, W. B., and G. A. Polis. 1998. Marine subsidies of island communities in the Gulf of California: Evidence from stable carbon and nitrogen isotopes. *Oikos* 81:75–80.

Ando, M., A. Itaya, S.-I. Yamamoto, and E. Shibata. 2006. Expansion of dwarf bamboo, *Sasa nipponica*, grassland under feeding pressure of sika deer, *Cervus nippon*, on subalpine coniferous forest in central Japan. *Journal of Forest Research* 11:51–55.

Andrew, N. L., Y. Agatsuma, E. Ballesteros, A. G. Bazhin, E. P. Creaser, D. K. A. Barnes, L. W. Botsford, et al. 2002. Status and management of world sea urchin fisheries. *Oceanography and Marine Biology: An Annual Review* 40:343–425.

Angelstam, P., P. E. Wikberg, P. Danilov, W. E. Faber, and K. Nygren. 2000. Effects of moose density on timber quality and biodiversity restoration in Sweden, Finland, and Russian Karelia. *Alces* 36:133–145.

Anouk Simard, M., S. D. Côté, R. B. Weladji, and J. Huot. 2008a. Feedback effects of chronic browsing on life history traits of a large herbivore. *Journal of Animal Ecology* 77:678–686.

Anouk Simard, M., S. D. Côté, R. B. Weladji, and J. Huot. 2008b. On being the right size: Food-limited feedback on optimal body size. *Journal of Animal Ecology* 77:635–637.

Anthony, R. G., J. A. Estes, M. A. Ricca, A. K. Miles, and E. D. Forsman. 2008. Bald eagles and sea otters in the Aleutian archipelago: Indirect effects of trophic cascades. *Ecology* 89:2725–2735.

Aponte, C., G. R. Barreto, and J. Terborgh. 2003. Consequences of habitat fragmentation on age structure and life history in a tortoise population. *Biotropica* 35:550–555.

Archibald, S. 2008. African grazing lawns: How fire, rainfall, and grazer numbers interact to affect grass community states. *Journal of Wildlife Management* 72:492–501.

Archibald, S., and W. J. Bond. 2003. Growing tall vs. growing wide: Tree architecture and allometry of *Acacia karroo* in forest, savanna, and arid environments. *Oikos* 102:3–14.

Archibald, S., W. J. Bond, W. D. Stock, and D. H. K. Fairbanks. 2005. Shaping the landscape: Fire–grazer interactions in an African savanna. *Ecological Applications* 15:96–109.

Arjo, W. M., and D. H. Pletscher. 1999. Behavioral responses of coyotes to wolf recolonization in northwestern Montana. *Canadian Journal of Zoology* 77:1919–1927.

Arnold, N. E. 1976. Fossil reptiles from the Aldabra Atoll, Indian Ocean. *Bulletin of the British Museum (Natural History), London* 29:6–116.

Arnold, N. E. 1979. Indian Ocean giant tortoises: Their systematics and island adaptations. *Philosophical Transactions of the Royal Society of London. Series B, Biological Sciences* 286:127–145.

Arthur, A. D., R. P. Pech, and C. R. Dickman. 2005. Effects of predation and habitat structure on the population dynamics of house mice in large outdoor enclosures. *Oikos* 108:562–572.

Asquith, N. M., J. Terborgh, A. E. Arnold, and C. M. Riveros. 1999. The fruits the agouti ate: *Hymenaea courbaril* seed fate when its disperser is absent. *Journal of Tropical Ecology* 15:229–235.

Asquith, N. M., S. J. Wright, and M. J. Clauss. 1997. Does mammal community composition control seedling recruitment in Neotropical forests? Evidence from Panama. *Ecology* 78:941–946.

Augustine, D. J., and S. J. McNaughton. 1998. Ungulate effects on the functional species composition of plant communities: Herbivore selectivity and plant tolerance. *Journal of Wildlife Management* 62:1165–1183.

Aunapuu, M., J. Dahlgren, T. Oksanen, D. Grellmann, L. Oksanen, J. Olofsson, Ü. Rammul, M. Schneider, B. Johansen, and H. O. Hygen. 2008. Spatial patterns and dynamic responses of arctic food webs corroborate the exploitation ecosystems hypothesis (EEH). *American Naturalist* 171:249–262.

Austin, J. J., E. N. Arnold, and R. Bour. 2003. Was there a second adaptive radiation of giant tortoises in the Indian Ocean? Using mitochondrial DNA to investigate speciation and biogeography of *Aldabrachelys* (Reptilia, Testudinidae). *Molecular Ecology* 12:1415–1424.

Ayal, Y. 2007. Trophic structure and the role of predation in shaping hot desert communities. *Journal of Arid Environments* 68:171–187.

Azam, F. 1998. Microbial control of organic carbon flux: The plot thickens. *Science* 280:694–696.

Babcock, R. C., S. Kelly, N. T. Shears, J. W. Walker, and T. J. Willis. 1999. Changes in community structure in temperate marine reserves. *Marine Ecology Progress Series* 189:125–134.

Baines, D., R. B. Sage, and M. M. Baines. 1994. The implications of red deer grazing to ground vegetation and invertebrate communities of Scottish native pinewoods. *Journal of Applied Ecology* 31:776–783.

Ball, I. J., R. L. Eng, and S. K. Ball. 1995. Population density and productivity of ducks on large grassland tracts in northcentral Montana. *Wildlife Society Bulletin* 23:767–773.

Bangert, R. K., and C. N. Slobodchikoff. 2004. Prairie dog engineering indirectly affects beetle movement behavior. *Journal of Arid Environments* 56:83–84.

Banks, P. B., K. Norrdahl, M. Nordström, and E. Korpimäki. 2004. Dynamic impacts of feral mink predation on vole metapopulations in the outer archipelago of the Baltic Sea. *Oikos* 105:79–88.

Banse, K. 1994. Grazing and zooplankton production as key controls of phytoplankton production in the open ocean. *Oceanography* 7:13–20.

Barber, N. A., R. J. Marquis, and W. P. Tori. 2008. Invasive prey impacts the abundance and distribution of native predators. *Ecology* 89:2678–2683.

Bardgett, R. D., and D. A. Wardle. 2003. Herbivore mediated linkages between aboveground and belowground communities. *Ecology* 84:2258–2268.

Barkai, A., and C. McQuaid. 1988. Predator–prey role reversal in a marine benthic ecosystem. *Science* 242:62–64.

Barmore, W. J. 2003. *Ecology of ungulates and their winter range in northern Yellowstone National Park: Research and synthesis 1962–1970.* Yellowstone Center for Resources, Yellowstone National Park, WY.

Barnes, R. F. W., and S. A. Lahm. 1997. An ecological perspective on human densities in the central African forest. *Journal of Applied Ecology* 34:245–260.

Barnosky, A. D., P. L. Koch, R. S. Feranec, S. L. Wing, and A. B. Shabel. 2004. Assessing the causes of Late Pleistocene extinctions on the continents. *Science* 306:70–75.

Barton, B. T. 2005. *Cascading effects of predator removal on the ecology of sea turtle nesting beaches.* University of Central Florida, Orlando.

Barton, R. A., and A. Whiten. 1994. Reducing complex diets to simple rules: Food selection by olive baboons. *Behavioral Ecology and Sociobiology* 35:283–293.

Bascompte, J., C. J. Melian, and E. Sala. 2005. Interaction strength combinations and the overfishing of a marine food web. *Proceedings of the National Academy of Sciences* 102:5443–5447.

Bates, H. W. 1862. Contributions to the insect fauna of the Amazon Valley. *Transactions of the Linnean Society of London* 23:495.

Baum, J. K., R. A. Myers, D. G. Kehler, B. Worm, S. J. Harley, and P. A. Doherty. 2003. Collapse and conservation of shark populations in the northwest Atlantic. *Science* 299:389–392.

Baum, J. K., and B. Worm. 2009. Cascading top-down effects of changing oceanic predator abundances. *Journal of Animal Ecology* 70:699–714.

Baxter, C. V., K. D. Fausch, M. Murakami, and P. L. Chapman. 2004. Fish invasion restructures stream and forest food webs by interrupting reciprocal prey subsidies. *Ecology* 85:2656–2663.

Baxter, C. V., K. D. Fausch, and W. C. Saunders. 2005. Tangled webs: Reciprocal flows of invertebrate prey link streams and riparian zones. *Freshwater Biology* 50:201–220.

Beauchamp, G. 2004. Reduced flocking by birds on islands with relaxed predation. *Proceedings of the Royal Society B: Biological Sciences* 271:1039–1042.

Beaugrand, G., K. M. Brander, J. A. Lindley, S. Souissi, and P. C. Reid. 2003. Plankton effect on cod recruitment in the North Sea. *Nature* 426:661–664.

Beckerman, A. P., M. Uriarte, and O. J. Schmitz. 1997. Experimental evidence for a behavior-mediated trophic cascade in a terrestrial food chain. *Proceedings of the National Academy of Sciences* 94:10735–10738.

Begon, M., J. L. Harper, and C. R. Townsend. 1996. *Ecology: Individuals, populations and communities*. Blackwell Science, Oxford.

Beguin, J., D. Pothier, and M. Prévost. 2009. Can the impact of deer browsing on tree regeneration be mitigated by shelterwood cutting and strip clearcutting. *Forest Ecology and Management* 257:38–45.

Bell, R. H. V. 1971. A grazing ecosystem in the Serengeti. *Scientific American* 225:86–93.

Bellingham, P. J., and C. N. Allan. 2003. Forest regeneration and the influences of white-tailed deer (*Odocoileus virginianus*) in cool temperate New Zealand rain forests. *Forest Ecology and Management* 175:71–86.

Bender, E. A., T. J. Case, and M. E. Gilpin. 1984. Perturbation experiments in community ecology: Theory and practice. *Ecology* 65:1–13.

Benndorf, J. 1990. Conditions for effective biomanipulation: Conclusions derived from whole-lake experiments in Europe. *Hydrobiologia* 200/201:187–203.

Benndorf, J., W. Boing, J. Koop, and I. Neubauer. 2002. Top-down control of phytoplankton: The role of time scale, lake depth and trophic state. *Freshwater Biology* 47:2282–2295.

Berger, J. 1993. Disassociations between black rhinoceros mothers and young calves: Ecologically variable or, as yet, undetected behaviour? *African Journal of Ecology* 31:261–264.

Berger, J. 1999. Anthropogenic extinctions of top carnivores and interspecific animal behaviour: Implications of the rapid decoupling of a web involving wolves, bears, moose and ravens. *Proceedings of the Royal Society B: Biological Sciences* 266:2261–2267.

Berger, J. 2005. Hunting by carnivores and by humans: Is functional redundancy possible and who really cares? Pp. 316–341 in J. Ray, K. H. Redford, R. Steneck, and J. Berger, eds. *Large carnivores and the conservation of biodiversity*. Island Press, Washington, DC.

Berger, J. 2007a. Carnivore repatriation and holarctic prey: Narrowing the deficit in ecological effectiveness. *Conservation Biology* 21:1105–1116.

Berger, J. 2007b. Fear, human shields, and the redistribution of prey and predators in protected areas. *Biology Letters* 3:620–623.

Berger, J. 2008. *The better to eat you with: Fear in wild animal societies*. University of Chicago Press, Chicago.

Berger, J., P. B. Stacey, L. Bellis, and M. P. Johnson. 2001a. A mammalian predator–prey imbalance: Grizzly bear and wolf extinction affect avian Neotropical migrants. *Ecological Applications* 11:947–960.

Berger, J., J. E. Swenson, and I.-L. Persson. 2001b. Recolonizing carnivores and naive prey: Conservation lessons from Pleistocene extinctions. *Science* 291:1036–1039.

Berger, K. M., and E. M. Gese. 2007. Does interference competition with wolves limit the distribution and abundance of coyotes? *Journal of Animal Ecology* 76:1075–1085.

Berger, K. M., E. M. Gese, and J. Berger. 2008. Indirect effects and traditional trophic cascades: A test involving wolves, coyotes, and pronghorn. *Ecology* 89:818–828.

Bergquist, A. M., and S. R. Carpenter. 1986. Limnetic herbivory: Effects on phytoplankton populations and primary production. *Ecology* 67:1351–1360.

Bertness, M. D., S. D. Gaines, and M. E. Hay. 2001. *Marine community ecology*. Sinauer, Sunderland, MA.

Beschta, R. L. 2003. Cottonwoods, elk, and wolves in the Lamar Valley of Yellowstone National Park. *Ecological Applications* 13:1295–1309.

Beschta, R. L. 2005. Reduced cottonwood recruitment following extirpation of wolves in Yellowstone's Northern Range. *Ecology* 86:391–403.

Beschta, R. L., and W. J. Ripple. 2006. River channel dynamics following extirpation of wolves in northeastern Yellowstone National Park, USA. *Earth Surface Processes and Landforms* 31:1525–1539.

Beschta, R. L., and W. J. Ripple. 2007a. Increased willow heights along northern Yellowstone's Blacktail Deer Creek following wolf reintroduction. *Western North American Naturalist* 67:613–617.

Beschta, R. L., and W. J. Ripple. 2007b. Wolves, elk, and aspen in the winter range of Jasper National Park, Canada. *Canadian Journal of Forest Research* 37:1873–1885.

Beschta, R. L., and W. J. Ripple. 2008. Wolves, trophic cascades, and rivers in western Olympic National Park. *Ecohydrology* 1:118–130.

Beyer, H. L., E. H. Merrill, N. Varley, and M. S. Boyce. 2007. Willow on Yellowstone's Northern Range: Evidence for a trophic cascade? *Ecological Applications* 17:1563–1571.

Bigger, D. S., and M. A. Marvier. 1998. How different would a world without herbivory be? A search for generality in ecology. *Integrative Biology* 1:60–67.

Biggs, B. J., S. N. Francoeur, A. D. Huryn, R. Young, C. J. Arbuckle, and C. R. Townsend. 2000. Trophic cascades in streams: Effects of nutrient enrichment on autotrophic and consumer benthic communities under two different fish predation regimes. *Canadian Journal of Fisheries and Aquatic Sciences* 57:1380–1394.

Binkley, D. 2008. Age distribution of aspen in Rocky Mountain National Park, USA. *Forest Ecology and Management* 255:797–802.

Binkley, D., M. M. Moore, W. H. Romme, and P. M. Brown. 2006. Was Aldo Leopold right about the Kaibab deer herd? *Ecosystems* 9:227–241.

Birkeland, C. 1988. The influence of echinoderms on coral-reef communities. *Echinoderm Studies* 3:1–79.

Birkeland, C., and J. S. Lucas. 1990. Acanthaster planci: *Major management problem of coral reefs.* CRC Press, West Palm Beach, FL.

Bjorndal, K., and J. B. C. Jackson. 2003. Roles of sea turtles in marine ecosystems: Reconstructing the past. Pp. 259–273 in P. L. Lutz, J. A. Musick, and J. Wyneken, eds. *The biology of sea turtles*, Vol. 2. CRC Press, Boca Raton, FL.

Bjørnstad, O. N., M. Peltonen, A. M. Liebhold, and W. Baltenweiler. 2002. Waves of larch budmoth outbreaks in the European Alps. *Science* 298:1020–1023.

Blumstein, D. T. 2006. Developing an evolutionary ecology of fear: How life history and natural history traits affect disturbance tolerance in birds. *Animal Behaviour* 71:389–399.

Boag, B. 2000. The impact of the New Zealand flatworm on earthworms and moles in agricultural land in Scotland. *Aspects of Applied Biology* 62:79–84.

Boitani, L. 2003. Wolf conservation and recovery. Pp. 317–340 in L. D. Mech and L. Boitani, eds. *Wolves: Behavior, ecology, and conservation.* University of Chicago Press, Chicago.

Bolker, B., M. Holyoak, V. Krivan, L. Rowe, and O. Schmitz. 2003. Connecting theoretical and empirical studies of trait-mediated interactions. *Ecology* 84:1101–1114.

Bond, W. J. 2005. Large parts of the world are brown or black: A different view on the "green world" hypothesis. *Journal of Vegetation Science* 16:261–266.

Bond, W. J., and D. Loffell. 2001. Introduction of giraffe changes acacia distribution in a South African savanna. *African Journal of Ecology* 39:286–294.

Bond, W. J., J. A. Silander, J. Ranaivonasy, and J. Ratsirarson. 2008. The antiquity of Madagascar's grasslands and the rise of C4 grassy biomes. *Journal of Biogeography* 35:1743–1758.

Bond, W. J., K. A. Smythe, and D. A. Balfour. 2001. *Acacia* species turnover in space and time in an African savanna. *Journal of Biogeography* 28:117–128.

Bond, W. J., F. I. Woodward, and G. G. Midgley. 2005. The global distribution of ecosystems in a world without fire. *New Phytologist* 165:525–538.

Borer, E. T., C. J. Briggs, W. W. Murdoch, and S. L. Swarbrick. 2003. Testing intraguild predation theory in a field system: Does numerical dominance shift along a gradient of productivity? *Ecology Letters* 6:929–935.

Borer, E. T., E. W. Seabloom, J. B. Shurin, K. E. Anderson, C. A. Blanchette, B. Broitman, S. D. Cooper, and B. S. Halpern. 2005. What determines the strength of a trophic cascade? *Ecology* 86:528–537.

Bourn, D., and M. J. Coe. 1979. Features of tortoise mortality and decomposition on Aldabra. *Philosophical Transactions of the Royal Society of London. Series B, Biological Sciences* 286:188–193.

Bourn, D., C. Gibson, D. Augeri, C. J. Wilson, J. Church, and S. I. Hay. 1999. The rise and fall of the Aldabran giant tortoise population. *Proceedings of the Royal Society B: Biological Sciences* 266:1091–1100.

Bourque, B. J., B. Johnson, and R. S. Steneck. 2007. Possible prehistoric hunter–gatherer impacts on food web structure in the Gulf of Maine. Pp. 165–187 in J. Erlandson and R. Torben, eds. *Human impacts on ancient marine environments.* University of California Press, Berkeley.

Bowyer, R. T., and J. G. Kie. 2004. Effects of foraging activity on sexual segregation in mule deer. *Journal of Mammalogy* 85:498–504.

Bowyer, R. T., V. van Ballenberghe, J. G. Kie, and J. A. K. Maier. 1999. Birth-site selection by Alaskan moose: Maternal strategies for coping with a risky environment. *Journal of Mammalogy* 80:1070–1083.

Boyle, R., and M. D. Dearing. 2003. Ingestion of juniper foliage reduces metabolic rates in woodrat (*Neotoma*) herbivores. *Zoology* 106:151–158.

Brainard, R., J. Maragos, R. Schroeder, J. Kenyon, P. Vroom, S. Godwin, R. Hoeke, et al. 2005. The state of coral reef ecosystems of the Pacific Remote Island Areas. Pp. 338–372 in J. E. Waddell, ed. *The state of coral reef ecosystems of the United States and Pacific freely associated states: 2005.* NOAA Technical Memorandum. NOS NCCOS, Honolulu.

Braithwaite, C. J. R., J. D. Taylor, and W. J. Kennedy. 1973. The evolution of an atoll: The depositional and erosional history of Aldabra. *Philosophical Transactions of the Royal Society of London, Series B. Biological Sciences* 266:307–340.

Branch, M. P., ed. 2001. *John Muir's last journey: South to the Amazon and east to Africa. Unpublished journals and selected correspondence.* Island Press, Washington, DC.

Brashares, J. S. 2003. Behavioral, ecological, and life-history correlates of mammal extinctions in West Africa. *Conservation Biology* 17:733–743.

Brashares, J. S., P. Arcese, and M. K. Sam. 2001. Human demography and reserve size predict wildlife extinction in West Africa. *Proceedings of the Royal Society B: Biological Sciences* 268:2473–2478.

Brashares, J. S., P. Arcese, M. K. Sam, P. B. Coppolillo, A. R. E. Sinclair, and A. Balmford. 2004. Bushmeat hunting, wildlife declines and fish supply in West Africa. *Science* 306:1180–1183.

Brashares, J. S., and M. K. Sam. 2005. How much is enough? Estimating the minimum sampling required for effective monitoring of African reserves. *Biodiversity and Conservation* 14:2709–2722.

Bråthen, K. A., R. A. Ims, N. G. Yoccoz, P. Fauchald, T. Tveraa, and V. H. Hausner. 2007. Induced shift in ecosystem productivity? Extensive scale effects of abundant large herbivores. *Ecosystems* 10:773–789.

Breen, P. A., T. A. Carson, J. B. Foster, and E. A. Steward. 1982. Changes in subtidal community structure associated with British Columbia sea otter transplants. *Marine Ecology Progress Series* 7:13–20.

Brett, M. T., and C. R. Goldman. 1996. A meta-analysis of the freshwater trophic cascade. *Proceedings of the National Academy of Sciences* 93:7723–7726.

Brochu, C. A. 2007. Morphology, relationships, and biogeographical significance of an extinct horned crocodile (Crocodylia, Crocodylidae) from the Quaternary of Madagascar. *Zoological Journal of the Linnean Society* 150:835–863.

Broitman, S. D. C., et al. 2005. What determines the strength of a trophic cascade? *Ecology* 86:528–537.

Brooks, J. L., and S. I. Dodson. 1965. Predation, body size, and composition of plankton. *Science* 150:28–35.

Brower, L. P. 1988. Avian predation on the monarch butterfly and its implications for mimicry theory. *American Naturalist* (Supplement) 131:S4–S6.

Brown, J. H., and E. J. Heske. 1990. Control of a desert–grassland transition by a keystone rodent guild. *Science* 250:1705–1707.

Brown, J. S., and B. P. Kotler. 2004. Hazardous duty pay and the foraging cost of predation. *Ecology Letters* 7:999–1014.

Brown, J. S., J. W. Laundre, and M. Gurung. 1999. The ecology of fear: Optimal foraging, game theory, and trophic interactions. *Journal of Mammalogy* 89:385–399.

Brown, T. L., D. J. Decker, S. J. Riley, J. W. Enck, T. B. Lauber, P. D. Curtis, and G. F. Mattfeld. 2000. The future of hunting as a mechanism to control white-tailed deer populations. *Wildlife Society Bulletin* 28:797–807.

Brown, V. C., and A. C. Gange. 1992. Secondary plant succession: How is it modified by insect herbivory? *Vegetatio* 101:3–13.

Brozovic, N., and W. Schlenker. 2007. *Optimal management of an ecosystem with an unknown threshold.* Social Science Research Network. Retrieved November 15, 2009 from ssrn.com/abstract=990613.

Bruno, J. F., and B. J. Cardinale. 2008. Cascading effects of predator richness. *Frontiers in Ecology and the Environment* 6:539–546.

Bryant, J. P., F. S. Chapin, and D. J. Klein. 1983. Carbon/nutrient balance of boreal plants in relation to vertebrate herbivory. *Oikos* 40:357–368.

Buchman, N., K. Cuddington, and J. Lambrinos. 2007. A historical perspective on ecosystem engineering. In K. Cuddington, J. E. Byers, W. G. Wilson, and A. Hastings, eds. *Ecosystem engineers, plants to protists.* Academic Press, Amsterdam.

Burkepile, D. E., and M. E. Hay. 2006. Herbivore vs. nutrient control of marine primary producers: Context-dependent effects. *Ecology* 87:3128–3139.

Burkepile, D. E., and M. E. Hay. 2007. Predator release of the gastropod *Cyphoma gibbosum* increases predation on gorgonian corals. *Oecologia* 154:167–173.

Burney, D. A., L. P. Burney, L. R. Godfrey, W. L. Jungers, S. J. Goodman, H. T. Wright, and A. J. T. Jull. 2004. A chronology for late prehistoric Madagascar. *Journal of Human Evolution* 27:25–63.

Burney, D. A., and T. F. Flannery. 2005. Fifty millennia of catastrophic extinctions after human contact. *Trends in Ecology and Evolution* 20:395–401.

Burney, D. A., G. S. Robinson, and L. P. Burney. 2003. *Sporormiella* and the late Holocene extinctions in Madagascar. *Proceedings of the National Academy of Sciences* 100:10800–10805.

Burton, R. A. 2008. *On being certain: Believing you are right even when you're not.* St. Martin's Press, New York.

Buss, L. W., and J. B. C. Jackson. 1979. Competitive networks: Nontransitive competitive relationships in cryptic coral reef environments. *American Naturalist* 113:223–234.

Butler, J. R. 2000. The economic costs of wildlife predation on livestock in Gokwe communal land, Zimbabwe. *African Journal of Ecology* 38:23–30.

Byrnes, J., J. J. Stachowicz, K. M. Hultgren, A. R. Hughes, S. V. Olyranik, and C. S. Thornber. 2006. Predator diversity strengthens trophic cascades in kelp forests by modifying herbivore behaviour. *Ecology Letters* 9:61–71.

Cabascon, C., and D. Pothier. 2007. Browsing of tree regeneration by white-tailed deer in large clearcuts on Anticosti Island, Quebec. *Forest Ecology and Management* 253:112–119.

Cabascon, C., and D. Pothier. 2008. Impact of deer browsing on plant communities in cut-over sites on Anticosti Island. *Écoscience* 15:389–397.

Caccone, A., J. P. Gibbs, V. Ketmaier, E. Suatoni, and J. R. Powell. 1999. Origin and evolutionary relationships of giant Galapagos tortoises. *Proceedings of the National Academy of Sciences* 96:13223–13228.

Caddy, J. F., and P. G. Rodhouse. 1998. Cephalopod and groundfish landings: Evidence for ecological change in global fisheries? *Reviews in Fish Biology and Fisheries* 8:431–444.

Caley, M. J., M. H. Carr, M. A. Hixon, T. P. Hughes, G. P. Jones, and B. A. Menge. 1996. Recruitment and the local dynamics of open marine populations. *Annual Review of Ecology and Systematics* 27:477–500.

Campbell, D. G., K. S. Lowell, and M. E. Lightbourn. 1991. The effect of introduced hutias (*Geocapromys ingrahami*) on the woody vegetation of Little Wax Cay, Bahamas. *Conservation Biology* 5:536–541.

Campbell, K., C. J. Donlan, F. Cruz, and V. Carrio. 2004. Eradication of feral goats *Capra hircus* from Pinta Island, Galapagos, Ecuador. *Oryx* 38:328–333.

Caraco, N. F., and J. J. Cole. 2004. When terrestrial organic matter is sent down the river: Importance of allochthonous C inputs to the metabolism in lakes and rivers. Pp. 301–316 in G. A. Polis, M. E. Power, and G. R. Huxel, eds. *Food webs at the landscape level.* University of Chicago Press, Chicago.

Cardillo, M., G. M. Mace, K. E. Jones, J. Bielby, O. R. P. Bininda-Emonds, W. Sechrest, C. D. L. Orme, and A. Purvis. 2005. Multiple causes of high extinction risk in large mammal species. *Science* 309:1239–1241.

Carlquist, S. 1974. *Island biology.* Columbia University Press, New York.

Caro, T. M. 2005. *Antipredator defenses in birds and mammals.* University of Chicago Press, Chicago.

Carpenter, R. C. 1984. Predator and population density control of homing behavior in the Caribbean echinoid *Diadema antillarum. Marine Biology* 82:101–108.

Carpenter, R. C. 1986. Partitioning herbivory and its effects on coral reef algal communities. *Ecological Monographs* 56:346–363.

Carpenter, S. R. 1992. Destabilization of planktonic ecosystems and blooms of blue-green algae. Pp. 461–482 in J. F. Kitchell, ed. *Food web management: A case study of Lake Mendota.* Springer-Verlag, New York.

Carpenter, S. R. 1996. Microcosm experiments have limited relevance for community and ecosystem ecology. *Ecology* 77:677–680.

Carpenter, S. R. 1998. The need for large-scale experiments to assess and predict the response of ecosystems to perturbation. Pp. 287–312 in M. L. Pace and P. M. Groffman, eds. *Successes, limitations and frontiers in ecosystem science.* Springer-Verlag, New York.

Carpenter, S. R. 2003. *Regime shifts in lake ecosystems: Pattern and variation.* Excellence in Ecology Series. Ecology Institute, Oldendorf/Luhe, Germany.

Carpenter, S. R., and W. A. Brock. 2006. Rising variance: A leading indicator of ecological transition. *Ecology Letters* 9:308–315.

Carpenter, S. R., W. A. Brock, J. J. Cole, J. F. Kitchell, and M. L. Pace. 2008. Leading indicators of trophic cascades. *Ecology Letters* 11:128–138.

Carpenter, S. R., J. J. Cole, J. R. Hodgson, J. F. Kitchell, M. L. Pace, D. Bade, K. L. Cottingham, T. E. Essington, J. N. Houser, and D. E. Schindler. 2001. Trophic cascades, nutrients, and lake productivity: Whole-lake experiments. *Ecological Monographs* 71:163–186.

Carpenter, S. R., T. M. Frost, J. F. Kitchell, T. K. Kratz, D. W. Schindler, J. Shearer, W. G. Sprules, M. J. Vanni, and A. P. Zimmerman. 1991. Patterns of primary production and herbivory in 25 North American lake ecosystems. Pp. 67–96 in J. Cole, S. Findlay, and G. Lovett, eds. *Comparative analyses of ecosystems: Patterns, mechanisms, and theories.* Springer-Verlag, New York.

Carpenter, S. R., and J. F. Kitchell. 1988. Consumer control of lake productivity. *BioScience* 38:764–769.

Carpenter, S. R., and J. F. Kitchell, eds. 1993. *The trophic cascade in lakes.* Cambridge University Press, Cambridge.

Carpenter, S. R., J. F. Kitchell, and J. R. Hodgson. 1985. Cascading trophic interactions and lake productivity. *BioScience* 35:634–649.

Carpenter, S. R., J. F. Kitchell, J. R. Hodgson, P. A. Cochran, J. J. Elser, M. M. Elser, D. M. Lodge, D. Kretchmer, X. He, and C. N. von Ende. 1987. Regulation of lake primary productivity by food web structure. *Ecology* 68:1863–1876.

Carr, M. H., and M. A. Hixon. 1995. Predation effects on early post-settlement survivorship of coral-reef fishes. *Marine Ecology Progress Series* 124:31–42.

Casini, M., J. Lovgren, J. Hjelm, M. Cardinale, J. C. Molinero, and G. Kornilovs. 2008. Multi-level trophic cascades in a heavily exploited open marine ecosystem. *Proceedings of the Royal Society B: Biological Sciences* 275:1793–1801.

Castilla, J. C. 1999. Coastal marine communities: Trends and perspectives from human-exclusion experiments. *Trends in Ecology and Evolution* 14:280–283.

Castilla, J. C., and L. R. Duran. 1985. Human exclusion from the rocky intertidal zone of central Chile: The effects on *Concholepas concholepas* (Gastropoda). *Oikos* 45:391–399.

Castro, K., and T. Angell. 2000. Prevalence and progression of shell disease in American lobster, *Homarus americanus*, from Rhode Island waters and the offshore canyons. *Journal of Shellfish Research* 19:691–700.

Catling, P. C., and R. J. Burt. 1995. Why are red foxes absent from some eucalypt forests in eastern New South Wales? *Wildlife Research* 22:535–546.

Caughley, G. 1983. *The deer wars: The story of deer in New Zealand*. Heinemann, Auckland, NZ.

Ceballos, G., and P. R. Ehrlich. 2002. Mammal population losses and the extinction crisis. *Science* 296:904–907.

Cebrian, J. 1999. Patterns in the fate of production in plant communities. *American Naturalist* 154:449–468.

Cerling, T. E., J. M. Harris, B. J. MacFadden, M. G. Leakey, J. Quade, V. Eisenmann, and J. R. Ehleringer. 1997. Global vegetation change through the Miocene/Pliocene boundary. *Nature* 389:153–158.

Certini, G. 2005. Effects of fire on properties of forest soils: A review. *Oecologia* 143:1–10.

Chalfoun, A. D., F. R. Thompson, and M. J. Ratnaswamy. 2002. Nest predators and fragmentation: A review and meta analysis. *Conservation Biology* 16:306–318.

Chamaille-Jammes, S., H. Fritz, M. Valeix, F. Murindagomo, and J. Clobert. 2008. Resource variability, aggregation and direct density dependence in an open context: The local regulation of an African elephant population. *Journal of Animal Ecology* 77:135–144.

Chapin, D. M., R. L. Beschta, and H. W. Shen. 2002. Relationships between flood frequencies and riparian plant communities in the upper Klamath Basin, Oregon. *Journal of the American Water Resources Association* 38:603–617.

Chase, J. M., M. A. Leibold, A. L. Downing, and J. B. Shurin. 2000. The effects of productivity, herbivory, and plant species turnover in grassland food webs. *Ecology* 81:2485–2497.

Chavez, F. P., J. Ryan, S. E. Lluch-Cota, and M. Niquen. 2003. From anchovies to sardines and back: Multidecadal change in the Pacific Ocean. *Science* 299:217–221.

Cherfas, J. 1973. Goats must go to save the Galapagos tortoises. *New Scientist* 146:9.

Chesher, R. H. 1969. Destruction of Pacific corals by the sea star *Acanthaster planci*. *Science* 165:280–283.

Chiappone, M., H. Dienes, D. W. Swanson, and S. L. Miller. 2003. Density and gorgonian host-occupation patterns by flamingo tongue snails (*Cyphoma gibbosum*) in the Florida Keys. *Caribbean Journal of Science* 39:116–127.

Chouinard, A., and L. Filion. 2005. Impact of introduced white-tailed deer and native insect defoliators on the density and growth of conifer saplings on Anticosti Island, Quebec. *Écoscience* 12:506–518.

Christianson, D. A., and S. Creel. 2007. A review of environmental factors affecting elk winter diets. *Journal of Wildlife Management* 71:164–176.

Clark, J. S., S. R. Carpenter, M. Barber, S. Collins, A. Dobson, J. A. Foley, D. M. Lodge, et al. 2001. Ecological forecasts: An emerging imperative. *Science* 293:657–660.

Cleaveland, S., C. Packer, K. Hampson, M. Kaare, R. Kock, M. Craft, T. Lembo, T. Mlengeya, and A. Dobson. 2008. The multiple roles of infectious diseases in the Serengeti ecosystem. In A. R. E. Sinclair, C. Packer, S. A. R. Mduma, and J. M. Fryxell, eds. *Serengeti III: Human impacts on ecosystem dynamics*. University of Chicago Press, Chicago.

Close, D. C., and C. McArthur. 2002. Rethinking the role of many plant phenolics: Protection from photodamage not herbivores? *Oikos* 99:166–172.

Coblentz, B. E. 1978. The effects of feral goats (*Capra hircus*) on island ecosystems. *Biological Conservation* 13:279–285.

Cocroft, R. B. 2002. Antipredator defense as a limited resource: Unequal predation risk in broods of an insect with maternal care. *Behavioral Ecology* 12:125–133.

Coe, M. J., D. H. Cumming, and J. Phillipson. 1976. Biomass and production of large African herbivores in relation to rainfall and primary production. *Oecologia* 22:341–354.

Coen, L. D., R. D. Brumbaugh, D. Bushek, R. Grizzle, M. W. Luckenbach, M. H. Posey, S. P. Powers, and S. G. Tolley. 2007. Ecosystem services related to oyster restoration. *Marine Ecology Progress Series* 341:303–307.

Cohen, J. E., T. Jonsson, and S. R. Carpenter. 2003. Ecological community description using the food web, species abundance, and body size. *Proceedings of the National Academy of Sciences* 100:1781–1786.

Cole, J. J., S. R. Carpenter, M. L. Pace, M. C. Van de Bogert, J. F. Kitchell, and J. R. Hodgson. 2006. Differential support of lake food webs by three types of terrestrial organic carbon. *Ecology Letters* 9:558–568.

Cole, J. J., Y. T. Prairie, N. F. Caraco, W. H. McDowell, L. J. Tranvik, R. G. Striegl, C. M. Duarte, et al. 2007. Plumbing the global carbon cycle: Integrating inland waters into the terrestrial carbon budget. *Ecosystems* 10:171–184.

Coley, P. D. 1983. Herbivory and defensive characteristics of tree species in a lowland tropical forest. *Ecological Monographs* 53:209–233.

Collette, B. B., and G. Klein-MacPhee, eds. 2002. *Bigelow and Schroeder's fishes of the Gulf of Maine*, 3rd ed. Smithsonian Institution Press, Washington, DC.

Collie, J., and A. K. DeLong. 1999. Multispecies interactions in the Georges Bank fish community. Pp. 187–210 in *Ecosystem approaches for fisheries management*. Alaska Sea Grant College Program, Fairbanks.

Collins, W. B., and P. J. Urness. 1983. Feeding behavior and habitat selection of mule deer and elk on northern Utah summer range. *Journal of Wildlife Management* 47:646–663.

Connell, J. H. 1971. On the role of natural enemies in preventing competitive exclusion in some marine animals and rain forest trees. Pp. 298–312 in P. J. den Boer and G. R. Gradwell, eds. *Dynamics of populations*. Centre for Agricultural Publishing and Documentation, Wageningen.

Connell, S. W., and B. M. Gillanders. 2007. *Marine ecology*. Oxford University Press, Oxford.

Connor, M. M. 2001. *Elk movement in response to early-season hunting in the White River area, Colorado*. Ph.D. dissertation, Colorado State University, Fort Collins.

Coomes, D. A., R. B. Allen, D. M. Forsyth, and W. G. Lee. 2003. Factors preventing the recovery of New Zealand forests following the control of invasive deer. *Conservation Biology* 17:450–459.

Coope, G. R. 1987. The response of late Quaternary insect communities to sudden climatic changes. Pp. 421–438 in J. H. R. Gee and P. S. Giller, eds. *Organization of communities: Past and present*. Blackwell, Oxford.

Corbett, G. N. 1995. Review of the history and present status of moose in the national parks of the Atlantic region: Management implications. *Alces* 31:225–267.

Cordeiro, N., and H. F. Howe. 2003. Forest fragmentation severs mutualism between seed dispersers and an endemic African tree. *Proceedings of the National Academy of Sciences* 100:14052–14054.

Corfield, T. F. 1973. Elephant mortality in Tsavo National Park, Kenya. *East African Wildlife Journal* 11:339–368.

Corkeron, P. J., and R. C. Connor. 1999. Why do baleen whales migrate? *Marine Mammal Science* 15:1228–1245.

Cosner, C., D. L. DeAngelis, J. S. Ault, and D. B. Olson. 1999. Effects of spatial grouping on the functional response of predators. *Theoretical Population Biology* 56:65–75.

Costanza, R., H. Daly, C. Folke, P. Hawken, C. S. Holling, A. J. McMichael, D. Pimentel, and D. Rapport. 2000. Managing our environmental portfolio. *BioScience* 50:149–155.

Côté, S. D. 2005. Extirpation of a large black bear population by introduced white-tailed deer. *Conservation Biology* 19:1668–1671.

Côté, S. D., T. P. Rooney, J. Tremblay, C. Dussault, and D. M. Waller. 2004. Ecological impacts of deer overabundance. *Annual Review of Ecology and Systematics* 35:113–147.

Courchamp, F., L. Berc, and J. Gascoigne. 2008. *Allee effects in ecology and conservation*. Oxford University Press, Oxford.

Courchamp, F., J. L. Chapuis, and M. Pascal. 2003a. Mammal invaders on islands: Impact, control, and control impact. *Biological Reviews* 78:347–383.

Courchamp, F., M. Langlais, and G. Sugihara. 1999. Cats protecting birds: Modelling the mesopredator release effect. *Journal of Animal Ecology* 68:282–292.

Courchamp, F., R. Woodroffe, and G. Roemer. 2003b. Removing protected populations to save endangered species. *Science* 302:1532.

Cousins, S. 1987. The decline of the trophic level concept. *Trends in Ecology and Evolution* 2:312–316.

Cowen, R. K. 1983. The effects of sheephead (*Semicossyphus pulcher*) predation on red sea urchin (*Strongylocentrotus franciscanus*) populations: An experimental analysis. *Oecologia* 58:249–255.

Crawley, M. J. 1989. The relative importance of vertebrate and invertebrate herbivores in plant populations dynamics. Pp. 45–71 in E. A. Bernays, ed. *Insect–plant interactions*, Vol. 1. CRC Press, Boca Raton, FL.

Crawley, M. J. 1997. Plant–herbivore dynamics. Pp. 401–475 in M. J. Crawley, ed. *Plant ecology*. Blackwell, London.

Crépin, A. 2007. Using fast and slow processes to manage resources with thresholds. *Environmental Resource Economics* 36:191–213.

Crête, M. 1999. The distribution of deer biomass in North America supports the hypothesis of exploitation ecosystems. *Ecology Letters* 2:223–227.

Crête, M., and C. Dangle. 1999. Management of indigenous North American deer at the end of the 20th century in relation to large predators and primary production. *Acta Veterinaria Hungarica* 47:1–16.

Crête, M., and D. J. Doucet. 1998. Persistent suppression in dwarf birch after release from heavy summer browsing by caribou. *Arctic and Alpine Research* 30:126–132.

Crête, M., and M. Manseau. 1996. Natural regulation of cervidae along a 1000 km latitudinal gradient: Change in trophic dominance. *Evolutionary Ecology* 10:51–62.

Croizat, L. 1958. *Panbiogeography*. L. Croizat, Caracas, Venezuela.

Croizat, L., G. Nelson, and D. E. Rosen. 1974. Centers of origin and related concepts. *Systematic Zoology* 23:265–287.

Croll, D. A., R. Kudela, and B. R. Tershy. 2006. Ecosystem impacts of the decline of large whales in the North Pacific. Pp. 202–214 in J. A. Estes, D. P. DeMaster, D. F. Doak, T. M. Williams and R. L. Brownell Jr., eds. *Whales, whaling and ocean ecosystems*. University of California Press, Berkeley.

Croll, D. A., J. L. Maron, J. A. Estes, E. M. Danner, and G. V. Byrd. 2005. Introduced predators transform subarctic islands from grassland to tundra. *Science* 307:1959–1961.

Crooks, K., and M. Sanjayan. 2006. *Connectivity conservation*. Cambridge University Press, Cambridge.

Crooks, K. R., and M. E. Soulé. 1999. Mesopredator release and avifaunal extinctions in a fragmented ecosystem. *Nature* 400:563–566.

Crowder, L. B., and W. E. Cooper. 1982. Habitat structural complexity and the interaction between bluegills and their prey. *Ecology* 63:1802–1813.

Crowell, K. 1961. The effects of reduced competition in birds. *Proceedings of the National Academy of Sciences* 47:240–243.

Croze, H., A. K. K. Hillman, and E. M. Lang. 1981. Elephants and their habitats: How do they tolerate each other? In S. A. R. Fowler and W. T. D. Smith, eds. *Dynamics of large mammal populations*. Wiley, New York.

Cumming, D. H., M. B. Fenton, I. L. Rautenbach, R. D. Taylor, G. S. Cumming, M. S. Cumming, K. M. Dunlop, et al. 1997. Elephants, woodlands and biodiversity in southern Africa. *South African Journal of Science* 93:231–236.

Curran, L. M., I. Caniago, G. D. Paoli, D. Astianti, M. Kusneti, M. Leighton, C. E. Niararita, and H. Haeruman. 1999. Impact of El Niño and logging on canopy tree recruitment in Borneo. *Science* 286:2184–2188.

Curran, L. M., and M. Leighton. 2000. Vertebrate responses to spatiotemporal variation in seed production of mast-fruiting Dipterocarpaceae. *Ecological Monographs* 70:101–128.

Currie, D. J., P. Dilworth-Christie, and F. Chapleau. 1999. Assessing the strength of top-down influences on plankton abundance in unmanipulated lakes. *Canadian Journal of Fisheries and Aquatic Sciences* 56:427–436.

Cushing, D. H. 1996. *Towards a science of recruitment in fish populations.* Ecology Institute, Luhe, Germany.

Dahlgren, J., L. Oksanen, T. Oksanen, J. Olofsson, P. A. Hambäck, and Å. Lindgren. 2009a. Plant defenses to no avail? Species specific responses of plants to food web manipulations in a low arctic scrubland. *Evolutionary Ecology Research* 11:1189–1202.

Dahlgren, J., L. Oksanen, J. Olofsson, and T. Oksanen. 2009b. Plant defenses at no cost? The recovery of tundra scrubland following heavy grazing by gray-sided voles (*Myodes rufocanus*). *Evolutionary Ecology Research* 11:1205–1216.

Damuth, J. 1993. Cope's rule, the island rule and the scaling of mammalian population density. *Nature* 365:784–750.

Danell, K., and R. Bergström. 2002. Mammalian herbivory in terrestrial environments. Pp. 107–131 in C. M. Herrera and O. Pellmyr, eds. *Plant–animal interactions, an evolutionary approach.* Blackwell, Oxford.

Darnell, R. M. 1961. Trophic spectrum of an estuarine community, based on studies of Lake Pontchartrain, Louisiana. *Ecology* 42:553–568.

Daskalov, G. M. 2002. Overfishing drives a trophic cascade in the Black Sea. *Marine Ecology Progress Series* 225:53–63.

Daskalov, G. M., A. N. Grishin, S. Rodionov, and V. Mihneva. 2007. Trophic cascades triggered by overfishing reveal possible mechanisms of ecosystem regime shifts. *Proceedings of the National Academy of Sciences* 104:10518–10523.

Davenport, A. C., and T. W. Anderson. 2007. Positive indirect effects of reef fishes on kelp performance: The importance of mesograzers. *Ecology* 88:1548–1561.

Davidson, D. W. 1993. The effects of herbivory and granivory on terrestrial plant succession. *Oikos* 68:23–35.

Davis, M. B., and R. G. Shaw. 2001. Range shifts and adaptive responses to Quaternary climate change. *Science* 292:673–679.

Dayton, P. K. 1971. Competition, disturbance and community organization: The provision and utilization of space in a rocky intertidal community. *Ecological Monographs* 41:351–389.

Dayton, P. K. 1972. Towards an understanding of community resilience and the potential effects of enrichment to the benthos at McMurdo Sound, Antarctica. Pp. 81–96 in B. C.

Parker, ed. *Proceedings of the Colloquium on Conservation Problems in Antarctica.* Allen Press, Lawrence, KS.

Dayton, P. K. 1975. Experimental studies of algal-canopy interactions in a sea otter–dominated kelp community at Amchitka Island, Alaska. *Fishery Bulletin* 73:230–237.

Dayton, P. K., M. J. Tegner, P. B. Edwards, and K. L. Riser. 1998. Sliding baselines, ghosts, and reduced expectations in kelp forest communities. *Ecological Applications* 8:309–322.

De Boer, S. G. 1947. The damage to forest reproduction survey. *Wisconsin Conservation Bulletin* 12:1–23.

de Calesta, D. S. 1994. Effect of white-tailed deer on songbirds within managed forests in Pennsylvania. *Journal of Wildlife Management* 58:711–718.

de Mazancourt, C., and M. Loreau. 2000. Effect of herbivory and plant species replacement on primary production. *American Naturalist* 155:735–754.

de Roos, A. M., L. Persson, and H. R. Thieme. 2003. Emergent Allee effects in top predators feeding on structured prey populations. *Proceedings of the Royal Society B: Biological Sciences* 270:611–618.

de Steven, D. 1991. Experiments on mechanisms of tree establishment in old-field succession: Seedling survival and growth. *Ecology* 72:1076–1088.

den Herder, M. R. Virtanen, and H. Roininen. 2006. Reindeer herbivory reduces willow growth and grouse forage in a forest–tundra ecotone. *Basic and Applied Ecology* 9:324–331.

DeBruyn, A. M. H., K. S. McCann, and J. B. Rasmussen. 2004. Migration supports uneven consumer control in a sewage-enriched river food web. *Journal of Animal Ecology* 73:737–746.

del Giorgio, P. A., and C. M. Duarte. 2002. Respiration in the open ocean. *Nature* 420:379–384.

DeMartini, E. E., A. M. Friedlander, and S. R. Holzwarth. 2005. Size at sex change in protogynous labroids, prey body size distributions, and apex predator densities at NW Hawaiian atolls. *Marine Ecology Progress Series* 297:259–271.

DeMartini, E. E., A. M. Friedlander, S. A. Sandin, and E. Sala. 2008. Differences in the structure of shallow-reef fish assemblages between fished and unfished atolls in the northern Line Islands, central Pacific. *Marine Ecology Progress Series* 365:190–215.

DeMelo, R., R. France, and D. J. McQueen. 1992. Biomanipulation: Hit or myth? *Limnology and Oceanography* 37:192–207.

Dempster, J. P. 2008. The population dynamics of grasshoppers and locusts. *Biological Reviews* 38:490–529.

Denno, R. F., D. S. Gruner, and I. Kaplan. 2008. Potential for entomopathogenic nematodes in biological control: A meta-analytical synthesis and insights from trophic cascade theory. *Journal of Nematology* 40:61–72.

Dial, R., and J. Roughgarden. 1995. Experimental removal of insectivores from rain forest canopy: Direct and indirect effects. *Ecology* 76:1821–1834.

Diamond, J. M. 1988. Urban extinction of birds. *Nature* 333:393–394.

Diamond, J. M. 2005. *Collapse: How societies choose to fail or succeed*. Penguin, New York.

Dicke, M., and L. E. M. Vet. 1999. Plant–carnivore interactions: Evolutionary and ecological consequences for plant, herbivore and carnivore. Pp. 483–520 in H. Olff, V. K. Brown, and R. H. Drent, eds. *Herbivores: Between plants and predators.* Blackwell Science, Oxford.

Diehl, S., and M. Feissel. 2000. Effects of enrichment on three-level food chains with omnivory. *American Naturalist* 155:200–218.

Dinsdale, E. A., O. Pantos, S. Smriga, R. A. Edwards, F. Angly, L. Wegley, M. Hatay, et al. 2008. Microbial ecology of four coral atolls in the northern Line Islands. *PLoS One* 3:e1584.

Dirzo, R., and A. Miranda. 1991. Altered patterns of herbivory and diversity in the forest understory: A case study of the possible consequences of contemporary defaunation. Pp. 273–287 in P. W. Price, T. M. Lewinsohn, G. W. Fernandes, and W. W. Benson, eds. *Plant–animal interactions, evolutionary ecology in tropical and temperate regions.* Wiley, New York.

Doak, D. F., J. Z. Estes, B. S. Halpern, U. Jacob, D. R. Lindberg, J. Lovvorn, D. H. Monson, et al. 2008. Understanding and predicting ecological dynamics: Are major surprises inevitable? *Ecology* 89:952–961.

Doak, D. K. 1992. Lifetime impacts of herbivory for a perennial plant. *Ecology* 73:2086–2099.

Dobson, A. 1995. The ecology and epidemiology of rinderpest virus in Serengeti and Ngorongoro Conservation Area. Pp. 485–505 in A. R. E. Sinclair and P. Arcese, eds. *Serengeti II: Dynamics, management, and conservation of an ecosystem.* University of Chicago Press, Chicago.

Dominy, N. I., P. J. Grubb, R. V. Jackson, P. W. Lucas, D. J. Metcalfe, J.-C. Svenning, and I. M. Turner. 2008. In tropical lowland rain forests monocots have tougher leaves than dicots, and include a new kind of tough leaf. *Annals of Botany* 101:1–15.

Done, T. J. 1992. Phase shifts in coral reef communities and their ecological significance. *Hydrobiologia* 247:121–132.

Donlan, C. J., J. Berger, C. E. Bock, J. H. Bock, D. A. Burney, J. A. Estes, D. Foreman, et al. 2006. Pleistocene rewilding: An optimistic agenda for twenty-first century conservation. *American Naturalist* 168:660–681.

Doroff, A. M., J. A. Estes, M. T. Tinker, D. M. Burn, and T. J. Evans. 2003. Sea otter population declines in the Aleutian archipelago. *Journal of Mammalogy* 84:55–64.

Downing, J. A., C. Plante, and S. Lalonde. 1990. Fish production correlated with primary productivity, not the morphoedaphic index. *Canadian Journal of Fisheries and Aquatic Sciences* 47:1929–1936.

Downing, J. A., Y. T. Prairie, J. J. Cole, C. M. Duarte, L. J. Tranvik, R. G. Striegl, W. H. McDowell, et al. 2006. The global abundance and size distribution of lakes, ponds, and impoundments. *Limnology and Oceanography* 51:2388–2397.

Drenner, R. W., and K. D. Hambright. 2002. Piscivores, trophic cascades and lake management. *Scientific World* 2:284–307.

Dublin, H. T. 1986. *Decline of the Mara woodlands: The role of fire and elephants.* University of British Columbia, Vancouver.

Dublin, H. T. 1995. Vegetation dynamics in the Serengeti–Mara ecosystem: The role of elephants, fire, and other factors. Pp. 71–90 in A. R. E. Sinclair and P. Arecese, eds. *Serengeti*

II: Dynamics, management, and conservation of an ecosystem. University of Chicago Press, Chicago.

Dublin, H. T., A. R. E. Sinclair, and J. McGlade. 1990. Elephants and fire as causes of multiple stable states in the Serengeti–Mara woodlands. *Journal of Animal Ecology* 59:1147–1164.

Duffy, J. E. 2002. Biodiversity and ecosystem function: The consumer connection. *Oikos* 99:201–219.

Duffy, J. E. 2006. Biodiversity and the functioning of seagrass ecosystems. *Marine Ecology Progress Series* 311:233–250.

Duffy, J. E., J. P. Richardson, and K. E. France. 2005. Ecosystem consequences of diversity depend on food chain length in estuarine vegetation. *Ecology Letters* 8:301–309.

Duggins, D. O. 1980. Kelp beds and sea otters: An experimental approach. *Ecology* 61:447–453.

Duggins, D. O. 1983. Starfish predation and the creation of mosaic patterns in a kelp-dominated community. *Ecology* 64:1610–1619.

Duggins, D. O., C. A. Simenstad, and J. A. Estes. 1989. Magnification of secondary production by kelp detritus in coastal marine ecosystems. *Science* 245:170–173.

Dulvy, N. K., R. P. Freckleton, and N. V. C. Polunin. 2004a. Coral reef cascades and the indirect effects of predator removal by exploitation. *Ecology Letters* 7:410–416.

Dulvy, N. K., and N. V. C. Polunin. 2004. Using informal knowledge to infer human-induced rarity of a conspicuous reef fish. *Animal Conservation* 7:365–374.

Dulvy, N. K., N. V. C. Polunin, A. C. Mill, and N. A. J. Graham. 2004b. Size structural change in lightly exploited coral reef fish communities: Evidence for weak indirect effects. *Canadian Journal of Fisheries and Aquatic Sciences* 61:466–475.

Dunham, A. E. 2008. Above and below ground impacts of terrestrial mammals and birds in a tropical forest. *Oikos* 117:571–579.

Dunlap, T. R. 1988. *Saving America's wildlife.* Princeton University Press, Princeton, NJ.

Dupont, L. M., S. Jahns, F. Marret, and S. Ning. 2000. Vegetation change in equatorial West Africa: Time-slices for the last 150 ka. *Palaeogeography, Palaeoclimatology, Palaeoecology* 155:95–122.

Duran, L. R., and J. C. Castilla. 1989. Variation and persistence of the middle rocky intertidal community of central Chile, with and without human harvesting. *Marine Biology* 103:555–562.

Durant, S. M. 1998. Competition refuges and coexistence: An example from Serengeti carnivores. *Journal of Animal Ecology* 67:370–386.

Dwyer, G., J. Dushoff, and S. H. Yee. 2004. The combined effects of pathogens and predators on insect outbreaks. *Nature* 430:341–345.

Dyer, L. A., and D. Letourneau. 2003. Top down and bottom up diversity cascades in detrital vs. living food webs. *Ecology Letters* 6:60–68.

Dyer, L. A., M. S. Singer, J. T. Lill, J. O. Stireman III, G. L. Gentry, R. J. Marquis, R. E. Ricklefs, et al. 2007. Host specificity of Lepidoptera in tropical and temperate forests. *Nature* 448:696–700.

East, R. 1984. Rainfall, soil nutrient status and biomass of large African savanna mammals. *African Journal of Ecology* 22:245–270.

Eckhardt, H. C., B. W. van Wilgen, and H. C. Biggs. 2000. Trends in woody vegetation cover in the Kruger National Park, South Africa, between 1940 and 1998. *African Journal of Ecology* 38:108–115.

Edmeades, B. 2006. *Megafauna: First victims of the human-caused extinction.* Retrieved November 15, 2009 from Megafauna.com.

Ehrlen, J. 1995. Demography of the perennial herb *Lathyrus vernus*. II. Herbivory and population dynamics. *Journal of Ecology* 83:297–308.

Ehrlen, J. 1996. Spatiotemporal variation in predispersal seed predation intensity. *Oecologia* 108:708–713.

Ehrlich, P., and L. C. Birch. 1967. Evolutionary history and population biology. *Nature* 214:349–352.

Ehrlich, P. R., and P. H. Raven. 1964. Butterflies and plants: A study in coevolution. *Evolution* 18:586–608.

Ekerholm, P., L. Oksanen, and T. Oksanen. 2001. Long-term dynamics of voles and lemmings at the timberline and above the willow limit as a test of hypotheses on trophic interactions. *Ecography* 24:555–568.

Ekerholm, P., L. Oksanen, T. Oksanen, and M. Schneider. 2004. The impact of short term predator removal on vole dynamics in a subarctic-alpine habitat complex. *Oikos* 106:457–468.

Ellenberg, H. 1988. *Vegetation ecology of Central Europe*, 4th ed. Cambridge University Press, Cambridge.

Ellison, A. M., M. S. Bank, B. D. Clinton, E. A. Colburn, K. Elliott, C. R. Ford, D. R. Foster, et al. 2005. Loss of foundation species: Consequences for the structure and dynamics of forested ecosystems. *Frontiers in Ecology and the Environment* 3:479–486.

Elmhagen, B., and S. P. Rushton. 2007. Trophic control of mesopredators in terrestrial ecosystems: Top-down or bottom-up? *Ecology Letters* 10:197–206.

Elser, J. J., T. H. Chrzanowski, R. W. Sterner, and K. H. Mills. 1998. Stoichiometric constraints on food-web dynamics: A whole-lake experiment on the Canadian shield. *Ecosystems* 1:120–136.

Elser, J. J., M. M. Elser, N. A. MacKay, and S. R. Carpenter. 1988. Zooplankton mediated transitions between N and P limited algal growth. *Limnology and Oceanography* 33:1–14.

Elser, J. J., W. F. Fagan, R. F. Denno, D. R. Dobberfuhl, A. Folarin, A. Huberty, S. Interlandi, et al. 2000a. Nutritional constraints in terrestrial and freshwater food webs. *Nature* 408:578–580.

Elser, J. J., and C. R. Goldman. 1991. Zooplankton effects on phytoplankton in lakes of contrasting trophic status. *Limnology and Oceanography* 36:64–90.

Elser, J. J., and R. P. Hassett. 1994. A stoichiometric analysis of the zooplankton–phytoplankton interaction in marine and fresh-water ecosystems. *Nature* 370:211–213.

Elser, J. J., R. W. Sterner, A. E. Galford, T. H. Chrzanowski, D. L. Findlay, K. H. Mills, M. J. Paterson, M. P. Stainton, and D. W. Schindler. 2000b. Pelagic C:N:P stoichiometry in a eutrophied lake: Responses to a whole-lake food-web manipulation. *Ecosystems* 3:293–307.

Elton, C. 1927. *Animal ecology.* Sidgwick and Jackson, London.

Elton, C. 2001. *Animal ecology*, 2nd ed., reissued. University of Chicago Press, Chicago.

Eltringham, S. K. 1980. A quantitative assessment of range use by large African mammals with particular reference to the effect of elephants on trees. *African Journal of Ecology* 18:53–71.

Emmerson, M. C., and D. Raffaelli. 2004. Predator–prey size, interaction strength, and the stability of real food webs. *Journal of Animal Ecology* 73:399–409.

Emmons, L. H. 1987. Comparative feeding ecology of felids in a Neotropical rainforest. *Behavioral Ecology and Sociobiology* 20:271–283.

Endean, R., and A. Cameron. 1990. *Acanthaster planci* population outbreaks. Pp. 419–437 in Z. Dubinsky, ed. *Ecosystems of the world: Coral reefs*. Elsevier, Amsterdam.

Eskildsen, L. I., J. M. Olesen, and C. G. Jones. 2004. Feeding response of the Aldabra giant tortoise (*Geochelone gigantea*) to island plants showing heterophylly. *Journal of Biogeography* 31:1785–1790.

Essington, T. E. 2006. Pelagic ecosystem response to a century of commercial whaling and fishing. Pp. 38–49 in J. A. Estes, D. P. DeMaster, D. F. Doak, T. M. Williams and R. L. Brownwell, eds. *Whales, whaling and ocean ecosystems*. University of California Press, Berkeley.

Essington, T. E., A. H. Beaudreau, and J. Wiedenmann. 2006. Fishing through marine food webs. *Proceedings of the National Academy of Sciences* 103:3171–3175.

Essington, T. E., and S. Hansson. 2004. Predator-dependent functional response and interaction strengths in a natural food web. *Canadian Journal of Fisheries and Aquatic Sciences* 61:2227–2236.

Estes, J., K. Crooks, and R. D. Holt. 2001. Ecological role of predators. Pp. 857–878 in S. Levin, ed. *Encyclopedia of biodiversity*. Vol. 4, Academic Press, San Diego.

Estes, J. A. 2005. Carnivory and trophic connectivity in kelp forests. Pp. 61–81 in C. Ray, K. H. Redford, R. S. Steneck, and J. Berger, eds. *Large carnivores and the conservation of biodiversity*. Island Press, Washington, DC.

Estes, J. A., E. M. Danner, D. F. Doak, B. Konar, A. M. Springer, P. D. Steinberg, M. T. Tinker, and T. M. Williams. 2004. Complex trophic interactions in kelp forest ecosystems. *Bulletin of Marine Science* 74:621–638.

Estes, J. A., D. P. DeMaster, D. F. Doak, T. M. Williams, and R. L. Brownell Jr., eds. 2006. *Whales, whaling and ocean ecosystems*. University of California Press, Berkeley.

Estes, J. A., and D. O. Duggins. 1995. Sea otters and kelp forests in Alaska: Generality and variation in a community ecological paradigm. *Ecological Monographs* 65:75–100.

Estes, J. A., D. O. Duggins, and G. B. Rathbun. 1989. The ecology of extinctions in kelp forest communities. *Conservation Biology* 3:252–264.

Estes, J. A., D. R. Lindberg, and C. Wray. 2005. Evolution of large body size in abalones (*Haliotis*): Patterns and implications. *Paleobiology* 31:591–606.

Estes, J. A., and J. F. Palmisano. 1974. Sea otters: Their role in structuring nearshore communities. *Science* 185:1058–1060.

Estes, J. A., N. S. Smith, and J. F. Palmisano. 1978. Sea otter predation and community organization in the western Aleutian Island, Alaska. *Ecology* 59:822–833.

Estes, J. A., M. T. Tinker, T. M. Williams, and D. F. Doak. 1998. Killer whale predation on sea otters linking oceanic and nearshore ecosystems. *Science* 282:473–476.

Fa, J. E., C. A. Peres, and J. Meeuwig. 2002. Bushmeat exploitation in tropical forests: An intercontinental comparison. *Conservation Biology* 16:232–237.

Fagan, W. F., and J. G. Bishop. 2000. Trophic interactions during primary succession: Herbivores slow a plant reinvasion at Mount St. Helens. *American Naturalist* 155:238–251.

Falkowski, P., R. J. Scholes, E. Boyle, J. Canadell, D. Canfield, J. Elser, N. Gruber, et al. 2000. The global carbon cycle: A test of our knowledge of earth as a system. *Science* 290:291–296.

Fan, M., Y. Kuang, and Z. Feng. 2005. Cats protecting birds revisited. *Bulletin of Mathematical Biology* 67:1081–1106.

Feeley, K. 2005. The role of clumped defecation in the spatial distribution of soil nutrients and the availability of nutrients for plant uptake. *Journal of Tropical Ecology* 21:99–102.

Feeley, K. J., and J. W. Terborgh. 2005. The effects of herbivore density on soil nutrients and tree growth in tropical forest fragments. *Ecology* 86:116–124.

Feeley, K. J., and J. W. Terborgh. 2006. Habitat fragmentation and effects of herbivore (howler monkey) abundances on bird species richness. *Ecology* 87:144–150.

Feeny, P. 1970. Seasonal changes in oak leaf tannins and nutrients as a cause of spring feeding by winter moth caterpillars. *Ecology* 51:565–581.

Fenton, A., and S. A. Rands. 2006. The impact of parasite manipulation and predator foraging behavior on predator–prey communities. *Ecology* 87:2832–2841.

Fey, K., P. B. Banks, L. Oksane, and E. Korpimäki. 2009. Does removal of an alien predator from small islands in the Baltic Sea induce a trophic cascade? *Ecography* 32:546–552.

Fine, P. V. A., I. Mesones, and P. D. Coley. 2004. Herbivores promote habitat specialization by trees in Amazonian forests. *Science* 305:663–665.

Finke, D. L., and R. F. Denno. 2004. Predator diversity dampens trophic cascades. *Nature* 429:473–476.

Flannery, T. F. 1994. *The future eaters: An ecological history of the Australasian lands and people.* Griffin Press, Adelaide, Australia.

Flannery, T. F. 2002. *The eternal frontier: An ecological history of North America and its peoples.* Grove, New York.

Flecker, A. S. 1992. Ecosystem engineering by a dominant detritivore in a diverse tropical stream. *Ecology* 77:1845–1854.

Flueck, W. T. 2000. Population regulation in large northern herbivores: Evolution, thermodynamics, and large predators. *Zeitschrift fuer Jagdwissenschaft* 46:139–166.

Forbes, S. A. 1887. The lake as a microcosm. Reprinted in *Bulletin of the Illinois Natural History Survey* 15:537–550.

Forkner, R. E., R. J. Marquis, J. T. Lill, and J. Le Corff. 2008. Timing is everything? Phenological synchrony and population variability in leaf-chewing herbivores of *Quercus*. *Ecological Entomology* 33:276–285.

Fortin, D., H. L. Beyer, M. S. Boyce, D. W. Smith, T. Duchesne, and J. S. Mao. 2005. Wolves influence elk movements: Behavior shapes a trophic cascade in Yellowstone National Park. *Ecology* 86:1320–1330.

Fraenkel, G. S. 1959. The raison d'etre of secondary plant substances. *Science* 129:1466–1470.

Frank, D. A. 2008. Evidence for top predator control of a grazing system. *Oikos* 117:1718–1724.

Frank, K. T., B. Petrie, J. S. Choi, and W. C. Leggett. 2005. Trophic cascades in a formerly cod-dominated ecosystem. *Science* 308:1621–1623.

Frank, K. T., B. Petrie, and N. L. Shackell. 2007. The ups and downs of trophic control in continental shelf ecosystems. *Trends in Ecology and Evolution* 22:236–242.

Frank, K. T., B. Petrie, N. L. Shackell, and J. S. Choi. 2006. Reconciling differences in trophic control in mid-latitude marine ecosystems. *Ecology Letters* 9:1096–1105.

Freedman, H. I., and P. Waltman. 1977. Mathematical analysis of some three species food chain models. *Mathematical Biosciences* 33:257–276.

Fretwell, S. D. 1977. The regulation of plant communities by food chains exploiting them. *Perspectives of Biology and Medicine* 20:169–185.

Fretwell, S. D. 1987. Food chain dynamics: The central theory of ecology? *Oikos* 50:291–301.

Friedlander, A. M., and E. E. DeMartini. 2002. Contrasts in density, size, and biomass of reef fishes between the northwestern and the main Hawaiian islands: The effects of fishing down apex predators. *Marine Ecology Progress Series* 230:253–264.

Friedlander, A. M., and J. D. Parrish. 1998. Temporal dynamics of fish communities on an exposed shoreline in Hawaii. *Environmental Biology of Fishes* 53:1–18.

Fritz, H., and P. Duncan. 1993. Large herbivores in rangelands. *Nature* 364:292–293.

Fritz, H., P. Duncan, I. J. Gordon, and A. W. Illius. 2002. Megaherbivores influence trophic guilds structure in African ungulate communities. *Oecologia* 131:620–625.

Fritz, H., and A. Loison. 2006. Large herbivores across biomes. Pp. 19–49 in K. Danell, P. Duncan, R. Bergström, and J. Pastor, eds. *Large herbivore ecology, ecosystem dynamics and conservation*. Cambridge University Press, Cambridge.

Fryxell, J., A. Mosser, A. R. E Sinclair, and C. Packer. 2007. Group formation and predator–prey dynamics in Serengeti. *Nature* 449:1041–1044.

Fryxell, J., and A. R. E. Sinclair. 1988a. Causes and consequences of migration by large herbivores. *Trends in Ecology and Evolution* 3:237–241.

Fryxell, J. M., J. Greever, and A. R. E. Sinclair. 1988. Why are migratory ungulates so abundant? *American Naturalist* 131:781–798.

Fryxell, J. M., and A. R. E. Sinclair. 1988b. Seasonal migration by white-eared kob in relation to resources. *African Journal of Ecology* 26:17–31.

Fuhlendorf, S. D., and D. M. Engle. 2004. Application of the grazing–fire interaction to restore a shifting mosaic on tallgrass prairie. *Journal of Applied Ecology* 41:604–614.

Fukami, T., D. A. Wardle, P. J. Bellingham, C. P. H. Mulder, D. R. Towns, G. W. Yeates, K. I. Bonner, M. S. Durrett, M. N. Grant-Hoffman, and W. M. Williamson. 2006. Above- and below-ground impacts of introduced predators in seabird-dominated island ecosystems. *Ecology Letters* 9:1299–1307.

Fuller, R. J., and R. M. A. Gill. 2001. Ecological impacts of increasing numbers of deer in British woodland. *Forestry* 74:193–199.

Gange, A. C., V. K. Brown, and D. M. Aplin. 2003. Multitrophic links between arbuscular mycorrhizal fungi and insect parasitoids. *Ecology Letters* 6:1051–1055.

Gange, A. C., and H. M. West. 1994. Interactions between arbuscular mycorrhizal fungi and foliar-feeding insects in *Plantago lanceolata* L. *New Phytologist* 128:79–87.

Gasaway, W. C., R. D. Boertje, D. V. Grangaard, D. G. Kelleyhouse, R. O. Stephenson, and D. G. Larsen. 1992. The role of predation in limiting moose at low densities in Alaska and Yukon and implications for conservation. *Wildlife Monographs* 120:1–59.

Gasaway, W. C., R. O. Stephenson, J. L. Davis, P. E. K. Shepherd, and O. E. Burris. 1983. Interrelationships of wolves, prey, and man in interior Alaska. *Wildlife Monographs* 84:1–50.

Gaston, A. J., S. A. Stockton, and J. L. Smith. 2006. Species–area relationship and the impact of deer browse in the complex phytogeography of the Haida Gwaii archipelago (Queen Charlotte Islands), British Columbia. *Ecoscience* 13:511–522.

Gautier-Hion, A., J. M. Duplantier, R. Quris, F. Feer, C. Sourd, J. P. Decoux, L. Emmons, et al. 1985. Fruit characters as a basis of fruit choice and seed dispersal in a tropical forest vertebrate community. *Oecologia* 65:324–337.

Gehring, C. A., and T. G. Whitham. 1994. Interactions between aboveground herbivores and the mycorrhizal mutualists of plants. *Trends in Ecology and Evolution* 9:251–255.

Gehrt, S. D., and W. R. Clark. 2003. Raccoons, coyotes, and reflections on the mesopredator release hypothesis. *Wildlife Society Bulletin* 31:836–842.

Gehrt, S. D., and S. Prange. 2007. Interference competition between coyotes and raccoons: A test of the mesopredator release hypothesis. *Behavioral Ecology* 18:204–214.

Georgiadis, N. J., F. Ihwagi, J. G. N. Olwero, and S. S. Romanach. 2007a. Savanna herbivore dynamics in a livestock-dominated landscape. II: Ecological, conservation, and management implications of predator restoration. *Biological Conservation* 137:473–482.

Georgiadis, N. J., J. G. N. Olwero, G. Ojwang, and S. S. Romanach. 2007b. Savanna herbivore dynamics in a livestock-dominated landscape: I. Dependence on land use, rainfall, density, and time. *Biological Conservation* 137:460–472.

Gerlach, J. 2004. *Giant tortoises of the Indian Ocean. The genus* Dipsochelys *inhabiting the Seychelles Islands and the extinct giants of Madagascar and the Mascarenes.* Chimaira, Frankfurt am Main.

Gerlach, J., and K. L. Canning. 1993. On the crocodiles of the western Indian Ocean. *Phelsuma* 2:54–58.

Gese, E. M., and S. Grothe. 1995. Analysis of coyote predation on deer and elk during winter in Yellowstone National Park, Wyoming. *American Midland Naturalist* 133:36–43.

Gese, E. M., R. L. Ruff, and R. L. Crabtree. 1996. Foraging ecology of coyotes (*Canis latrans*): The influence of extrinsic factors and a dominance hierarchy. *Canadian Journal of Zoology* 74:769–783.

Gibson, C. W. D., T. C. Guilford, C. Hambler, and P. H. Sterling. 1983. Transition matrix models after release from grazing on Aldabra Atoll. *Vegetatio* 52:151–159.

Gibson, C. W. D., and J. Phillipson. 1983. The vegetation map of the Aldabra Atoll: Preliminary analysis and explanation of the vegetation map. *Philosophical Transactions of the Royal Society of London. Series B, Biological Sciences* 302:201–235.

Gilbert, L. E. 1980. Food web organization and conservation of Neotropical diversity. Pp.

11–33 in M. E. Soulé and B. A. Wilcox, eds. *Conservation biology*. Sinauer, Sunderland, MA.

Gill, R. M. A., and R. J. Fuller. 2007. The effects of deer browsing on woodland structure and songbirds in lowland Britain. *Ibis* 149:117–127.

Gillson, L. 2004. Evidence of hierarchical patch dynamics in an East African savanna? *Landscape Ecology* 19:883–894.

Gimingham, C. H. 1972. *Ecology of heathlands*. Chapman & Hall, London.

Glanz, W. E. 1990. Neotropical mammal densities: How unusual is the community on Barro Colorado Island, Panama? Pp. 287–313 in A. H. Gentry, ed. *Four Neotropical rainforests*. Yale University Press, New Haven, CT.

Godwin, H., and A. G. Tansley. 1941. Prehistoric charcoals as evidence of former vegetation, soil and climate. *Journal of Ecology* 29:117–126.

Goheen, J. R., F. Keesing, B. F. Allan, D. L. Ogada, and R. S. Ostfeld. 2004. Net effects of large mammals on *Acacia* seedling survival in an African savanna. *Ecology* 85:1555–1556.

Gomez, J. M., and A. Gonzalez-Megias. 2002. Asymmetrical interactions between ungulates and phytophagous insects: Being different matters. *Ecology* 83:201–211.

Goodrich, J. M., and S. W. Buskirk. 1995. Control of abundant native vertebrates for conservation of endangered species. *Conservation Biology* 9:1357–1364.

Gosselin, L. A., and P.-Y. Qian. 1997. Juvenile mortality in benthic marine invertebrates. *Marine Ecology Progress Series* 146:265–282.

Grabowski, J. H., and C. H. Peterson. 2007. Restoring oyster reefs to recover ecosystem services. Pp. 281–298 in K. Cuddington, J. E. Byers, W. G. Wilson, and A. Hastings, eds. *Ecosystem engineers, plants to protists*. Elsevier/Academic Press, Amsterdam.

Graham, N. A. J., R. D. Evans, and G. R. Russ. 2003. The effects of marine reserve protection on the trophic relationships of reef fishes on the Great Barrier Reef. *Environmental Conservation* 30:200–208.

Graham, R. W. Jr., E. L. Lundelius, M. A. Graham, E. K. Schroeder, R. S. Toomey III, E. Anderson, A. D. Barnosky, et al. 1996. Spatial response of mammals to Late Quaternary environmental fluctuations. *Science* 272:1601–1606.

Grange, S., P. Duncan, J. M. Gaillard, A. R. E. Sinclair, P. J. P. Gogan, C. Packer, H. Hofer, and M. East. 2004. What limits the Serengeti zebra population? *Oecologia* 140:523–532.

Green, E. P., and A. W. Bruckner. 2000. The significance of coral disease epizootiology for coral reef conservation. *Biological Conservation* 96:347–361.

Green, P. T., D. J. O'Dowd, and P. S. Lake. 1997. Control of seedling recruitment by land crabs in rain forest on a remote oceanic island. *Ecology* 78:2474–2486.

Greene, C. H., and A. J. Pershing. 2007. Oceans: Climate drives sea change. *Science* 315:1084–1085.

Griffin, P. C., S. C. Griffin, C. Waroquiers, and L. S. Mills. 2005. Mortality by moonlight: Predation risk and the snowshoe hare. *Behavioral Ecology* 16:938–944.

Grubb, P. 1971. The growth, ecology and population structure of giant tortoises on Aldabra. *Philosophical Transactions of the Royal Society of London. Series B, Biological Sciences* 260:327–372.

Grubb, P. J., R. V. Jackson, I. M. Barberis, J. N. Bee, D. A. Coomes, N. I. Dominy, M. A. S. de La Fuente, et al. 2008. Monocot leaves are eaten less than dicot leaves in tropical lowland rain forests: Correlations with toughness and leaf presentation. *Annals of Botany* 101:1379–1389.

Gruner, D. S. 2004. Attenuation of top-down and bottom-up forces in a complex terrestrial community. *Ecology* 85:3010–3022.

Gruner, D. S., J. E. Smith, E. W. Seabloom, S. A. Sandin, J. T. Ngai, H. Hillebrand, W. S. Harpole, et al. 2008. A cross-system synthesis of consumer and nutrient resource control on producer biomass. *Ecology Letters* 11:740–755.

Guthrie, R. D. 1990. *Frozen fauna of the mammoth steppe: The story of Blue Babe.* University of Chicago Press, Chicago.

Hacker, S. D., and M. D. Bertness. 1999. Experimental evidence for factors maintaining plant species diversity in a New England salt marsh. *Ecology* 80:2064–2073.

Haidt, J. 2006. *The happiness hypothesis: Finding modern truth in ancient wisdom.* Basic Books, New York.

Haines, B. 1983. Leaf-cutting ants bleed mineral elements out of rainforest in southern Venezuela. *Tropical Ecology* 24:85–93.

Hairston, N. G. Sr. 1991. The literature glut: Causes and consequences: Reflections of a dinosaur. *Bulletin of the Ecological Society of America* 72:171–174.

Hairston, N. G., F. E. Smith, and L. B. Slobodkin. 1960. Community structure, population control, and competition. *American Naturalist* 94:421–425.

Häkkinen, I., and M. Jokinen. 1974. Jäniksen talviekologiasta ulkosaaristossa. (In Finnish; English summary: On the winter ecology of the snow hare (*Lepus timidus*) in the outer archipelago.) *Suomen Riista* 25:5–14.

Halaj, J., and D. H. Wise. 2001. Terrestrial trophic cascades: How much do they trickle? *American Naturalist* 157:262–281.

Hall, M. O., M. J. Durako, J. W. Fourqurean, and J. C. Zieman. 1999. Decadal changes in seagrass distribution and abundance in Florida Bay. *Estuaries* 22:445–459.

Hall, S. J., and D. Raffaelli. 1991. Food-web patterns: Lessons from a species-rich web. *Journal of Animal Ecology* 60:823–842.

Hall, S. R., J. B. Shurin, S. Diehl, and R. M. Nisbet. 2007. Food quality, nutrient limitation of secondary production, and the strength of trophic cascades. *Oikos* 116:1128–1143.

Hallam, T. G. 1986. Community dynamics in a homogeneous environment. Pp. 61–94 in T. G. Hallam and S. A. Levin, eds. *Mathematical ecology: An introduction.* Springer-Verlag, Berlin.

Halofsky, J. S., and W. J. Ripple. 2008a. Fine-scale predation risk on elk after wolf-reintroduction in Yellowstone National Park, USA. *Oecologia* 155:869–877.

Halofsky, J. S., and W. J. Ripple. 2008b. Linkages between wolf presence and aspen recruitment in the Gallatin elk winter range of southwestern Montana, USA. *Forestry* 81:195–207.

Halofsky, J. S., W. J. Ripple, and R. L. Beschta. 2008. Recoupling fire and aspen recruitment after wolf reintroduction in Yellowstone National Park, USA. *Forest Ecology and Management* 256:1004–1008.

Halpern, B. S., S. Walbridge, K. A. Selkoe, C. V. Kappel, F. Micheli, C. D'Agrosa, J. F. Bruno, et al. 2008. A global map of human impact on marine ecosystems. *Science* 319:948–952.

Hamann, O. 2004. Vegetation changes over three decades on Santa Fe Island, Galapagos, Ecuador. *Nordic Journal of Botany* 23:143–152.

Hambäck, P. A. 1998. Seasonality, optimal foraging, and prey coexistence. *American Naturalist* 152:881–895.

Hambäck, P. A., and P. Ekerholm. 1997. Mechanisms of apparent competition in seasonal environments: An example with vole herbivory. *Oikos* 80:276–288.

Hambäck, P. A., L. Oksanen, P. Ekerholm, Å. Lindgren, T. Oksanen, and M. Schneider. 2004. Predators indirectly protect tundra plants by reducing herbivore abundance. *Oikos* 106:85–92.

Hamilton, W. D. 1971. Geometry for the selfish herd. *Journal of Theoretical Biology* 31:295–311.

Hamilton, W. D., and C. D. Busse. 1978. Primate carnivory and its significance to human diets. *BioScience* 28:761–766.

Hanby, J. P., J. D. Bygott, and C. Packer. 1995. Ecology, demography and behaviour of lions in two contrasting habitats: Ngorongoro Crater and Serengeti plains. Pp. 315–331 in A. R. E. Sinclair and P. Arcese, eds. *Serengeti II: Dynamics, management, and conservation of an ecosystem.* University of Chicago Press, Chicago.

Hanks, J. 1981. Characterization of population condition. Pp. 47–73 in C. W. Fowler and T. D. Smith, eds. *Dynamics of large mammal populations.* Wiley, New York.

Hansen, A. J., R. Knight, J. Marzluff, S. Powell, K. Brown, P. Hernandez, and K. Jones. 2005. Effects of exurban development on biodiversity: Patterns, mechanisms, research needs. *Ecological Applications* 15:1893–1905.

Hansen, B. B., S. Henriksen, R. Aanes, and B.-E. Saether. 2007. Ungulate impact on vegetation in a two trophic level system. *Polar Biology* 30:549–558.

Hanski, I., L. Hansson, and H. Henttonen. 1991. Specialist predators, generalist predators, and the microtine rodent cycle. *Journal of Animal Ecology* 60:353–367.

Hanski, I., J. Poyry, T. Pakkala, and M. Kuussaari. 1995. Multiple equilibria in metapopulation dynamics. *Nature* 377:618–621.

Hansson, L. 1985a. *Clethrionomys* food: Generic, specific and regional characteristics. *Annales Zoologici Fennici* 22:315–318.

Hansson, L. 1985b. Damage by wildlife, especially small rodents, to North American *Pinus contorta* provenances introduced into Sweden. *Canadian Journal of Forest Research* 15:1167–1671.

Hansson, L. A., H. Annadotter, E. Bergman, S. F. Hamrin, E. Jeppesen, T. Kairesalo, E. Luokkanen, P. A. Nilsson, M. Sondergaard, and J. Strand. 1998. Biomanipulation as an application of food-chain theory: Constraints, synthesis and recommendations for temperate lakes. *Ecosystems* 1:558–574.

Harrington, L. M., K. Fabricius, G. De'ath, and A. Negri. 2004. Habitat selection of settlement substrata determines post-settlement survival in corals. *Ecology* 85:3428–3437.

Harrington, R. N., N. Owen-Smith, P. Viljoen, H. Biggs, and D. Mason. 1999. Establishing the causes of the roan antelope decline in the Kruger National Park, South Africa. *Biological Conservation* 90:69–78.

Harrold, C., and D. C. Reed. 1985. Food availability, sea urchin grazing, and kelp forest community structure. *Ecology* 66:1160–1169.

Hartley, S. E., and C. G. Jones. 1997. Plant chemistry and herbivory, or why the world is green. Pp. 284–324 in M. J. Crawley, ed. *Plant ecology*. Blackwell, Oxford.

Hartvigsen, G., and S. J. McNaughton. 1995. Tradeoff between height and relative growth rate in a dominant grass from the Serengeti ecosystem. *Oecologia* 102:273–276.

Harvell, C. D., E. Jordan-Dahlgren, S. Merkel, E. Roscnberg, L. Raymundo, G. Smith, E. Weil, and B. Willis. 2007. Coral disease, environmental drivers and the balance between coral and microbial associates. *Oceanography* 20:58–81.

Hastings, A., and T. Powell. 1991. Chaos in a three-species food chain. *Ecology* 72:896–903.

Hatcher, B. G., and A. W. D. Larkum. 1983. An experimental analysis of factors controlling the standing crop of the epilithic algal community on a coral reef. *Journal of Experimental Marine Biology and Ecology* 69:61–84.

Hawkins, S. J., and R. G. Hartnoll. 1983. Grazing of intertidal algae by marine invertebrates. *Oceanography and Marine Biology* 21:195–282.

Haxeltine, A., and I. C. Prentice. 1996. BIOME3: An equilibrium terrestrial biosphere model based on ecophysiological constraints, resource availability, and competition among plant functional types. *Global Biogeochemical Cycles* 10:693–709.

Hay, M. E., J. E. Duffy, V. J. Paul, P. E. Renaud, and W. Fenical. 1990. Specialist herbivores reduce their susceptibility to predation by feeding on the chemically defended seaweed *Avrainvillea longicaulis*. *Limnology and Oceanography* 35:1734–1743.

He, X., and J. F. Kitchell. 1990. Direct and indirect effects of predation on a fish community: A whole lake experiment. *Transaction of the American Fisheries Society* 119:825–835.

He, X., R. A. Wright, and J. F. Kitchell. 1993. Fish behavioral and community responses to manipulation. Pp. 69–84 in S. R. Carpenter and J. F. Kitchell, eds. *The trophic cascade in lakes*. Cambridge University Press, Cambridge.

Hebblewhite, M., and E. H. Merrill. 2007. Multiscale wolf predation risk for elk: Does migration reduce risk? *Oecologia* 152:377–387.

Hebblewhite, M., and D. H. Pletscher. 2002. Effects of elk group size on predation by wolves. *Canadian Journal of Zoology* 80:800–809.

Hebblewhite, M., C. A. White, C. G. Nietvelt, J. A. McKenzie, T. E. Hurd, J. M. Fryxell, S. E. Bayley, and P. C. Paquet. 2005. Human activity mediates a trophic cascade caused by wolves. *Ecology* 86:2135–2144.

Heck, K. L. Jr., G. Hays, and R. J. Orth. 2003. Critical evaluation of the nursery role hypothesis for seagrass meadows. *Marine Ecology Progress Series* 253:123–136.

Heck, K. L. Jr., J. R. Pennock, J. F. Valentine, L. D. Coen, and S. A. Sklenar. 2000. Effects of nutrient enrichment and small predator density on seagrass ecosystems: An experimental assessment. *Limnology and Oceanography* 45:1041–1057.

Heck, K. L. Jr., and J. F. Valentine. 2006. Plant–herbivore interactions in seagrass meadows. *Journal of Experimental Marine Biology and Ecology* 330:420–436.

Heck, K. L. Jr., and J. F. Valentine. 2007. The primacy of top-down effects in shallow benthic ecosystems. *Estuaries and Coasts* 30:371–381.

Hedges, L. V., J. Gurevich, and P. S. Curtis. 1999. The meta-analysis of response ratios in experimental ecology. *Ecology* 80:1150–1156.

Hedlund, K., and M. S. Öhrn. 2000. Tritrophic interactions in a soil community enhance decomposition rates. *Oikos* 88:585–591.

Heinrich, B. 1991. *Ravens in winter.* Vintage, New York.

Heithaus, M. R., A. Frid, A. J. Wirsing, L. M. Dill, J. W. Fourqurean, D. Burkholder, J. Thomson, and L. Bejder. 2007. State-dependent risk-taking by green sea turtles mediates top-down effects of tiger shark intimidation in a marine ecosystem. *Journal of Animal Ecology* 76:837–844.

Heithaus, M. R., A. Frid, A. J. Wirsing, and B. Worm. 2008. Ecological consequences of declines in apex predators in the sea. *Trends in Ecology and Evolution* 23:202–210.

Helldin, J. O., O. Liberg, and G. Gloersen. 2006. Lynx (*Lynx lynx*) killing red foxes (*Vulpes vulpes*) in boreal Sweden: Frequency and population effects. *Journal of Zoology* 270:657–663.

Hendrix, S. D. 1988. Herbivory and its impact on plant reproduction. Pp. 246–263 in J. L. Doust and L. L. Doust, eds. *Plant reproductive ecology: Patterns and strategies.* Oxford Press, Oxford.

Hendrix, S. D., V. K. Brown, and H. Dingle. 1988a. Arthropod guild structure during early old field succession in a new and old world site. *Journal of Animal Ecology* 57:1053–1066.

Hendrix, S. D., V. K. Brown, and A. C. Gange. 1988b. Effects of insect herbivory on early plant succession: Comparison of an English UK site and an American site. *Biological Journal of the Linnean Society* 35:206–216.

Henke, S. E., and F. C. Bryant. 1999. Effects of coyote removal on the faunal community in western Texas. *Journal of Wildlife Management* 63:1066–1081.

Henneman, M. L., and J. Memmott. 2001. Infiltration of a Hawaiian community by introduced biological control agents. *Science* 293:1314–1316.

Hill, C. M. 2000. Conflict of interest between people and baboons: Crop raiding in Uganda. *International Journal of Primatology* 21:299–315.

Hillebrand, H., and J. B. Shurin. 2005. Biodiversity and aquatic food webs. Pp. 184–197 in A. Belgrano, U. M. Scharler, J. Dunne, and R. E. Ulanowicz, eds. *Aquatic food webs: An ecosystem approach.* Oxford University Press, Oxford.

Hilty, J. W., W. Lidicker Jr., and A. Merenlender. 2006. *Corridor ecology: The science and practice of linking landscapes for biodiversity conservation.* Island Press, Washington, DC.

Hinke, J. T., I. C. Kaplan, K. Aydin, G. M. Watters, R. J. Olson, and J. F. Kitchell. 2004. Visualizing the food-web effects of fishing for tunas in the Pacific Ocean. *Ecology and Society* 9:10.

Hixon, M. A. 1991. Predation as a process structuring coral reef fish communities. Pp. 475–508 in P. F. Sale, ed. *The ecology of fishes on coral reefs.* Academic Press, San Diego, CA.

Hixon, M. A., and J. P. Beets. 1993. Predation, prey refuges, and the structure of coral-reef fish assemblages. *Ecological Monographs* 63:77–101.

Hixon, M. A., and M. H. Carr. 1997. Synergistic predation, density dependence, and population regulation in marine fish. *Science* 277:946–949.

Hixon, M. A., S. W. Pacala, and S. A. Sandin. 2002. Population regulation: Historical context and contemporary challenges of open vs. closed systems. *Ecology* 83:1490–1508.

Hnatiuk, S. 1978. Plant dispersal by the Aldabran giant tortoise, *Geochelone gigantea* (Schweigger). *Oecologia* 36:345–350.

Hnatiuk, S., S. R. J. Woodell, and D. M. Bourn. 1976. Giant tortoise and vegetation interaction on Aldabra atoll. II. Coastal. *Biological Conservation* 9:305–316.

Hobbs, N. T. 1996. Modification of ecosystems by ungulates. *Journal of Wildlife Management* 60:695–713.

Hochberg, M. E., M. P. Hassell, and R. M. May. 1990. The dynamics of host–parasitoid–pathogen interactions. *American Naturalist* 135:74–94.

Hodgson, J. R., and E. M. Hansen. 2005. Terrestrial prey items in the diet of largemouth bass, *Micropterus salmoides*, in a small, north temperate lake. *Journal of Freshwater Ecology* 20:793–794.

Hodgson, J. R., and J. F. Kitchell. 1987. Opportunistic foraging by largemouth bass (*Micropterus salmoides*). *American Midland Naturalist* 118:323–336.

Holdo, R. M., R. D. Holt, and J. M. Fryxell. 2009a. Grazers, browsers, and fire influence the extent and spatial pattern of tree cover in the Serengeti. *Ecological Applications* 19:95–109.

Holdo, R. M., R. D. Holt, K. A. Galvin, S. Polasky, and E. Knapp. In press. Responses to alternative rainfall regimes and antipoaching enforcement in a migratory system. *Ecological Applications*.

Holdo, R. M., A. R. E. Sinclair, K. L. Metzger, B. M. Bolker, A. P. Dobson, M. E. Ritchie, and R. D. Holt. 2009b. A disease-mediated trophic cascade in the Serengeti and its implications for ecosystem C. *PLoS Biology* 7(9):1–12, e1000210.

Holdridge, L. 1967. *Life zone ecology*. Tropical Science Center, San José, Costa Rica.

Hollenbeck, J. P., and W. J. Ripple. 2008. Aspen snag dynamics, cavity-nesting birds, and trophic cascades in Yellowstone's northern range. *Forest Ecology and Management* 255:1095–1103.

Holling, C. S. 1973. Resilience and stability of ecological systems. *Annual Review of Ecology and Systematics* 4:1–23.

Holmgren, M., and M. Scheffer. 2001. El Niño as a window of opportunity for the restoration of degraded arid ecosystems. *Ecosystems* 4:151–159.

Holt, R. D. 1977. Predation, apparent competition, and the structure of prey communities. *Theoretical Population Biology* 12.197–229.

Holt, R. D. 1984. Spatial heterogeneity, indirect interactions, and the coexistence of prey species. *American Naturalist* 124:377–406.

Holt, R. D. 1996. Food webs in space. Pp. 313–323 in G. A. Polis and K. O. Winemiller, eds. *Food webs: Integration of patterns and dynamics*. Chapman and Hall, New York.

Holt, R. D. 1997. Community modules. Pp. 333–349 in A. C. Gange and V. K. Brown, eds. *Multitrophic interactions in terrestrial ecosystems, 36th symposium of the British Ecological Society*. Blackwell Science, Oxford.

Holt, R. D. 2000. Trophic cascades in terrestrial systems: Reflections on Polis et al. *Trends in Ecology and Evolution* 15:444–445.

Holt, R. D. 2002. Food webs in space: On the interplay of dynamic instability and spatial processes. *Ecological Research* 17:261–273.

Holt, R. D. 2008a. The community context of disease emergence: Could changes in predation be a key driver? In R. S. Ostfeld, F. Keesing, and V. T. Eviner, eds. *Infectious disease ecology: Effects of ecosystems on disease and of disease on ecosystems.* Princeton University Press, Princeton, NJ.

Holt, R. D. 2008b. Theoretical perspectives on resource pulses. *Ecology* 89:671–681.

Holt, R. D. 2009. Toward a trophic island biogeography: Reflections on the interface of island biogeography and food web ecology. Pp. 143–185 in J. Losos and R. Ricklefs, eds. *Island biogeography at 40.* Princeton University Press, Princeton, NJ.

Holt, R. D., M. Barfield, R. Holdo, and S. Bhotika. In preparation. *Vegetation structure and predation risk: Implications for food chain dynamics.*

Holt, R. D., and G. R. Huxel. 2007. Alternative prey and the dynamics of intraguild predation: Theoretical perspectives. *Ecology* 88:2706–2712.

Holt, R. D., and G. A. Polis. 1997. A theoretical framework for intraguild predation. *American Naturalist* 149:745–764.

Holt, R. D., and M. Roy. 2007. Predation can increase the prevalence of infectious disease. *American Naturalist* 169:690–699.

Holyoak, M., M. A. Leibold, and R. D. Holt, eds. 2005. *Metacommunities: Spatial dynamics and ecological communities.* University of Chicago Press, Chicago.

Homer-Dixon, T., and D. Keith. 2008. Blocking the sky to save the earth. *New York Times*, September 20, p. A19.

Hooijer, D. A. 1974. Giant land tortoise, *Geochelone-Atlas* (Falconer + Cautley), from Pleistocene of Timor. *Proceedings of the Koninklije Nederlandse Akademie van Wetenschappen. Series B: Physical Sciences.* 74: 504–525.

Hopcraft, J. G. C., A. R. E Sinclair, and C. Packer. 2005. Prey accessibility outweighs prey abundance for the location of hunts in Serengeti lions. *Journal of Animal Ecology* 74:559–566.

Horn, H. S. 1971. *The adaptive geometry of trees.* Princeton University Press, Princeton, NJ.

Hörnberg, S. 2001. Changes in population density of moose (*Alces alces*) and damage to forests in Sweden. *Forest Ecology and Management* 149:141–151.

Horsley, S. B., S. L. Stout, and D. S. DeCalesta. 2003. White-tailed deer impact on the vegetation dynamics of a northern hardwood forest. *Ecological Applications* 13:98–118.

Hosper, S. H. 1998. Stable states, buffers and switches: An ecosystem approach to the restoration and management of shallow lakes in the Netherlands. *Water Science and Technology* 37:151–164.

Hough, A. F. 1965. A twenty-year record of understory vegetational change in a virgin Pennsylvania forest. *Ecology* 46:370–373.

Howard, J. J., J. Cazin, and D. F. Wiemer. 1988. Toxicity of terpenoid deterrents to the leaf-cutting ant *Atta cephalotes* and its mutualistic fungus. *Journal of Chemical Ecology* 14:59–69.

Howe, H. F., and M. N. Miriti. 2000. No question: Seed dispersal matters. *Trends in Ecology and Evolution* 15:434–436.

Howe, H. F., B. Zorn-Arnold, A. Sullivan, and J. S. Brown. 2006. Massive and distinctive effects of meadow vole on grassland ecosystems. *Ecology* 87:3007–3013.

Hrbáček, J., M. Dvorakova, V. Korinek, and L. Prochazkova. 1961. Demonstration of the effects of the fish stock on the species composition of zooplankton and the intensity of metabolism of the whole plankton assemblage. *Verhandlungen: Internationale Vereinigung für Theoretische und Angewandte Limnologie* 14:192–195.

Hudson, P. J., A. P. Dobson, and D. Newborn. 1992. Do parasites make prey vulnerable to predation? Red grouse and parasites. *Journal of Animal Ecology* 61:681–692.

Hughes, A. R., K. J. Bando, L. F. Rodrigues, and S. L. Williams. 2004. Relative effects of grazers and nutrients on seagrasses: A meta-analysis approach. *Marine Ecology Progress Series* 282:87–89.

Hughes, T. P. 1994. Catastrophes, phase shifts, and large-scale degradation of a Caribbean coral reef. *Science* 265:1547–1551.

Hughes, T. P., A. H. Baird, D. R. Bellwood, M. Card, S. R. Connolly, C. Folke, R. Grosberg, et al. 2003. Climate change, human impacts, and the resilience of coral reefs. *Science* 301:929–933.

Hughes, T. P., D. C. Reed, and M. J. Boyle. 1987. Herbivory on coral reefs: Community structure following mass mortalities of sea urchins. *Journal of Experimental Marine Biology and Ecology* 113:39–59.

Hughes, T. P., M. J. Rodrigues, D. R. Bellwood, D. Ceccarelli, O. Hoegh-Guldberg, L. McCook, N. Moltschaniwskyj, M. S. Pratchett, R. S. Steneck, and B. Willis. 2007. Phase shifts, herbivory, and the resilience of coral reefs to climate change. *Current Biology* 17:360–365.

Hulme, P. E. 1994. Seedling herbivory in grassland: Relative impact of vertebrate and invertebrate herbivores. *Journal of Ecology* 82:873–880.

Hulme, P. E. 1996. Herbivores and the performance of grassland plants: A comparison of arthropod mollusc, and rodent herbivory. *Journal of Ecology* 84:43–51.

Hunt, G. L., P. Stabeno, G. Walters, E. Sinclair, R. D. Brodeur, J. M. Napp, and N. A. Bond. 2002. Climate change and control of the southeastern Bering Sea pelagic ecosystem. *Deep-Sea Research Part II: Topical Studies in Oceanography* 49:5821–5853.

Hunt, H. L., and R. E. Scheibling. 1997. Role of early post-settlement mortality in recruitment of benthic marine invertebrates. *Marine Ecology Progress Series* 155:269–301.

Hunte, W., and D. Younglao. 1988. Recruitment and population recovery of *Diadema antillarum* (Echinodermata; Echinoidea) in Barbados. *Marine Ecology Progress Series* 45:109–119.

Hunter, A. F. 1995. Ecology, life history, and phylogeny of outbreak and nonoutbreak species. Pp. 41–64 in N. Cappuccino and P. W. Price, eds. *Population dynamics: New approaches and synthesis.* Academic Press, San Diego, CA.

Hunter, A. F. 2000. Gregariousness and repellent defenses in the survival of phytophagous insects. *Oikos* 91:213–224.

Hunter, J., and T. Caro. 2008. Interspecific competition and predation in American carnivore families. *Ethology, Ecology & Evolution* 20:295–324.

Hunter, M. D., and R. E. Forkner. 1999. Hurricane damage influences foliar polyphenolics and subsequent herbivory on surviving trees. *Ecology* 80:2676–2682.

Hurlbert, S. H. 1997. Functional importance of keystoneness: Reformulating some questions in theoretical ecology. *Australian Journal of Ecology* 22:369–382.

Husheer, S. W., D. A. Coomes, and A. W. Robertson. 2003. Long-term influences of introduced deer on the composition of New Zealand *Nothofagus* forests. *Forest Ecology and Management* 181:99–117.

Huxel, G. R., and K. McCann. 1998. Food web stability: The influence of trophic flows across habitats. *American Naturalist* 152:460–469.

Ickes, K., S. J. Dewalt, and S. Appanah. 2001. Effects of native pigs (*Sus scrofa*) on woody understorey vegetation in a Malaysian lowland rain forest. *Journal of Tropical Ecology* 17:191–206.

Inamdar, A. 1996. *The ecological consequences of elephant depletion.* Ph.D. thesis, Cambridge University, Cambridge, UK.

Irons, D. B., R. G. Anthony, and J. A. Estes. 1986. Foraging strategies of glaucous-winged gulls in rocky intertidal communities. *Ecology* 67:1460–1474.

Itô, H., and T. Hino. 2007. Dwarf bamboo as an ecological filter for forest regeneration. *Ecological Research* 22:706–711.

Ives, A. R., and S. R. Carpenter. 2007. Stability and diversity in ecosystems. *Science* 317:58–62.

Jackson, J. B. C. 1997. Reefs since Columbus. *Coral Reefs* 16:S23–S32.

Jackson, J. B. C. 2001. What was natural in the coastal oceans? *Proceedings of the National Academy of Sciences* 98:5411–5418.

Jackson, J. B. C. 2008. Ecological extinction and evolution in the brave new ocean. *Proceedings of the National Academy of Sciences* 105:11458–11465.

Jackson, J. B. C., M. X. Kirby, W. H. Berger, K. A. Bjorndal, L. W. Botsford, B. J. Borque, R. H. Bradbury, et al. 2001. Historical overfishing and the recent collapse of coastal ecosystems. *Science* 293:629–638.

Jameson, R. J., K. W. Kenyon, A. M. Johnson, and H. M. Wight. 1982. History and status of translocated sea otter populations in North America. *Wildlife Society Bulletin* 10:100–107.

Jansen, R., R. M. Little, and T. M. Crowe. 1999. Implications of grazing and burning of grasslands on the sustainable use of francolins (*Francolinus* spp.) and on overall bird conservation in the highlands of Mpumalanga province, South Africa. *Biodiversity and Conservation* 8:587–602.

Jansson, B., and H. Velner. 1995. The Baltic: The sea of surprises. Pp. 292–372 in L. H. Gunderson, C. S. Holling, and S. S. Light, eds. *Barriers and bridges to the renewal of ecosystems and institutions.* Columbia University Press, New York.

Jansson, M., L. Persson, A. M. de Roos, R. I. Jones, and L. J. Tranvik. 2007. Terrestrial carbon and intraspecific size-variation shape lake ecosystems. *Trends in Ecology and Evolution* 22:316–322.

Janzen, D. H. 1970. Herbivores and the number of tree species in tropical forests. *American Naturalist* 104:501–528.

Janzen, D. H. 1983. Insects. Pp. 619–645 in D. H. Janzen, ed. *Costa Rican natural history.* University of Chicago Press, Chicago.

Janzen, D. H., and P. S. Martin. 1982. Neotropical anachronisms: The fruits the gomphotheres ate. *Science* 215:19–27.

Jennings, S., E. M. Grandcourt, and N. V. C. Polunin. 1995. The effects of fishing on the diversity, biomass and trophic structure of Seychelles' reef fish communities. *Coral Reefs* 14:225–235.

Jennings, S., and N. V. C. Polunin. 1997. Impacts of predator depletion by fishing on the biomass and diversity of non-target reef fish communities. *Coral Reefs* 16:71–82.

Jensen, O. P., T. R. Hrabik, S. J. D. Martell, C. J. Walters, and J. F. Kitchell. 2006. Diel migration in the Lake Superior pelagic community: Modeling trade-offs at an intermediate trophic level. *Canadian Journal of Fisheries and Aquatic Sciences* 63:2296–2307.

Jeppesen, E., J. P. Jensen, M. Sondergaard, T. Lauridsen, L. J. Pedersen, and L. Jensen. 1997. Top-down control in freshwater lakes: The role of nutrient state, submerged macrophytes and water depth. *Hydrobiologia* 342:151–164.

Jeppesen, E., M. Søndergaard, and K. Christofferson. 1998. *The structuring role of submerged macrophytes in lakes*. Springer-Verlag, Berlin.

Jisaka, M., H. Ohigashi, K. Takegawa, M. Hirota, R. Irie, M. A. Huffman, and K. Koshimizu. 1993. Steroid glycosides from *Vernonia amygdalina*, a possible chimpanzee medicinal plant. *Phytochemistry* 34:409–413.

Johnson, C. N., J. L. Isaac, and D. O. Fisher. 2007. Rarity of a top predator triggers continent-wide collapse of mammal prey: Dingoes and marsupials in Australia. *Proceedings of the Royal Society B: Biological Sciences* 274:341–346.

Jones, C. G., J. H. Lawton, and M. Shachak. 1994. Organisms as ecosystem engineers. *Oikos* 69:373–386.

Jones, C. G., J. H. Lawton, and M. Shachak. 1997. Positive and negative effects of organisms as physical ecosystem engineers. *Ecology* 78:1946–1957.

Jones, M. T., I. Castellanos, and M. R. Weiss. 2002. Do leaf shelters always protect caterpillars from invertebrate predators? *Ecological Entomology* 27:753–757.

Kajak, A., K. Chmielewski, M. Kaczmarek, and E. Rembialkowska. 1993. Experimental studies on the effect of epigeic predators on matter decomposition process in managed peat grassland. *Polish Ecological Studies* 17:289–310.

Kalela, O. 1957. Regulation of reproduction rate in subarctic populations of the vole *Clethrionomys rufocanus* (Sund.). *Annales Academiae Scientarum Fenniae, Serial A* 55:1–72.

Karanth, K. U., and J. D. Nichols. 1998. Estimation of tiger densities in India using photographic captures and recaptures. *Ecology* 79:2852–2862.

Karanth, K. U., J. D. Nichols, N. S. Kumar, W. A. Link, and J. E. Hines. 2004. Tigers and their prey: Predicting carnivore densities from prey abundance. *Proceedings of the National Academy of Sciences* 101:4854–4858.

Karanth, K. U., and M. E. Sunquist. 1995. Prey selection by tiger, leopard and dhole in tropical forests. *Journal of Animal Ecology* 64:439–450.

Karban, R., and I. T. Baldwin. 1997. *Induced responses to herbivory*. University of Chicago Press, Chicago.

Kauffman, S. A. 2008. *Reinventing the sacred: A new view of science, reason, and religion*. Basic Books, New York.

Kaunzinger, C. M. K., and P. J. Morin. 1998. Productivity controls food-chain properties in microbial communities. *Nature* 395:495–497.

Kay, C. E. 1990. *Yellowstone's northern elk herd: A critical evaluation of the "natural regulation" paradigm.* Dissertation, Utah State University, Logan.

Kay, C. E. 2001. Long-term aspen exclosures in Yellowstone ecosystem. Pp. 225–240 in W. D. Shepperd, D. Binkley, D. L. Bartos, and T. J. Stohlgren, eds. *Sustaining Aspen in western landscapes: Symposium proceedings, June 13–15, 2000, Grand Junction, Colorado,* RMRS-P-18. U.S. Forest Service, Fort Collins, CO.

Keane, R. M., and M. J. Crawley. 2002. Exotic plant invasions and the enemy release hypothesis. *Trends in Ecology and Evolution* 17:164–170.

Kearns, C. A., and D. W. Inouye. 1997. Pollinators, flowering plants, and conservation biology. *BioScience* 47:297–307.

Keeling, H. C., and O. L. Phillips. 2007. The global relationship between forest productivity and biomass. *Global Ecology and Biogeography* 16:618–631.

Keller, M., D. S. Schimel, W. W. Hargrove, and F. M. Hoffman. 2008. A continental strategy for the national ecological observatory network. *Frontiers in Ecology and the Environment* 6:282–284.

Kelly, C. A., and R. J. Dyer. 2002. Demographic consequences of inflorescence-feeding insects for *Lyatris cylindracea*, an iteroparous perennial. *Oecologia* 132:350–360.

Kelly, D., and V. Sork. 2002. Mast seeding in perennial plants: Why, how, where? *Annual Review of Ecology and Systematics* 33:427–447.

Kenyon, K. W. 1969. The sea otter in the eastern Pacific Ocean. *North American Fauna* 68:1–352.

Kenyon, K. W. 1977. Caribbean monk seal extinct. *Journal of Mammalogy* 58:97–98.

Kie, J. G., R. T. Bowyer, and K. M. Stewart. 2003. Ungulates in western forests: Habitat requirements, population dynamics, and ecosystem processes. Pp. 296–340 in C. J. Zabel and R. G. Anthony, eds. *Mammal community dynamics: Management and conservation in the coniferous forests of western North America.* Cambridge University Press, Cambridge.

Kinley, T. A., and C. D. Apps. 2001. Mortality patterns in a subpopulation of endangered mountain caribou. *Wildlife Society Bulletin* 29:158–164.

Kirby, K. J. 2001. The impact of deer on the ground flora of British broadleaved woodland. *Forestry* 74:219–229.

Klemola, T., M. Koivula, E. Korpimäki, and K. Norrdahl. 2000. Experimental tests of predation and food hypotheses for population cycles of voles. *Proceedings of the Royal Society B: Biological Sciences* 267:351–356.

Kline, D. I., N. M. Kuntz, M. Breitbart, N. Knowlton, and F. Rohwer. 2006. Role of elevated organic carbon levels and microbial activity in coral mortality. *Marine Ecology Progress Series* 314:119–125.

Klironomos, J. N., and W. B. Kendrick. 1995. Stimulative effects of arthropods on endomycorrhizas of sugar maple in the presence of decaying litter. *Functional Ecology* 9:528–536.

Knight, T. M. 2004. The effect of herbivory and pollen limitation on a declining population of *Trillium grandiflorum*. *Ecological Applications* 14:915–928.

Knight, T. M., J. M. Chase, H. Hillebrand, and R. D. Holt. 2006. Predation on mutualists can reduce the strength of trophic cascades. *Ecology Letters* 9:1173–1178.

Knight, T. M., M. W. McCoy, J. M. Chase, K. A. McCoy, and R. D. Holt. 2005. Trophic cascades across ecosystems. *Nature* 437:880–884.

Knowlton, N. 1992. Thresholds and multiple stable states in coral reef community dynamics. *American Zoologist* 32:674–682.

Knowlton, N. 2004. Multiple "stable" states and the conservation of marine ecosystems. *Progress in Oceanography* 60:387–396.

Knowlton, N., and J. B. C. Jackson. 2008. Shifting baselines, local impacts, and global change on coral reefs. *PLoS Biology* 6:e54.

Koch, P. L., and A. D. Barnosky. 2006. Late Quaternary extinctions: State of the debate. *Annual Review of Ecology, Evolution and Systematics* 37:215–250.

Konar, B. 2000. Seasonal inhibitory effects of marine plants on sea urchins: Structuring communities the algal way. *Oecologia* 125:208–217.

Kortello, A. D., T. E. Hurd, and D. L. Murray. 2007. Interactions between cougars (*Puma concolor*) and gray wolves (*Canis lupus*) in Banff National Park, Alberta. *Ecoscience* 14:214–222.

Kotanen, P. M., and R. L. Jeffries. 1997. Long-term destruction of sub-arctic wetland vegetation by lesser snow geese. *Ecoscience* 4:179–182.

Kotler, B. P., J. S. Brown, and O. Hasson. 1991. Factors affecting gerbil foraging behavior and rates of owl predation. *Ecology* 72:2249–2260.

Krebs, C. J., S. Boutin, R. Boonstra, A. R. E. Sinclair, J. N. M. Smith, M. R. T. Dale, K. Martin, and R. Turkington. 1995. Impact of food and predation on the snowshoe hare cycle. *Science* 269:1112–1115.

Krook, K., W. J. Bond, and P. Hockey. 2007. The effect of grassland shifts on the avifauna of a South African savanna. *Ostrich: Journal of African Ornithology* 78:271–279.

Kunert, G., S. Otto, U. S. R. Rose, J. Gershenzon, and W. W. Weisser. 2005. Alarm pheromone mediates production of winged dispersal morphs in aphids. *Ecology Letters* 8:596–603.

Kuntz, N. M., D. I. Kline, S. A. Sandin, and F. Rohwer. 2005. Pathologies and mortality rates caused by organic carbon and nutrient stressors in three Caribbean coral species. *Marine Ecology Progress Series* 294:173–180.

Kurtén, B. 1969. *The Ice Age*. International Book Production, Stockholm.

Kurtén, B. 1971. *The age of mammals*. Weidenfeld & Nicholson, London.

Kuznetsov, Y. A., O. De Feo, and S. Rinaldi. 2001. Belyakov homoclinic bifurcations in a tritrophic food chain model. *SIAM Journal on Applied Mathematics* 62:462–487.

Kvitek, R. G., P. Iampietro, and C. E. Bowlby. 1998. Sea otters and benthic prey communities: A direct test of the sea otter as keystone predator in Washington State. *Marine Mammal Science* 14:895–902.

Kvitek, R. G., J. S. Oliver, A. R. DeGange, and B. S. Anderson. 1992. Changes in Alaskan soft-bottom prey communities along a gradient in sea otter predation. *Ecology* 73:413–428.

Laakso, J., and H. Setälä. 1999a. Population- and ecosystem-effects of predation on microbial-feeding nematodes. *Oecologia* 120:279–286.

Laakso, J., and H. Setälä. 1999b. Sensitivity of primary production to changes in the architecture of belowground food webs. *Oikos* 87:57–64.

Lafferty, K. D. 2004. Fishing for lobsters indirectly increases epidemics in sea urchins. *Ecological Applications* 14:1566–1573.

Lai, C. H., and A. T. Smith. 2003. Keystone status of plateau pikas (*Ochotona curzoniae*): Effect of control on biodiversity of native birds. *Biodiversity and Conservation* 12:1901–1912.

Laliberte, A. S., and W. J. Ripple. 2003. Wildlife encounters by Lewis and Clark: A spatial analysis of interactions between Native Americans and wildlife. *BioScience* 53:994–1003.

Laliberte, A. S., and W. J. Ripple. 2004. Range contractions of North American carnivores and ungulates. *BioScience* 54:123–138.

Lalli, C. M., and T. R. Parsons. 1997. *Biological oceanography. An introduction*. Elsevier, Amsterdam.

Lambert, T. D., G. H. Adler, C. M. Riveros, L. Lopez, R. Ascanio, and J. Terborgh. 2003. Rodents on tropical land-bridge islands. *Journal of Zoology* 260:179–187.

Landis, D. A., M. M. Gardinder, W. van der Werf, and S. M. Swinton. 2008. Increasing corn for biofuel production reduces biocontrol services in agricultural landscapes. *Proceedings of the National Academy of Sciences* 105:20552–20557.

Larkum, A. W. D., R. J. Orth, and C. M. Duarte. 2006. *Seagrasses: Biology, ecology and conservation*. Dordrecht, the Netherlands: Springer.

Laslett, G. M., A. Baynes, M. A. Smith, R. Jones, and B. L. Smith. 2001. New ages for the last Australian megafauna: Continent-wide extinction about 46,000 years ago. *Science* 292:1888–1892.

Lathrop, R. C., B. M. Johnson, T. B. Johnson, M. T. Vogelsang, S. R. Carpenter, T. R. Hrabik, J. F. Kitchell, J. J. Magnuson, L. G. Rudstam, and R. S. Stewart. 2002. Stocking piscivores to improve fishing and water clarity: A synthesis of the Lake Mendota biomanipulation project. *Freshwater Biology* 47:2410–2424.

Laurenson, M. K. 1995. Implications for high offspring mortality for cheetah population dynamics. Pp. 385–399 in A. R. E. Sinclair and P. Arcese, eds. *Serengeti II: Dynamics, management, and conservation of an ecosystem*. University of Chicago Press, Chicago.

Lavelle, P. 1997. Faunal activities and soil processes: Adaptive strategies that determine ecosystem function. *Advances in Ecological Research* 27:93–132.

Lawrence, K. L., and D. H. Wise. 2004. Unexpected indirect effect of spiders on the rate of litter disappearance in a deciduous forest. *Pedobiologia* 48:149–157.

Laws, R. M. 1970. Elephants as agents of habitat and landscape change in East Africa. *Oikos* 21:1–15.

Laws, R. M., C. Parker, and R. C. B. Johnstone. 1975. *Elephants and their habitats*. Clarendon, London.

Lawton, J. H. 1999. Are there general laws in ecology? *Oikos* 84:177–192.

Lawton, J. H., and S. McNeil. 1979. Between the devil and the deep blue sea: On the problem of being a herbivore. *Symposium of the British Ecological Society* 20:223–244.

Le Corff, J., R. J. Marquis, and J. B. Whitfield. 2000. Temporal and spatial variation in a parasitoid community associated with the herbivores that feed on Missouri *Quercus*. *Environmental Entomology* 29:181–194.

Lehodey, P., M. Bertignac, J. Hampton, A. Lewis, and J. Picaut. 1997. El Niño southern os-cillation and tuna in the western Pacific. *Nature* 389:715–718.

Leibold, M. A. 1996. A graphical model of keystone predators in food webs: Trophical reg-ulation of abundance, indices, and diversity patterns in communities. *American Natural-ist* 147:784–812.

Leibold, M. A., J. M. Chase, J. B. Shurin, and A. L. Downing. 1997. Species turnover and the regulation of trophic structure. *Annual Review of Ecology and Systematics* 28:467–494.

Leigh, E. G. Jr., R. T. Paine, J. F. Quinn, and T. H. Suchanek. 1987. Wave energy and intertidal productivity. *Proceedings of the National Academy of Sciences* 84:1314–1318.

Leimu, R., and J. Koricheva. 2006. A meta-analysis of tradeoffs between plant tolerance and resistance to herbivores: Combining the evidence from ecological and agricultural studies. *Oikos* 112:1–9.

Leland, A. 2002. *A new apex predator in the Gulf of Maine: Large, mobile Jonah crabs (*Cancer bo-realis*) control benthic community structure*. University of Maine, Orono.

Lenihan, H. S., C. H. Peterson, J. H. Grabowski, J. E. Byers, G. W. Thayer, and D. R. Colby. 2001. Cascading of habitat degradation: Oyster reefs invaded by refugee fishes escaping stress. *Ecological Applications* 11:764–782.

Lenoir, L., T. Persson, J. Bengtsson, H. Wallander, and A. Wiren. 2007. Bottom-up or top-down control in forest soil microcosms? Effects of soil fauna on fungal biomass and C/N mineralization. *Biology and Fertility of Soils* 43:281–294.

Lensing, J. R., and D. H. Wise. 2006. Predicted climate change alters the indirect effect of predators on an ecosystem process. *Proceedings of the National Academy of Sciences* 103:15502–15505.

Leopold, A. 1936. Deer and Dauerwald in Germany II. Ecology and policy. *Journal of Forestry* 34:460–466.

Leopold, A. 1937. Conservationist in Mexico. *American Forests* 43:118–119, 146.

Leopold, A. 1943. Deer irruptions. *Wisconsin Academy of Sciences, Arts, and Letters* 35:351–366.

Leopold, A. 1949. *A Sand County almanac and sketches here and there*. Oxford University Press, New York.

Leopold, A., L. K. Sowls, and D. L. Spencer. 1947. A survey of over-populated deer ranges in the United States. *Journal of Wildlife Management* 11:162–183.

Leptich, D. J., and J. R. Gilbert. 1989. Summer home range and habitat use by moose in northern Maine. *Journal of Wildlife Management* 53:880–885.

Leroux, S. J., and M. Loreau. 2008. Subsidy hypothesis and strength of trophic cascades across ecosystems. *Ecology Letters* 11:1147–1156.

Lessios, H. A., D. R. Robertson, and J. D. Cubit. 1984. Spread of *Diadema* mass mortality through the Caribbean. *Science* 226:335–337.

Letourneau, D. K., and L. A. Dyer. 1998. Experimental tests in lowland tropical forest shows top-down effects through four trophic levels. *Ecology* 79:1678–1687.

Levin, S. A. 1992. The problem of pattern and scale in ecology. *Ecology* 73:1943–1967.

Lewis, S. M. 1986. The role of herbivorous fishes in the organization of a Caribbean reef community. *Ecological Monographs* 56:183–200.

Lewontin, R. C. 1969. The meaning of stability. *Brookhaven Symposium in Biology* 22:13–24.

Libralato, S., V. Christensen, and D. Pauly. 2006. A method for identifying keystone species in food web models. *Ecological Modelling* 195:153–171.

Lill, J. T., and R. J. Marquis. 2003. Ecosystem engineering by caterpillars increases insect herbivore density on white oak. *Ecology* 84:682–690.

Lill, J. T., and R. J. Marquis. 2007. Microhabitat manipulation: Ecosystem engineering by shelter-building insects. Pp. 107–138 in K. Cuddington, J. E. Byers, W. G. Wilson, and A. Hastings, eds. *Ecosystem engineers, plants to protists.* Academic Press, Amsterdam.

Lill, J. T., R. J. Marquis, R. E. Forkner, J. Le Corff, N. Holmberg, and N. A. Barber. 2006. Leaf pubescence affects distribution and abundance of generalist slug caterpillars (Lepidoptera: Limacodidae). *Environmental Entomology* 35:797–806.

Lima, S. L. 1998. Stress and decision-making under the risk of predation: Recent developments from behavioral, reproductive, and ecological perspectives. *Advances in the Study of Behavior* 27:215–290.

Lima, S. L. 2002. Putting predators back into behavioral predator–prey interactions. *Trends in Ecology and Evolution* 17:70–75.

Lima, S. L., and L. M. Dill. 1990. Behavioral decisions made under the risk of predation: A review and prospectus. *Canadian Journal of Zoology* 68:619–640.

Lindeman, R. L. 1942. The trophic-dynamic aspect of ecology. *Ecology* 23:399–418.

Lindroth, R. L. 1989. Mammalian herbivore–plant interactions. Pp. 163–206 in W. G. Abrahamson, ed. *Plant–animal interactions.* McGraw-Hill, New York.

Lister, A. M. 1989. Rapid dwarfing of red deer on Jersey in the last interglacial. *Nature* 342:539–542.

Lister, A. M. 1993. Mammoth in miniature. *Nature* 362:288–289.

Livaitis, J. A., and R. Villafuerte. 1996. Intraguild predation, mesopredator release, and prey stability. *Conservation Biology* 10:676–677.

Lloyd, P. 2007. Predator control, mesopredator release, and impacts on bird nesting success: A field test. *African Zoology* 42:180–186.

Logan, J. A., and J. A. Powell. 2001. Ghost forests, global warming, and the mountain pine beetle (Coleoptera: Scolytidae). *American Entomologist* 47:160–172.

Longhurst, A. R. 1998. *Ecological geography of the sea.* Academic Press, San Diego.

Lopez, L., and J. Terborgh. 2007. Seed predation and seedling herbivory as factors in tree recruitment failure on predator-free forested islands. *Journal of Tropical Ecology* 23:129–137.

Lotze, H., K., H. S. Lenihan, B. J. Bourque, R. H. Bradbury, R. G. Cooke, M. C. Kay, S. M. Kidwell, M. X. Kirby, C. H. Peterson, and J. B. C. Jackson. 2006. Depletion, degradation and recovery potential of estuaries and coastal seas. *Science* 312:1806–1809.

Louda, S. M. 1982. Distribution ecology: Variation in plant recruitment over a gradient in relation to insect seed predation. *Ecological Monographs* 52:25–42.

Louda, S. M., K. H. Keeler, and R. D. Holt. 1990. Herbivore influences on plant performance and competitive interactions. Pp. 413–444 in J. B. Grace and D. Tilman, eds. *Perspectives on plant competition.* Academic Press, New York.

Louda, S. M., and M. A. Potvin. 1995. Effect of inflorescence-feeding insects on the demography and fitness of a native plant. *Ecology* 76:229–245.

Lovaas, A. L. 1970. *People and the Gallatin elk herd.* Montana Fish and Game Department, Helena.

Lovett, G. M., and A. E. Ruesink. 1995. Carbon and nitrogen mineralization from decomposing gypsy moth frass. *Oecologia* 104:133–138.

Lubchenco, J. 1978. Plant species diversity in a marine intertidal community: Importance of herbivore food preferences and algal competitive abilities. *American Naturalist* 112:23–29.

Lubchenco, J., and S. D. Gaines. 1981. A unified approach to marine plant–herbivore interactions. I. Populations and communities. *Annual Review of Ecology and Systematics* 12:405–437.

Luttbeg, B., and J. L. Kerby. 2005. Are scared prey as good as dead? *Trends in Ecology and Evolution* 20:416–418.

Lutz, H. J. 1930. The vegetation of Heart's Content, a virgin forest in northwestern Pennsylvania. *Ecology* 11:1–29.

Lynch, A. H., J. Beringer, P. Kershaw, A. Marshall, S. Mooney, N. Tapper, C. Turney, and S. van der Kaars. 2007. Using the paleorecord to evaluate climate and fire interactions in Australia. *Annual Review of Earth Planetary Sciences* 35:215–239.

MacArthur, R. H., and E. R. Pianka. 1966. On optimal use of a patchy environment. *American Naturalist* 100:603–609.

MacArthur, R. H., and E. O. Wilson. 1967. *The theory of island biogeography.* Princeton University Press, Princeton, NJ.

Mack, R. N., and J. N. Thompson. 1982. Evolution in steppe with few large, hooved mammals. *American Naturalist* 119:757–773.

Maezono, Y., R. Kobayashi, M. Kusahara, and T. Miyashita. 2005. Direct and indirect effects of exotic bass and bluegill on exotic and native organisms in farm ponds. *Ecological Applications* 15:638–650.

Magurran, A. E. 1990. The adaptive significance of schooling as an antipredator defense in fish. *Annales Zoologici Fennici* 27:51–66.

Maina, G. G., and W. M. Jackson. 2003. Effects of fragmentation on artificial nest predation in a tropical forest in Kenya. *Biological Conservation* 111:161–169.

Mann, K. H. 2000. *Ecology of coastal waters.* Blackwell Science, Malden, MA.

Mann, K. H., and P. A. Breen. 1972. Relation between lobster abundance, sea-urchins, and kelp beds. *Journal of the Fisheries Research Board of Canada* 29:603–609.

Maron, J. L., J. A. Estes, D. A. Croll, E. M. Danner, S. C. Elmendorf, and S. Buckalew. 2006. An introduced predator transforms Aleutian Island plant communities by thwarting nutrient subsidies. *Ecological Monographs* 76:3–24.

Marquis, R. J. 1992. Selective impact of herbivores. Pp. 301–325 in R. S. Fritz and E. L. Simms, eds. *Ecology and plant resistance to herbivores and pathogens.* University of Chicago Press, Chicago.

Marquis, R. J. 2005. Impacts of herbivores on tropical plant diversity. Pp. 328–346 in D. Burslem, M. Pinard, and S. Hartley, eds. *Biotic interactions in the tropics: Their role in maintenance of species diversity.* Cambridge University Press, Cambridge.

Marquis, R. J., and J. T. Lill. 2006. Effects of herbivores as physical ecosystem engineers on plant-based trophic interaction webs. Pp. 246–274 in T. Ohgushi, T. P. Craig, and P. W.

Price, eds. *Ecological communities: Plant mediation in indirect interaction webs.* Columbia University Press, New York.

Marquis, R. J., J. T. Lill, and A. Piccini. 2002. Effect of plant architecture on colonization and damage by leaf-tying caterpillars of *Quercus alba. Oikos* 99:531–537.

Marquis, R. J., and C. J. Whelan. 1994. Insectivorous birds increase plant growth through their impact on herbivore communities of white oak. *Ecology* 75:2007–2014.

Marshall, K. N. 2007. *Integrating energetics and interaction strengths in natural ecosystems.* University of Washington, Seattle.

Martin, P. S. 1984. Prehistoric overkill: The global model. Pp. 354–403 in P. S. Martin and R. G. Kline, eds. *Quaternary extinctions.* University of Arizona Press, Tucson.

Martin, P. S. 2005. *Twilight of the mammoths.* University of California Press, Berkeley.

Martin, P. S., and D. W. Steadman. 1999. Prehistoric extinctions on islands and continents. Pp. 17–55 in R. D. E. MacPhee, ed. *Extinctions in near time: Causes, contexts, and consequences.* Kluwer Academic/Plenum, Amsterdam.

Martin, P. S., and C. R. Szuter. 1999. War zones and game sinks in Lewis and Clark's West. *Conservation Biology* 13:36–45.

Marzluff, J. M., and T. Angell. 2007. *In the company of crows and ravens.* Yale University Press, New Haven, CT.

Masson, D. G. 1984. Evolution of the Mascarene Basin, western Indian Ocean, and the significance of the Amirante Arc. *Marine Geophysical Researches* 4:365–382.

Masters, G. J., V. K. Brown, and A. C. Gange. 1993. Plant mediated interactions between above- and below-ground herbivores. *Oikos* 66:148–151.

Masters, G. J., T. H. Jones, and M. Rogers. 2001. Host-plant mediated effects of root herbivory on insect seed predators and their parasitoids. *Oecologia* 127:246–250.

Mattson, D. J. 1990. Human impacts on bear habitat use. *International Conference of Bear Research and Management* 8:35–56.

May, R. M. 1973a. *Stability and complexity in model ecosystems.* Princeton University Press, Princeton, NJ.

May, R. M. 1973b. Time-delay versus stability in population models with two and three trophic levels. *Ecology* 54:315–325.

May, R. M. 1974. Biological populations with nonoverlapping generations: Stable points, stable cycles, and chaos. *Science* 186:645–647.

May, R. M. 1977. Thresholds and breakpoints in ecosystems with a multiplicity of stable states. *Nature* 269:471–477.

Mayle, F. E., D. J. Beerling, W. D. Gosling, and M. B. Bush. 2004. Responses of Amazonian ecosystems to climatic and atmospheric carbon dioxide changes since the last glacial maximum. *Philosophical Transactions of the Royal Society of London. Series B, Biological Sciences* 359:499–514.

Mazumder, A. 1994. Patterns of algal biomass in dominant odd- vs. even-link lake ecosystems. *Ecology* 75:1141–1149.

McBride, R. S., and A. K. Richardson. 2007. Evidence of size-selective fishing mortality from an age and growth study of a hogfish (Labridae: *Lachnolaimus maximus*), a hermaphroditic reef fish. *Bulletin of Marine Science* 80:401–417.

McCann, K. S., A. Hastings, and D. R. Strong. 1998. Trophic cascades and trophic trickles in pelagic food webs. *Proceedings of the Royal Society of London Series B: Biological Sciences* 265:205–209.

McCann, K., J. Rasmussen, and J. Umbanhowar. 2005. The dynamics of spatially coupled webs. *Ecology Letters* 8:513–523.

McCauley, E., R. M. Nisbet, W. W. Murdoch, A. M. de Roos, and W. S. C. Gurney. 1999. Large amplitude cycles of *Daphnia* and its algal prey in enriched environments. *Nature* 402:653–656.

McClanahan, T., N. Polunin, and T. Done. 2002. Ecological states and the resilience of coral reefs. *Conservation Ecology* 6:18.

McClanahan, T. R. 1997. Primary succession of coral-reef algae: Differing patterns on fished versus unfished reefs. *Journal of Experimental Marine Biology and Ecology* 218:77–102.

McClanahan, T. R. 2000. Recovery of a coral reef keystone predator, *Balistapus undulatus*, in East African marine parks. *Biological Conservation* 94:191–198.

McClanahan, T. R. 2005. Recovery of carnivores, trophic cascades, and diversity in coral reef marine parks. Pp. 247–267 in J. C. Ray, K. H. Redford, R. S. Steneck, and J. Berger, eds. *Large carnivores and the conservation of biodiversity*. Island Press, Washington, DC.

McClanahan, T. R., and G. M. Branch. 2008. *Food webs and the dynamics of marine reefs*. Oxford University Press, New York.

McClanahan, T. R., N. A. J. Graham, J. M. Calnan, and M. A. MacNeil. 2007. Toward pristine biomass: Reef fish recovery in coral reef marine protected areas in Kenya. *Ecological Applications* 17:1055–1067.

McClanahan, T. R., E. Sala, P. A. Stickels, B. A. Cokos, A. C. Baker, C. J. Starger, and S. H. Jones IV. 2003. Interaction between nutrients and herbivory in controlling algal communities and coral condition on Glover's Reef, Belize. *Marine Ecology Progress Series* 261:135–147.

McClanahan, T. R., and S. H. Shafir. 1990. Causes and consequences of sea urchin abundance and diversity in Kenyan coral reef lagoons. *Oecologia* 83:362–370.

McClenachan, L., and A. B. Cooper. 2008. Extinction rate, historical population structure and ecological role of the Caribbean monk seal. *Proceedings of the Royal Society B: Biological Sciences* 275:1351–1358.

McClenachan, L., J. B. C. Jackson, and M. J. H. Newman. 2006. Conservation implications of historic sea turtle nesting beach loss. *Frontiers in Ecology and the Environment* 4:290–296.

McCoy, M. W., M. Barfield, and R. D. Holt. 2009. Predator shadows: Complex life histories as generators of spatially patterned indirect interactions across ecosystems. *Oikos* 118:87–100.

McFarland, W. N. 1991. The visual world of coral reef fishes. Pp. 16–38 in P. F. Sale, ed. *The ecology of fishes on coral reefs*. Academic Press, San Diego.

McLaren, B. E., and R. O. Peterson. 1994. Wolves, moose, and tree rings on Isle Royale. *Science* 266:1555–1558.

McNaughton, S. J. 1979a. Grassland–herbivore dynamics. Pp. 46–81 in A. R. E. Sinclair and M. Norton-Griffiths, eds. *Serengeti: Dynamics of an ecosystem.* University of Chicago Press, Chicago.

McNaughton, S. J. 1979b. Grazing as an optimization process: Grass–ungulate relationships in the Serengeti. *American Naturalist* 113:691–703.

McNaughton, S. J. 1984. Grazing lawns: Animals in herds, plant form, and coevolution. *American Naturalist* 124:863–886.

McNaughton, S. J. 1985. Ecology of a grazing ecosystem: The Serengeti. *Ecological Monographs* 55:259–294.

McNaughton, S. J. 1992. The propagation of disturbance in savannas through food webs. *Journal of Vegetation Science* 3:301–314.

McNaughton, S. J., W. Ruess, and S. W. Seagle. 1988. Large mammals and process dynamics in African ecosystems. *BioScience* 38:794–800.

McNaughton, S. J., G. Zuniga, M. M. McNaughton, and F. F. Banyikwa. 1997. Ecosystem catalysis: Soil urease activity and grazing in the Serengeti ecosystem. *Oikos* 80:467–469.

McQueen, D. J., J. R. Post, and E. L. Mills. 1986. Trophic relationships in freshwater pelagic ecosystems. *Canadian Journal of Fisheries and Aquatic Sciences* 43:1571–1581.

McShea, W. J., H. B. Underwood, and J. H. Rappole. 1997. *The science of overabundance.* Smithsonian Institution Press, Washington, DC.

Mduma, S. A. R., A. R. E. Sinclair, and R. Hilborn. 1999. Food regulates the Serengeti wildebeest: A 40–year record. *Journal of Animal Ecology* 68:1101–1122.

Mech, L. D. 1977. Wolf-pack buffer zones as prey reservoirs. *Science* 198:320–321.

Menge, B. A. 1995. Indirect effects in marine rocky intertidal interaction webs: Patterns and importance. *Ecological Monographs* 65:21–74.

Menge, B. A., E. L. Berlow, C. A. Blanchette, S. A. Navarrete, and S. B. Yamada. 1994. The keystone species concept: Variation in interaction strength in a rocky intertidal habitat. *Ecological Monographs* 64:249–287.

Menge, B. A., and G. M. Branch. 2001. Rocky intertidal communities. Pp. 221–251 in M. D. Bertness, S. D. Gaines, and M. E. Hay, eds. *Marine community ecology.* Sinauer, Sunderland, MA.

Menge, B. A., B. A. Daley, J. Lubchenco, E. Sanford, E. Dahlhoff, P. A. Halpin, G. Hudson, and J. L. Burnaford. 1999. Top-down and bottom-up regulation of New Zealand rocky intertidal communities. *Ecological Monographs* 69:297–330.

Menge, B. A., and T. M. Farrell. 1989. Community structure and interaction webs in shallow marine hard-bottom communities: Tests of an environmental stress model. *Advances in Ecological Research* 19:189–262.

Menge, B. A., and J. P. Sutherland. 1987. Community regulation: Variation in disturbance, competition, and predation in relation to environmental stress and recruitment. *American Naturalist* 130:730–757.

Merton, L. F., D. M. Bourn, and R. J. Hnatiuk. 1976. Giant tortoise and vegetation interaction on Aldabra atoll. I. Inland. *Biological Conservation* 9:293–304.

Messier, F. 1994. Ungulate population models with predation: A case study with the North American moose. *Ecology* 75:478–488.

Micheli, F. 1999. Eutrophication, fisheries, and consumer–resource dynamics in marine pelagic ecosystems. *Science* 285:1396–1398.

Mikola, J., and H. Setälä. 1998. No evidence of trophic cascades in an experimental microbial-based soil food web. *Ecology* 79:153–164.

Mikola, J., G. W. Yeates, G. M. Barker, D. A. Wardle, and K. I. Bonner. 2001. Effects of defoliation intensity on soil food web properties in an experimental grassland community. *Oikos* 92:333–343.

Millennium Ecosystem Assessment. 2005. *Ecosystems and human well-being: Summary for decision-makers*. Island Press, Washington, DC.

Miller, G. H., M. L. Fogel, J. W. Magee, M. K. Gagan, S. J. Clarke, and B. J. Johnson. 2005. Ecosystem collapse in Pleistocene Australia and a human role in megafaunal extinction. *Science* 309:287–290.

Miller, M. W., M. E. Hay, S. L. Miller, D. Malone, E. E. Sotka, and A. M. Szmant. 1999. Effects of nutrients versus herbivores on reef algae: A new method for manipulating nutrients on coral reefs. *Limnology and Oceanography* 44:1847–1861.

Mills, M. G. L., H. C. Biggs, and I. J. Whyte. 1995. The relationship between rainfall, lion predation and population trends in African herbivores. *Wildlife Research* 22:75–88.

Mills, S. M., M. E. Soulé, and D. F. Doak. 1993. The keystone-species concept in ecology and conservation. *BioScience* 43:219–224.

Mittelbach, G. G., A. M. Turner, D. J. Hall, J. E. Rettig, and C. W. Osenberg. 1995. Perturbation and resilience: A long-term, whole-lake study of predator extinction and reintroduction. *Ecology* 76:2347–2360.

Moe, S. R., L. P. Rutina, H. Hyttenborn, and J. T. du Toit. 2009. What controls woodland regeneration after elephants have killed big trees? *Journal of Applied Ecology* 46:223–230.

Moen, J., H. Gardfjell, L. Oksanen, and L. Ericson. 1993. Grazing by food-limited microtine rodents on a productive, experimental plant community: Does the green desert exist? *Oikos* 68:401–413.

Möllmann, C., and F. W. Köster. 2002. Population dynamics of calanoid copepods and the implications of their predation by clupeid fish in the central Baltic Sea. *Journal of Plankton Research* 10:959–977.

Moore, J. C., and H. W. Hunt. 1988. Resource compartmentation and the stability of real ecosystems. *Nature* 333:261–263.

Mopper, S., and S. Strauss. 1998. *Genetic structure and local adaptation in natural insect populations: Effects of ecology, life history, and behavior*. Chapman & Hall, New York.

Morales, L. C., and S. Gorzula. 1986. The interrelations of the Caroní River Basin ecosystems and hydroelectric power projects. *Interciencia* 11:272–277.

Morin, P. 1999. Productivity, intraguild predation, and population dynamics in experimental food webs. *Ecology* 80:752–760.

Morrison, L. W. 1997. The insular biogeography of small Bahamian cays. *Journal of Ecology* 85:441–454.

Morrison, L. W. 2002. Island biogeography and metapopulation dynamics of Bahamian ants. *Journal of Biogeography* 29:387–394.

Morse, D. E., N. Hooker, A. N. C. Morse, and R. A. Jensen. 1988. Control of larval metamorphosis and recruitment in sympatric agariciid corals. *Journal of Experimental Marine Biology and Ecology* 116:193–217.

Moser, J. C. 1963. Contents and structure of *Atta texana* nest in summer. *Annals of the Entomological Society of America* 56:256–291.

Moulton, G. E., ed. 1986–1996. *The journals of the Lewis and Clark expedition, by Lewis M, Clark W*, Vols. 2–10. University of Nebraska Press, Lincoln.

Moutinho, P., D. C. Nepstad, and E. A. Davidson. 2003. Influence of leaf-cutting ant nests on secondary forest growth and soil properties in Amazonia. *Ecology* 84:1265–1276.

Mulder, C. P. H., M. N. Grant-Hoffman, D. R. Towns, P. J. Bellingham, D. A. Wardle, M. S. Durrett, T. Fukami, and K. I. Bonner. 2009. Direct and indirect effects of rats: Will their eradication restore ecosystem functioning of New Zealand seabird islands? *Biological Invasions* 11:1671–1688.

Müller, F. 1879. *Ituna* and *Thyridia*: A remarkable case of mimicry in butterflies. *Transactions of the Entomological Society* 1879:xx–xxix.

Muller-Navarra, D. C., M. T. Brett, A. M. Liston, and C. R. Goldman. 2000. A highly unsaturated fatty acid predicts carbon transfer between primary producers and consumers. *Nature* 403:74–77.

Mumby, P. J., C. P. Dahlgren, A. R. Harborne, C. V. Kappel, F. Micheli, D. R. Brumbaugh, K. E. Holmes, et al. 2006. Fishing, trophic cascades, and the process of grazing on coral reefs. *Science* 311:98–101.

Mumby, P. J., A. Hastings, and H. J. Edwards. 2007. Thresholds and the resilience of Caribbean coral reefs. *Nature* 450:98–101.

Munsche, P. B. 1981. The gamekeeper and English rural society, 1660–1830. *Journal of British Studies* 20:82–105.

Murdoch, W. W. 1966. Community structure, population control, and competition: A critique. *American Naturalist* 100:219–226.

Musick, J. A. 1993. Trends in shark abundance 1974–1991 for the Chesapeake Bight of the U.S. mid-Atlantic coast. In S. Branstetter, ed. *Conservation biology of elasmobranchs.* NOAA Technical Report NMFS 115:1–18.

Myers, N. 1973. Tsavo National Park, Kenya, and its elephants: An interim appraisal. *Biological Conservation* 5:123–132.

Myers, R. A., J. K. Baum, T. D. Shepherd, S. P. Powers, and C. H. Peterson. 2007. Cascading effects of the loss of apex predatory sharks from a coastal ocean. *Science* 315:1846–1850.

Myers, R. A., K. G. Bowen, and N. J. Barrowman. 1999. The maximum reproductive rate of fish at low population sizes. *Canadian Journal of Fisheries and Aquatic Sciences* 56:2404–2419.

Myers, R. A., and B. Worm. 2003. Rapid worldwide depletion of predatory fish communities. *Nature* 423:280–283.

Mylius, S. D., K. Klumpers, A. M. de Roos, and L. Persson. 2001. Impact of intraguild predation and stage structure on simple communities along a productivity gradient. *American Naturalist* 158:259–276.

Nachman, G. 2006. A functional response model of a predator population foraging in a patchy habitat. *Journal of Animal Ecology* 75:948–958.

Nakano, S., H. Miyasaka, and N. Kuhara. 1999. Terrestrial–aquatic linkages: Riparian arthropod inputs alter trophic cascades in a stream food web. *Ecology* 80:2435–2441.

National Research Council. 2004. *Nonnative oysters in Chesapeake Bay.* The National Academies Press, Washington, DC.

Nelson, B. V., and R. R. Vance. 1979. Diel foraging patterns of the sea urchin *Centrostephanus coronatus* as a predator avoidance strategy. *Marine Biology* 51:251–258.

Newbery, D. M. 1980. Interactions between the coccid, *Icerya seychellarum* (Westw.), and its host tree species on Aldabra Atoll. *Oecologia* 46:171–179.

Newell, R. I. E., and E. W. Koch. 2004. Modeling seagrass density and distribution in response to changes in turbidity stemming from bivalve filtration and seagrass sediment stabilization. *Estuaries* 27:793–806.

Newman, M. J. H., G. A. Paredes, E. Sala, and J. B. C. Jackson. 2006. Structure of Caribbean coral reef communities across a large gradient of fish biomass. *Ecology Letters* 9:1216–1227.

Nilsen, E. G., E. J. Milner-Gulland, L. Schofield, A. Mysterud, N. C. Senseth, and T. Coulson. 2007. Wolf reintroduction to Scotland: Public attitudes and consequences for red deer management. *Proceedings of the Royal Society B: Biological Sciences* 274:995–1002.

Nilsson, M.-C., O. Zackrisson, O. Sterner, and A. Wallstedt. 2000. Characterisation of the differential interference effects of two boreal dwarf shrub species. *Oecologia* 123:122–128.

Nizinski, M. S. 2007. Predation in subtropical soft-bottom systems: Spiny lobster and molluscs in Florida Bay. *Marine Ecology Progress Series* 345:185–197.

Nkwabi, A. K. 2007. *Influence of wildfire on the avian community of the Serengeti National Park, Tanzania.* University of Dar es Salaam, Dar es Salaam, Tanzania.

Norrdahl, K., T. Klemola, E. Korpimäki, and M. Koivula. 2002. Strong seasonality may attenuate trophic cascades: Vertebrate predator exclusion in boreal grassland. *Oikos* 99:419–430.

Norse, E. A., and L. B. Crowder. 2005. *Marine conservation biology: The science of maintaining the sea's biodiversity.* Island Press, Washington, DC.

Norton, G. W., R. J. Rhine, G. W. Wynn, and R. D. Wynn. 1987. Baboon diet: A five-year study of stability and variability in the plant feeding and habitat on the yellow baboons (*Papio cynocephalus*) of Mikumi National Park, Tanzania. *Folia Primatologica* 48:78–120.

Norton-Griffiths, M. 1979. The influence of grazing, browsing, and fire on vegetation dynamics of the Serengeti. Pp. 310–352 in A. R. E. Sinclair and M. Norton-Griffiths, eds. *Serengeti: Dynamics of an ecosystem.* University of Chicago Press, Chicago.

Novotny, V., Y. Basset, S. E. Miller, G. D. Weiblen, B. Bremer, L. Cizek, and P. Drozd. 2002. Low host specificity of herbivorous insects in a tropical forest. *Nature* 416:841–844.

Novotny, V., P. Drozd, S. E. Miller, M. Kulfan, M. Janda, Y. Bassett, and G. D. Weiblen. 2006. Why are there so many species of herbivorous insects in tropical rainforests? *Science* 313:1115–1118.

Noy-Meir, I. 1975. Stability of grazing systems: An application of predator–prey graphs. *Journal of Ecology* 63:459–481.

O'Dowd, D. J., P. T. Green, and P. S. Lake. 2003. Invasional "meltdown" on an oceanic island. *Ecology Letters* 6:812–817.

O'Dowd, D. J., and P. S. Lake. 1990. Red crabs in rainforest, Christmas Island: Differential herbivory of seedlings. *Oikos* 58:289–292.

Odum, H. T. 1957. Trophic structure and productivity of Silver Springs, Florida. *Ecological Monographs* 27:55–112.

Odum, H. T., and E. P. Odum. 1955. Trophic structure and productivity of a windward coral reef community on Eniwetok Atoll. *Ecological Monographs* 25:291–320.

Ogden, J. C., R. A. Brown, and N. Salesky. 1973. Grazing by the echinoid *Diadema antillarum* Philippi: Formation of halos around West Indian patch reefs. *Science* 182:715–717.

Oguz, T., and D. Gilbert. 2007. Abrupt transitions of the top-down controlled Black Sea pelagic ecosystem during 1960–2000: Evidence for regime-shifts under strong fishery exploitation and nutrient enrichment modulated by climate-induced variations. *Deep-Sea Research Part I: Oceanographic Research Papers* 54:220–242.

Oksanen, L. 1983. Trophic exploitation and arctic phytomass patterns. *American Naturalist* 122: 45–52.

Oksanen, L. 1990. Exploitation ecosystems in seasonal environments. *Oikos* 57:14–24.

Oksanen, L. 1992. Evolution of exploitation ecosystems I. Predation, foraging ecology and population dynamics in herbivores. *Evolutionary Ecology* 6:15–33.

Oksanen, L., S. D. Fretwell, J. Arruda, and P. Niemelä. 1981. Exploitation ecosystems in gradients of primary productivity. *American Naturalist* 118:240–261.

Oksanen, L., and T. Oksanen. 1981. Lemmings (*Lemmus lemmus*) and grey-sided voles (*Clethrionomys rufocanus*) in interaction with their resources and predators on Finnmarksvidda, northern Norway. *Rep. Kevo Subarctic Research Station* 17:7–31.

Oksanen, L., and T. Oksanen. 2000. The logic and realism of the hypothesis of exploitation ecosystems. *American Naturalist* 155:703–723.

Oksanen, L., and J. Olofsson. 2009. Vertebrate herbivory and its ecosystem consequences. In *Encyclopaedia of life sciences*. Wiley, Chichester, UK. DOI: 10.1002/9780470015902 .a0003283.

Oksanen, T. 1990. Exploitation ecosystems in heterogeneous habitat complexes. *Evolutionary Ecology* 4:220–234.

Oksanen, T., L. Oksanen, J. Dahlgren, and J. Olofsson. 2008. Arctic lemmings, *Lemmus* spp. and *Dicrostonyx* spp.: Integrating ecological and evolutionary perspectives. *Evolutionary Ecology Research* 10:415–434.

O'Leary, W. M. 1996. *Maine sea fisheries: The rise and fall of a native industry 1830–1890*. Northeastern University Press, Boston.

Olofsson, J., P. E. Hulme, O. Suominen, and L. Oksanen. 2004. Importance of large and small herbivores for the plant community structure in the forest–tundra ecotone. *Oikos* 106:324–334.

Olofsson, J., H. Kitti, P. Rautiainen, S. Stark, and L. Oksanen. 2001. Impact of summer grazing by reindeer on vegetation structure, productivity and nutrient cycling in the north Fennoscandian tundra. *Ecography* 24:13–24.

Olofsson, J., L. Oksanen, T., Callaghan, P. E., Hulme, T. Oksanen, and O. Suominen. 2009. Herbivores inhibit climate driven shrub expansion on the tundra. *Global Change Biology* 15:2681–2693.

O'Reilly, D., L. Ogada, T. M. Palmer, and F. Keesing. 2006. Effects of fire on bird diversity and abundance in an East African savanna. *African Journal of Ecology* 44:165–170.

Ormond, R. F. G., R. H. Bradbury, S. J. Bainbridge, K. Fabricius, J. Keesing, L. M. DeVantier, P. Medlay, and A. Steven. 1990. Test of a model of regulation of crown-of-thorns starfish by fish predators. In R. H. Bradbury, ed. *Acanthaster and the coral reef: A theoretical perspective.* Springer-Verlag, Berlin.

Orth, R. J. 1975. Destruction of eelgrass, *Zostera marina*, by the cownose ray, *Rhinoptera bonasus*, in the Chesapeake Bay. *Chesapeake Science* 16:205–208.

Osterblom, H., S. Hansson, U. Larsson, O. Hjerne, F. Wulff, R. Elmgren, and C. Folke. 2007. Human-induced trophic cascades and ecological regime shifts in the Baltic Sea. *Ecosystems* 10:877–889.

Ostfeld, R. S., and R. D. Holt. 2004. Are predators good for your health? Evaluating evidence for top-down regulation of zoonotic disease reservoirs. *Frontiers in Ecology and the Environment* 2:13–20.

Ottoson, I. 1971. Något om skogshararnas inverkan på ön Jungfrun i Kalmarsund (On the impact of mountain hares on the island of Jungfrun in Kalmarsund, in Swedish). *Fauna och Flora* 66:229–240.

Owen-Smith, N. 1981. The white rhino overpopulation problem, and a proposed solution. Pp. 129–150 in J. A. Jewell, S. Holt, and D. Hart, eds. *Problems in management of locally abundant wild mammals.* Academic Press, New York.

Owen-Smith, N. 1987. Pleistocene extinctions: The pivotal role of megaherbivores. *Paleobiology* 13:351–362.

Owen-Smith, N. 1988. *Megaherbivores: The influence of very large body size on ecology.* Cambridge University Press, Cambridge.

Owen-Smith, N. 1990. Demography of a large herbivore, the greater kudu *Tragelaphus strepsiceros*, in relation to rainfall. *Journal of Animal Ecology* 59:893–913.

Owen-Smith, N., D. R. Mason, and J. O. Ogutu. 2005. Correlates of survival rates for 10 African ungulate populations: Density, rainfall and predation. *Journal of Animal Ecology* 74:774–788.

Owen-Smith, N., and M. G. L. Mills. 2008. Predator–prey size relationships in an African large-mammal food web. *Journal of Animal Ecology* 77:173–183.

Owen-Smith, N., and J. O. Ogutu. 2003. Rainfall influences on ungulate population dynamics in the Kruger National Park. Pp. 310–331 in J. T. du Toit, K. H. Rogers, and H. C. Biggs, eds. *The Kruger experience. Ecology and management of savanna heterogeneity.* Island Press, Washington, DC.

Oyama, M. D., and C. A. Nobre. 2003. A new climate–vegetation equilibrium state for tropical South America. *Geophysical Research Letters* 30:2199 doi:10.1029/2003GL018600.

Pacala, S. W., and J. Roughgarden. 1984. Control of arthropod abundance by *Anolis* lizards on St. Eustatius (Neth. Antilles). *Oecologia* 64:160–162.

Pace, M. L. 1984. Zooplankton community structure, but not biomass, influences the phosphorus–chlorophyll a relationship. *Canadian Journal of Fisheries and Aquatic Sciences* 41:1089–1096.

Pace, M. L. 2001. Getting it right and wrong: Extrapolations across experimental scales. Pp.

157–177 in R. H. Gardner, W. M. Kemp, V. S. Kennedy, and J. E. Peterson, eds. *Scaling relations in experimental ecology*. Columbia University Press, New York.

Pace, M. L., J. J. Cole, S. R. Carpenter, and J. F. Kitchell. 1999. Trophic cascades revealed in diverse ecosystems. *Trends in Ecology and Evolution* 14:483–488.

Packer, C., R. Hilborn, A. Mosser, B. Kissui, M. Borner, G. Hopcraft, J. Wilmhurst, S. Mduma, and A. R. E. Sinclair. 2005. Ecological change, group territoriality, and population dynamics in Serengeti lions. *Science* 307:390–393.

Packer, C., R. D. Holt, A. P. Dobson, and P. J. Hudson. 2003. Keeping the herds healthy and alert: Impacts of predation upon prey with specialist pathogens. *Ecology Letters* 6:797–802.

Packer, C., and L. M. Ruttan. 1988. The evolution of cooperative hunting. *American Naturalist* 132:159–198.

Paine, R. T. 1966. Food web complexity and species diversity. *American Naturalist* 100:65–75.

Paine, R. T. 1969a. A note on trophic complexity and community stability. *American Naturalist* 103:91–93.

Paine, R. T. 1969b. The *Pisaster–Tegula* interaction: Prey patches, predator food preference, and intertidal community structure. *Ecology* 50:950–961.

Paine, R. T. 1971. A short-term experimental investigation of resource partitioning in a New Zealand rocky intertidal habitat. *Ecology* 52:1096–1106.

Paine, R. T. 1976. Size-limited predation: An observational and experimental approach with the *Mytilus–Pisaster* interaction. *Ecology* 57:858–873.

Paine, R. T. 1980. Food webs: Linkage, interaction strength, and community infrastructure. *Journal of Animal Ecology* 49:667–685.

Paine, R. T. 1992. Food-web analysis through field measurement of per capita interaction strength. *Nature* 355:73–75.

Paine, R. T. 2002. Trophic control of production in a rocky intertidal community. *Science* 296:736–739.

Paine, R. T., and S. A. Levin. 1981. Intertidal landscapes: Disturbance and the dynamics of pattern. *Ecological Monographs* 51:145–178.

Paine, R. T., and T. H. Suchanek. 1983. Convergence of ecological processes between independently evolved competitive dominants: A tunicate–mussel comparison. *Evolution* 37:821–831.

Paine, R. T., and A. C. Trimble. 2004. Abrupt community change on a rocky shore: Biological mechanisms contributing to the potential formation of an alternative state. *Ecology Letters* 7:441–445.

Paine, R. T., and R. L. Vadas. 1969. The effects of grazing by sea urchins, *Strongylocentrotus* spp., on benthic algal populations. *Limnology and Oceanography* 14:710–719.

Palkovacs, E. P., J. Gerlach, and A. Caccone. 2002. The evolutionary origin of Indian Ocean tortoises (*Dipsochelys*). *Molecular Phylogenetics and Evolution* 24:216–227.

Palkovacs, E. P., M. Marschner, C. Ciofi, J. Gerlach, and A. Caccone. 2003. Are the native giant tortoises from the Seychelles really extinct? A genetic perspective based on mtDNA and microsatellite data. *Molecular Ecology* 12:1403–1413.

Palma, A. T., R. S. Steneck, and C. Wilson. 1999. Settlement-driven, multiscale demographic patterns of large benthic decapods in the Gulf of Maine. *Journal of Experimental Marine Biology and Ecology* 241:107–136.

Palmer, S. C. F., J. E. Broadhead, I. Ross, and D. E. Smith. 2007. Long-term habitat use and browsing by deer in a Caledonian pinewood. *Forest Ecology and Management* 242:273–280.

Palmer, T. M., M. L. Stanton, T. P. Young, J. R. Goheen, R. M. Pringle, and R. Karban. 2008. Breakdown of an ant–plant mutualism follows the loss of large herbivores from an African savanna. *Science* 319:192–195.

Palmisano, S., and L. R. Fox. 1997. Effects of mammal and insect herbivory on population dynamics of a native Californian thistle, *Cirsium occidentale*. *Oecologia* 111:413–421.

Palomares, E., P. Gaona, P. Ferraras, and M. Delibes. 1995. Positive effects on game species of top predators by controlling smaller predator populations: An example with lynx, mongooses and rabbits. *Conservation Biology* 9:295–306.

Palomares, F., and T. M. Caro. 1999. Interspecific killing among mammalian carnivores. *American Naturalist* 153:492–508.

Pandolfi, J. M., R. H. Bradbury, E. Sala, T. P. Hughes, K. A. Bjorndal, R. G. Cooke, D. McArdle, et al. 2003. Global trajectories of the long-term decline of coral reef ecosystems. *Science* 301:955–958.

Pandolfi, J. M., J. B. C. Jackson, N. Baron, R. H. Bradbury, H. M. Guzman, T. P. Hughes, C. V. Kappel, et al. 2005. Ecology: Are US coral reefs on the slippery slope to slime? *Science* 307:1725–1726.

Parker, B. R., and D. W. Schindler. 2006. Cascading trophic interactions in an oligotrophic species-poor alpine lake. *Ecosystems* 9:157–166.

Parr, C. L., and S. L. Chown. 2003. Burning issues for conservation: A critique of faunal fire research in southern Africa. *Austral Ecology* 28:384–395.

Parrish, F. A., and R. C. Boland. 2004. Habitat and reef-fish assemblages of banks in the northwestern Hawaiian Islands. *Marine Biology* 144:1065–1073.

Parsons, T. R., M. Takahashi, and B. Hargrave. 1983. *Biological oceanographic processes*. Pergamon, Oxford.

Pastor, J. B., R. J. Naiman, B. Dewey, and P. McInnes. 1988. Moose, microbes, and the boreal forest. *BioScience* 38:770–777.

Patten, D. T. 1968. Dynamics of the shrub continuum along the Gallatin River in Yellowstone National Park. *Ecology* 49:1107–1112.

Patterson, J. H. 1907. *The maneaters of Tsavo*. Fontan, London.

Pauly, D. 1982. Studying single-species dynamics in a multispecies context. Pp. 33–70 in D. Pauly and G. I. Murphy, eds. *Theory and management of tropical fisheries*. ICLARM Conference Proceedings, Manila.

Pauly, D. 1985. The population dynamics of short-lived species, with emphasis on squids. *NAFO Scientific Council Studies* 9:143–154.

Pauly, D. 1995. Anecdotes and the shifting baseline syndrome of fisheries. *Trends in Ecology and Evolution* 10:430.

Pauly, D., and V. Christensen. 1995. Primary production required to sustain global fisheries. *Nature* 374:255–257.

Pauly, D., V. Christensen, J. Dalsgaard, R. Froese, and F. Torres Jr. 1998. Fishing down the food web. *Science* 279:860–863.

Pauly, D., V. Christensen, and C. Walters. 2000. Ecopath, Ecosim, and Ecospace as tools for evaluating ecosystem impact of fisheries. *ICES Journal of Marine Science* 57:697–706.

Pech, R. P., A. R. E. Sinclair, A. E. Newsome, and P. C. Catling. 1992. Limits to predator regulation of rabbits in Australia: Evidence from predator-removal experiments. *Oecologia* 89:102–112.

Peckarsky, B. L., P. A. Abrams, D. I. Bolnick, L. M. Dill, J. H. Grabowski, B. Luttbeg, J. L. Orrock, et al. 2008. Revisiting the classics: Considering non-consumptive effects in textbook examples of predator–prey interactions. *Ecology* 89:2416–2425.

Pejchar, L., R. M. Pringle, J. Ranganathan, J. R. Zook, G. Duran, F. Oviedo, and G. C. Daily. 2008. Birds as agents of seed dispersal in a human-dominated landscape in southern Costa Rica. *Biological Conservation* 141:536–544.

Pellew, R. A. P. 1983. The impacts of elephant, giraffe and fire upon the *Acacia tortilis* woodlands of the Serengeti. *African Journal of Ecology* 21:41–74.

Peres, C. A., and E. Palacios. 2007. Basin-wide effects of game harvest on vertebrate population densities in Amazonian forests: Implications for animal-mediated seed dispersal. *Biotropica* 39:304–315.

Perlman, D. 2008. How humans, vanishing cougars changed Yosemite. *San Francisco Chronicle*, May 2, p. A-1.

Persson, L. 1999. Trophic cascades: Abiding heterogeneity and the trophic level concept at the end of the road. *Oikos* 85:385–397.

Persson, L., A. M. De Roos, D. Claessen, P. Bystrom, J. Lovgren, S. Sjogren, R. Svanback, E. Wahlstrom, and E. Westman. 2003. Gigantic cannibals driving a whole-lake trophic cascade. *Proceedings of the National Academy of Sciences* 100:4035–4039.

Peterken, G. F., and C. R. Tubbs. 1965. Woodland regeneration in the New Forest, Hampshire, since 1650. *Journal of Applied Ecology* 2:159–170.

Peters, C. R., R. J. Blumenschine, R. L. Hay, D. A. Livingstone, C. W. Marean, T. Harrison, M. Armour-Chelu, et al. 2008. Paleoecology of the Serengeti–Mara ecosystem. Pp. 47–94 in A. R. E. Sinclair, C. Packer, S. A. R. Mduma, and J. M. Fryxell, eds. *Serengeti III: Human impacts on ecosystem dynamics*. University of Chicago Press, Chicago.

Peterson, C. H. 1982. Clam predation by whelks (*Busycon* spp.): Experimental tests of the importance of prey size, prey density, and seagrass cover. *Marine Biology* 66:159–170.

Peterson, C. H., F. J. Fodrie, and S. P. Powers. 2001. Site-specific and density-dependent extinction of prey by schooling rays: Generation of a population sink in top-quality habitat for bay scallops. *Oecologia* 129:349–356.

Peterson, C. H., H. C. Summerson, S. R. Fegley, and R. C. Prescott. 1989. Timing, intensity, and sources of autumn mortality of adult bay scallops, *Argopecten irradians concentricus* Say. *Journal of Experimental Marine Biology and Ecology* 127:121–140.

Peterson, R. O., J. A. Vucetich, R. E. Page, and A. Chouinard. 2003. Temporal and spatial dynamics of predator–prey dynamics. *Alces* 39:215–232.

Pimm, S. L. 1979. Complexity and stability: Another look at MacArthur's original hypothesis. *Oikos* 33:351–357.

Pimm, S. L. 1982. *Food webs.* Chapman & Hall, New York.

Pinker, S. 2006. *The stuff of thought.* Penguin, New York.

Pinnegar, J. K., N.V. C. Polunin, P. Francour, F. Badalamenti, R. Chemello, M.-L. Harmelin-Vivien, B. Hereu, et al. 2000. Trophic cascades in benthic marine ecosystems: Lessons for fisheries and protected-area management. *Environmental Conservation* 27:179–200.

Pinto da Silveira, E. K. 1969. O lobo-guara (*Chrysocyon brachyurus*): Possivel acao inibidora de certas Solanaceas sobre o nematode renal [The maned wolf: Possible inhibitory action of certain Solanaccas on renal function]. *Vellozia* 7:58–60.

Pokki, J. 1981. Distribution, demography and dispersal of the field vole, *Microtus agrestis* (L.), in the Tvärminne archipelago, Finland. *Acta Zoologica Fennica* 164:1–48.

Polis, G. A. 1991. Complex trophic interactions in deserts: An empirical critique of food web theory. *American Naturalist* 138:123–155.

Polis, G. A. 1999. Why are parts of the world green? Multiple factors control productivity and the distribution of biomass. *Oikos* 86:3–15.

Polis, G. A., W. B. Anderson, and R. D. Holt. 1997. Toward an integration of landscape and food web ecology: The dynamics of spatially subsidized food webs. *Annual Review of Ecology and Systematics* 28:289–316.

Polis, G. A., and S. Hurd. 1996. Allochtonous inputs across habitats, subsidized consumers, and apparent trophic cascades: Examples from the ocean–land interface. Pp. 275–285 in G. A. Polis and K. O. Winemiller, eds. *Food webs: Integration of patterns and dynamics.* Chapman & Hall, London.

Polis, G. A., S. D. Hurd, C. T. Jackson, and F. Sanchez-Pinero. 1998. Multifactor population limitation: Variable spatial and temporal control of spiders on Gulf of California islands. *Ecology* 79:490–502.

Polis, G. A., and S. J. McCormick. 1986. Scorpions, spiders and solpugids: Predation and competition among distantly related taxa. *Oecologia* 71:111–116.

Polis, G. A., M. E. Power, and G. R. Huxel. 2004. *Food webs at the landscape level.* University of Chicago Press, Chicago.

Polis, G. A., A. L. W. Sears, G. R. Huxel, D. R. Strong, and J. Maron. 2000. When is a trophic cascade a trophic cascade? *Trends in Ecology and Evolution* 15:473–475.

Polis, G. A., and D. R. Strong. 1996. Food web complexity and community dynamics. *American Naturalist* 147:813–846.

Pollard, A. J. 1992. The importance of deterrence: Responses of grazing animals to plant variation. Pp. 216–239 in R. S. Fritz and E. L. Simms, eds. *Plant resistance to herbivores and pathogens.* University of Chicago Press, Chicago.

Polley, H. W., and S. L. Collins. 1984. Relationships of vegetation and environment in buffalo wallows. *American Midland Naturalist* 112:178–186.

Polls, G. A., M. E. Power, and G. R. Huxel, eds. 2004. *Food webs at the landscape level.* University of Chicago Press, Chicago.

Post, D. M. 2002. The long and short of food chain length. *Trends in Ecology and Evolution* 17:269–277.

Post, E., R. O. Peterson, N. C. Stenseth, and B. E. McLaren. 1999. Ecosystem consequences of wolf behavioral response to climate. *Nature* 401:905–907.

Potvin, F., P. Beaupré, and G. Laprise. 1997. *Déplacements et survie hivernale des biches d'Anticosti de 1989 à 1990: Une étude télémétrique.* Ministère de l'Environnement et de la Faune, Quebec.

Potvin, F., P. Beaupré, and G. Laprise. 2003. The eradication of balsam fir stands by white-tailed deer on Anticosti Island, Québec: A 150 year process. *Ecoscience* 10:487–495.

Poveda, K., I. Steffan-Dewenter, S. Scheu, and T. Tscharntke. 2005. Effects of decomposers and herbivores on plant performance and aboveground plant–insect interactions. *Oikos* 108:503–510.

Power, M. E. 1990. Effect of fish in river food webs. *Science* 250:811–815.

Power, M. E. 1992. Top-down and bottom-up forces in food webs: Do plants have primacy? *Ecology* 73:733–746.

Power, M. E., N. Brozovic, C. Bode, and D. Zilberman. 2005. Spatially explicit tools for understanding and sustaining inland water ecosystems. *Frontiers in Ecology and the Environment* 3:47–55.

Power, M. E., M. S. Parker, and W. E. Dietrich. 2008. Seasonal reassembly of river food webs under a Mediterranean hydrologic regime: Floods, droughts, and impacts of fish. *Ecological Monographs* 78:263–282.

Power, M. E., D. Tilman, J. A. Estes, B. A. Menge, W. J. Bond, L. S. Mills, G. Daily, J. C. Castilla, J. Lubchenco, and R. T. Paine. 1996. Challenges in the quest for keystones. *BioScience* 46:609–620.

Powers, J. S., P. Sollins, M. E. Harmon, and J. A. Jones. 1999. Plant–pest interactions in time and space: A Douglas-fir bark beetle outbreak as a case study. *Landscape Ecology* 14:105–120.

Prairie, Y. T., D. F. Bird, and J. J. Cole. 2002. The summer metabolic balance in the epilimnion of southeastern Quebec lakes. *Limnology and Oceanography* 47:316–321.

Prasad, R. P., and W. E. Snyder. 2006. Diverse trait-mediated interactions in a multi-predator, multi-prey community. *Ecology* 87:1131–1137.

Pressier, E. L., D. I. Bolnick, and M. F. Benard. 2005. Scared to death? The effects of intimidation and consumption in predator–prey interactions. *Ecology* 86:501–509.

Pringle, E. G., P. Álvarez-Loayza, and J. Terborgh. 2007a. Seed characteristics and susceptibility to pathogen attack in tree seeds of the Peruvian Amazon. *Plant Ecology* 193:211–222.

Pringle, R. M. 2008. Elephants as agents of habitat creation for small vertebrates at the patch scale. *Ecology* 89:26–33.

Pringle, R. M., T. P. Young, D. I. Rubenstein, and D. J. McCauley. 2007b. Herbivore-initiated interaction cascades and their modulation by productivity in an African savanna. *Proceedings of the National Academy of Sciences* 104:193–197.

Prins, H. H. T., and H. P. van der Jeugd. 1993. Herbivore population crashes and woodland structure in East Africa. *Journal of Ecology* 81:305–314.

Pritchard, P. C. H. 1996. The Galapagos tortoises: Nomenclature and survival status. *Chelonian Research Monographs* 1:1–85.

Prittinen, K., J. Pusenius, K. Koiunoro, M. Rousi, and H. Roininen. 2003. Mortality in seedling populations of silver birch: Genotype variation and herbivore effects. *Functional Ecology* 17:658–663.

Proulx, M., and A. Mazumder. 1998. Reversal of grazing impact on plant species richness in nutrient-poor vs. nutrient-rich ecosystems. *Ecology* 79:2581–2592.

Putman, R. J. 1996. Ungulates in temperate forest ecosystems: Perspectives and recommendations for future research. *Forest Ecology and Management* 88:205–214.

Pyare, S., and J. Berger. 2003. Beyond demography and delisting: Ecological recovery for Yellowstone's grizzly bears and wolves. *Biological Conservation* 113:63–73.

Radloff, F. G. T., and J. T. du Toit. 2004. Large predators and their prey in a southern African savanna: A predator's size determines its prey size range. *Journal of Animal Ecology* 73:410–423.

Raffaelli, D., and S. J. Hall. 1992. Compartments and predation in an estuarine food web. *Journal of Animal Ecology* 61:551–560.

Ramankutty, N., and J. A. Foley. 1999. Estimating historical changes in global land cover: Croplands from 1700 to 1992. *Global Biogeochemical Cycles* 13:997–1027.

Rammul, Ü., T. Oksanen, L. Oksanen, J. Lehtelä, R. Virtanen, J. Olofsson, I. Strengbom, I. Rammul, and L. Ericson. 2007. Vole–vegetation interactions in an experimental, enemy-free taiga floor system. *Oikos* 116:1501–1513.

Randall, J. E. 1965. Grazing effect on sea grasses by herbivorous reef fishes in the West Indies. *Ecology* 46:255–260.

Randall, J. E. 1967. Food habits of reef fishes of the West Indies. *Studies in Tropical Oceanography* 5:665–847.

Randall, J. E., and G. L. Warmke. 1967. The food habits of the hogfish (*Lachnolaimus maximus*), a labrid fish from the western Atlantic. *Canadian Journal of Science* 7:141–144.

Rao, M. 2000. Variation in leaf-cutter ant (*Atta* sp.) densities in forest isolates: The potential role of predation. *Journal of Tropical Ecology* 16:209–225.

Rausher, M. D., and P. Feeny. 1980. Herbivory, plant density, and plant reproductive success: The effect of *Battus philenor* on *Aristolochia reticulata*. *Ecology* 61:905–917.

Ravier, C., and J. Fromentin. 2001. Long-term fluctuations in the eastern Atlantic and Mediterranean bluefin tuna population. *ICES Journal of Marine Science* 58:1299–1317.

Ray, J., K. H. Redford, R. Steneck, and J. Berger, eds. 2005. *Large carnivores and the conservation of biodiversity*. Island Press, Washington, DC.

Rayner, M. J., M. E. Hauber, M. J. Imber, R. K. Stamp, and M. N. Clout. 2007. Spatial heterogeneity of mesopredator release within an oceanic island system. *Proceedings of the National Academy of Sciences* 104:20862–20865.

Redding, J. 1995. History of deer population trends and forest cutting in the Allegheny National Forest. *Proceedings of the Central Hardwood Conference* 10:214–224.

Redford, K. 1992. The empty forest. *BioScience* 42:412–422.

Reeves, R. R., J. Berger, and P. J. Clapham. 2006. Killer whales as predators of large baleen whales and sperm whales. Pp. 174–190 in J. A. Estes, ed. *Whales, whaling, and ocean ecosystems*. University of California Press, Berkeley.

Reichelt, R. E., R. H. Bradbury, and P. J. Moran. 1990. Distribution of *Acanthaster planci* outbreaks on the Great Barrier Reef between 1966 and 1989. *Coral Reefs* 9:97–103.

Reichman, O. J., and E. W. Seabloom. 2002. The role of pocket gophers as subterranean ecosystem engineers. *Trends in Ecology and Evolution* 17:44–49.

Reisewitz, S. E., J. A. Estes, and C. A. Simenstad. 2005. Indirect food web interactions: Sea otters and kelp forest fishes in the Aleutian archipelago. *Oecologia* 146:623–631.

Renvoize, S. A. 1971. Origin and distribution of flora of Aldabra. *Philosophical Transactions of the Royal Society of London, Series B, Biological Sciences* 260:227–236.

Renvoize, S. A. 1975. A floristic analysis of western Indian Ocean coral islands. *Kew Bulletin* 30:133–152.

Revkin, A. C. 2009. Environmental issues slide in poll of public's concerns. *The New York Times*, January 22, p. A13.

Reynolds, C. S. 1994. The ecological basis for the successful biomanipulation of aquatic communities. *Archiv für Hydrobiologie* 130:1–33.

Richards, L. A., and P. D. Coley. 2008. Combined effects of host plant quality and predation on a tropical lepidopteran: A comparison between treefall gaps and the understory in Panama. *Biotropica* 40:736–741.

Rico-Gray, V., and P. S. Oliveira. 2007. *The ecology and evolution of ant–plant interactions*. University of Chicago Press, Chicago.

Riley, J., J. M. Winch, A. F. Stimson, and R. D. Pope. 1986. The association of *Amphisbaena alba* (Reptilia: Amphisbaenia) with the leaf-cutting ant *Atta cephalotes* in Trinidad. *Journal of Natural History* 20:459–470.

Rinaldi, S., A. Gragnani, and S. De Monte. 2004. Remarks on antipredator behavior and food chain dynamics. *Theoretical Population Biology* 66:277–286.

Ripple, W. J., and R. L. Beschta. 2003. Wolf reintroduction, predation risk, and cottonwood recovery in Yellowstone National Park. *Forest Ecology and Management* 184:299–313.

Ripple, W. J., and R. L. Beschta. 2004a. Wolves and the ecology of fear: Can predation risk structure ecosystems? *BioScience* 54:755–766.

Ripple, W. J., and R. L. Beschta. 2004b. Wolves, elk, willows, and trophic cascades in the upper Gallatin range of southwestern Montana, USA. *Forest Ecology and Management* 200:161–181.

Ripple, W. J., and R. L. Beschta. 2005. Linking wolves and plants: Aldo Leopold on trophic cascades. *BioScience* 55:613–621.

Ripple, W. J., and R. L. Beschta. 2006a. Linking a cougar decline, trophic cascade, and catastrophic regime shift in Zion National Park. *Biological Conservation* 133:397–408.

Ripple, W. J., and R. L. Beschta. 2006b. Linking wolves to willows via risk-sensitive foraging by ungulates in the northern Yellowstone ecosystem. *Forest Ecology and Management* 230:96–106.

Ripple, W. J., and R. L. Beschta. 2007a. Hardwood tree decline following large carnivore loss on the Great Plains, USA. *Frontiers in Ecology and the Environment* 5:241–246.

Ripple, W. J., and R. L. Beschta. 2007b. Restoring Yellowstone's aspen with wolves. *Biological Conservation* 138:514–519.

Ripple, W. J., and R. L. Beschta. 2008. Trophic cascades involving cougar, mule deer, and black oaks in Yosemite National Park. *Biological Conservation* 141:1249–1256.

Ripple, W. J., and E. J. Larsen. 2000. Historic aspen recruitment, elk, and wolves in northern Yellowstone National Park, USA. *Biological Conservation* 95:361–370.

Ripple, W. J., E. J. Larsen, R. A. Renkin, and D. W. Smith. 2001. Trophic cascades among wolves, elk and aspen on Yellowstone National Park's Northern Range. *Biological Conservation* 102:227–234.

Roberts, C. 2007. *The unnatural history of the sea.* Island Press, Washington, DC.

Roberts, C. M. 1995. Effects of fishing on the ecosystem structure of coral reefs. *Conservation Biology* 9:988–995.

Roberts, E. G., T. F. Flannery, L. K. Ayliffe, G. H. Yoshida, J. M. Olley, G. J. Prideaux, G. M. Laslett, et al. 2001. New ages for the last Australian megafauna: Continent-wide extinction about 46,000 years ago. *Science* 292:1888–1892.

Robinson, G. S., P. R. Ackery, I. J. Kitching, G. W. Beccaloni, and L. M. Hernandez. 2002. Hostplants of the moth and butterfly caterpillars of America north of Mexico. *Memoirs of the American Entomological Institute* 69:1–824.

Robinson, G. S., L. P. Burney, and D. A. Burney. 2005. Landscape paleoecology and megafaunal extinction in southeastern New York State. *Ecological Monographs* 75:295–315.

Robles, C. 1987. Predator foraging characteristics and prey population structure on a sheltered shore. *Ecology* 68:1502–1514.

Robles, C., R. Sherwood-Stephens, and M. Alvarado. 1995. Responses of a key intertidal predator to varying recruitment of its prey. *Ecology* 76:565–579.

Roemer, G. W., T. J. Coonan, D. K. Garcelon, J. Bascompte, and L. Laughrin. 2001. Feral pigs facilitate hyperpredation by golden eagles and indirectly cause the decline of the island fox. *Animal Conservation* 4:307–318.

Rogers, C. M., and M. J. Caro. 1998. Song sparrows, top carnivores, and nest predation: A test of the mesopredator release hypothesis. *Oecologia* 116:227–233.

Romanuk, T. N., B. E. Beisner, N. D. Martinez, and J. Kolasa. 2006. Non-omnivorous generality promotes population stability. *Biology Letters* 2:374–377.

Romare, P., and L.-A. Hansson. 2003. A behavioral cascade: Top-predator induced behavioral shifts in planktivorous fish and zooplankton. *Limnology and Oceanography* 48:1956–1964.

Rooney, T. P., and W. J. Dress. 1997. Species loss over sixty-six years in the ground-layer vegetation of Heart's Content, an old-growth forest in Pennsylvania, USA. *Natural Areas Journal* 17:297–305.

Rooney, T. P., and K. Gross. 2003. A demographic study of deer browsing impacts on *Trillium grandiflorum*. *Plant Ecology* 168:267–277.

Rooney, T. P., R. J. McCormick, S. L. Solheim, and D. M. Waller. 2000. Regional variation in recruitment of hemlock seedlings and saplings in the upper Great Lakes, USA. *Ecological Applications* 10:1119–1132.

Rooney, T. P., S. M. Wiegmann, D. A. Rogers, and D. M. Waller. 2004. Biotic impoverishment and homogenization in unfragmented forest understory communities. *Conservation Biology* 18:787–798.

Rosén, E. 1982. Vegetation development and sheep grazing in limestone grasslands of south Öland, Sweden. *Acta Phytogeographica Suecica* 72:1–104.

Rosenberg, A. A., W. J. Bolster, K. E. Alexander, W. B. Leavenworth, A. B. Cooper, and M. G. McKenzie. 2005. The history of ocean resources: Modeling cod biomass using historical records. *Frontiers in Ecology and the Environment* 3:78–84.

Rosenthal, G. A., and M. R. Berenbaum, eds. 1992. *Herbivores, their interactions with secondary plant metabolites.* Academic Press, San Diego, CA.

Rosenzweig, M. L. 1968. Net primary productivity of terrestrial communities: Prediction from climatological data. *American Naturalist* 102:67–74.

Rosenzweig, M. L. 1971. The paradox of enrichment: Destabilization of exploitation ecosystems in ecological time. *Science* 171:385–387.

Rosenzweig, M. L. 1973. Exploitation in three trophic levels. *American Naturalist* 107:275–294.

Rosenzweig, M. L. 1996. *Species diversity in space and time.* Cambridge University Press, Cambridge.

Rowell-Rahier, M., J. M. Pasteels, A. Alonso-Mejia, and L. P. Brower. 1995. Relative unpalatability of leaf-beetles with either biosynthesized or sequestered chemical defense. *Animal Behaviour* 49:709–714.

Roy, M., and R. D. Holt. 2008. Effects of predation on host–pathogen dynamics in SIR models. *Theoretical Population Biology* 73:319–331.

Royo, A., and W. P. Carson. 2006. On the formation of dense understory layers in forests worldwide: Consequences and implications for forest dynamics, biodiversity, and succession. *Canadian Journal of Forest Research* 36:1345–1362.

Russ, G. R. 1991. Coral reef fisheries: Effects and yields. Pp. 601–635 in P. F. Sale, ed. *The ecology of fishes on coral reefs.* Academic Press, San Diego, CA.

Russell, F. L., D. B. Zippin, and N. L. Fowler. 2001. Effects of white-tailed deer (*Odocoileus virginianus*) on plants, plant populations, and communities: A review. *American Midland Naturalist* 146:1–26.

Ruth, T. K., D. W. Smith, M. A. Haroldson, P. C. Buotte, C. C. Schwartz, H. B. Quigley, S. Cherry, K. M. Murphy, D. Tyers, and K. Frey. 2003. Large-carnivore response to recreational big-game hunting along the Yellowstone National Park and Absaroka–Beartooth Wilderness boundary. *Wildlife Society Bulletin* 31:1150–1161.

Rutina, L. P., S. R. Moe, and J. E. Swenson. 2005. Elephant *Loxodonta africana* driven woodland conversion to shrubland improves dry-season browse availability for impalas *Aepyceros melampus*. *Wildlife Biology* 11:207–213.

Ruzzante, D. E., C. T. Taggart, D. Cook, and S. Goddard. 1996. Genetic differentiation between inshore and offshore Atlantic cod (*Gadua morhua*) off Newfoundland: Microsatellite DNA variations and antifreeze level. *Canadian Journal of Fisheries and Aquatic Sciences* 53:634–645.

Saap, J. 1999. *What is natural? Coral reef crisis.* Oxford University Press, New York.

Sale, P. F., ed. 1991. *The ecology of fishes on coral reefs.* Academic Press, San Diego, CA.

Salo, P., M. Nordstrom, R. L. Thomson, and E. Korpimaki. 2008. Risk induced by a native top predator reduces alien mink movements. *Journal of Animal Ecology* 77:1092–1098.

Salomon, A. K., N. M. Tanape Sr., and H. P. Huntington. 2007. Serial depletion of marine invertebrates leads to the decline of a strongly interacting grazer. *Ecological Applications* 17:1752–1770.

Sammarco, P. W. 1980. *Diadema* and its relationship to coral spat mortality: Grazing, competition, and biological disturbance. *Journal of Experimental Marine Biology and Ecology* 45:245–272.

Sammarco, P. W. 1982. Echinoid grazing as a structuring force in coral communities: Whole reef manipulations. *Journal of Experimental Marine Biology and Ecology* 61:31–55.

Sandin, S. A., and S. W. Pacala. 2005. Fish aggregation results in inversely density-dependent predation on continuous coral reefs. *Ecology* 86:1520–1530.

Sandin, S. A., J. E. Smith, E. E. DeMartini, E. A. Dinsdale, S. D. Donner, A. M. Friedlander, T. Konotchick, et al. 2008. Baselines and degradation of coral reefs in the northern Line Islands. *PLoS One* 3:e1548.

Santos, P. F., J. Phillips, and W. G. Whitford. 1981. The role of mites and nematodes in early stages of buried litter decomposition in a desert. *Ecology* 63:664–669.

Sarnelle, O. 1992. Nutrient enrichment and grazer effect on phytoplankton in lakes. *Ecology* 73:551–560.

Sass, G. G., J. F. Kitchell, S. R. Carpenter, T. R. Hrabik, A. E. Marburg, and M. G. Turner. 2006. Fish community and food web responses to a whole-lake removal of coarse woody habitat. *Fisheries Research* 31:321–330.

Sauvé, D. G., and S. D. Côté. 2007. Winter forage selection in white-tailed deer at high density: Balsam fir is the best of a bad choice. *Journal of Wildlife Management* 71:911–914.

Schall, J. J., and E. R. Pianka. 1978. Geographical trends in numbers of species. *Science* 201:679–686.

Scheel, D., and C. Packer. 1991. Group hunting behaviour of lions: A search for cooperation. *Animal Behaviour* 41:697–709.

Scheel, D., and C. Packer. 1995. Variation in predation by lions: Tracking a movable feast. Pp. 299–314 in A. R. E. Sinclair and P. Arcese, eds. *Serengeti II: Dynamics, management, and conservation of an ecosystem.* University of Chicago Press, Chicago.

Scheffer, M. 1998. *Ecology of shallow lakes.* Chapman & Hall, London.

Scheffer, M. 2009. *Critical transitions in nature and society.* Princeton University Press, Princeton, NJ.

Scheffer, M., and J. Beets. 1994. Ecological models and the pitfalls of causality. *Hydrobiologia* 275/276:115–124.

Scheffer, M., R. C. Carpenter, J. A. Foley, C. Folke, and B. Walker. 2001. Catastrophic shifts in ecosystems. *Nature* 413:591–596.

Scheffer, M., and S. R. Carpenter. 2003. Catastrophic regime shifts in ecosystems: Linking theory to observation. *Trends in Ecology and Evolution* 18:648–656.

Scheffer, M., M. Holmgren, V. Brovkin, and M. Claussen. 2005. Synergy between small- and large-scale feedbacks of vegetation on the water cycle. *Global Change Biology* 11:1003–1012.

Scheffer, M., S. H. Hosper, M. L. Meijer, B. Moss, and E. Jeppesen. 1993. Alternative equilibria in shallow lakes. *Trends in Ecology and Evolution* 8:275–279.

Scheffer, M., R. Portielje, and L. Zambrano. 2003a. Fish facilitate wave erosion. *Limnology and Oceanography* 48:1920–1926.

Scheffer, M., S. Rinaldi, A. Gragnani, L. R. Mur, and E. H. Van Nes. 1997. On the dominance of filamentous cyanobacteria in shallow, turbid lakes. *Ecology* 78:272–282.

Scheffer, M., S. Szabó, A. Gragnani, E. H. van Nes, S. Rinaldi, N. Kautsky, J. Norberg, R. M. M. Roijackers, and R. J. M. Franken. 2003b. Floating plant dominance as a stable state. *Proceedings of the National Academy of Sciences* 100:4040–4045.

Scheffer, M., and E. H. Van Nes. 2007. Shallow lakes theory revisited: Various alternative regimes driven by climate, nutrients, depth and lake size. *Hydrobiologia* 584:455–466.

Scheu, S., A. Theenhaus, and T. H. Jones. 1999. Links between the detritivore and herbivore system: Effects of earthworms and Collembola on plant growth and aphid development. *Oecologia* 119:541–551.

Schindler, D. E., S. R. Carpenter, J. J. Cole, J. F. Kitchell, and M. L. Pace. 1997. Food web structure alters carbon exchange between lakes and the atmosphere. *Science* 277:248–251.

Schindler, D. E., P. R. Leavitt, S. P. Johnson, and C. S. Brock. 2006. A 500-year context for the recent surge in sockeye salmon (*Oncorhynchus nerka*) abundance in the Alagnak River, Alaska. *Canadian Journal of Fisheries and Aquatic Sciences* 63:1439–1444.

Schindler, D. E., and M. D. Scheuerell. 2002. Habitat coupling in lake ecosystems. *Oikos* 98:177–189.

Schindler, D. W. 1998. Replication versus realism: The need for ecosystem-scale experiments. *Ecosystems* 1:323–334.

Schindler, D. W. 2006. Recent advances in the understanding and management of eutrophication. *Limnology and Oceanography* 51:356–363.

Schmidt, K. A. 2003. Nest predation and population declines in Illinois songbirds: A case for mesopredator effects. *Conservation Biology* 17:1141–1150.

Schmitz, O. J. 1994. Resource edibility and trophic exploitation in an old field food web. *Proceedings of the National Academy of Sciences* 91:5364–5367.

Schmitz, O. J. 2006. Predators have large effects on ecosystem properties by changing plant diversity, not plant biomass. *Ecology* 87:1432–1437.

Schmitz, O. J. 2008a. Effects of predator hunting mode on grassland ecosystem function. *Science* 319:952–954.

Schmitz, O. J. 2008b. Herbivory from individuals to ecosystems. *Annual Review of Ecology and Systematics* 39:133–152.

Schmitz, O. J., P. A. Hambäck, and A. P. Beckerman. 2000. Trophic cascades in terrestrial systems: A review of the effects of carnivore removals on plants. *American Naturalist* 155:141–153.

Schmitz, O. J., V. Krivan, and O. Ovadia. 2004. Trophic cascades: The primacy of trait-mediated indirect interactions. *Ecology Letters* 7:153–163.

Schoener, T. W. 1987. Leaf pubescence in buttonwood: Community variation in a putative defense against defoliation. *Proceedings of the National Academy of Sciences* 84:7992–7995.

Schoener, T. W. 1988. Leaf damage in island buttonwood, *Conocarpus erectus*: Correlations with pubescence, island area, isolation and the distribution of major carnivores. *Oikos* 53:253–266.

Schoener, T. W. 1989. Food webs from the small to the large. *Ecology* 70:1559–1589.

Schoener, T. W. 1993. On the relative importance of direct versus indirect effects in ecological communities. Pp. 365–411 in H. Kawanabe, J. E. Cohen, and K. Iwasaki, eds. *Mutu-*

alism and community organization: Behavioral, theoretical and food-web approaches. Oxford University Press, New York.

Schoener, T. W. 2008. Encyclopedia of ecology. Pp. 2040–2050 in B. Ronan, ed. *Island biogeography.* Elsevier, Oxford.

Schoener, T. W. 2009. The MacArthur–Wilson equilibrium model: What it said and how it was tested. Pp. 52–87 in J. B. Losos and R. E. Ricklefs, eds. *The theory of island biogeography revisited.* Princeton University Press, Princeton, NJ.

Schoener, T. W., and D. A. Spiller. 1987. Effect of lizards on spider populations: Manipulative reconstruction of a natural experiment. *Science* 236:949–952.

Schoener, T. W., and D. A. Spiller. 1996. Devastation of prey diversity by experimentally introduced predators in the field. *Nature* 381:691–694.

Schoener, T. W., and D. A. Spiller. 1999a. Indirect effects in an experimentally staged invasion by a major predator. *American Naturalist* 153:347–358.

Schoener, T. W., and D. A. Spiller. 1999b. Variation in the magnitude of a predator's effect from small to large islands. *Ecologia de les Illes. Monografies de la Societat d'Història Natural de les Balears* 6:35–66.

Schoener, T. W., D. A. Spiller, and J. B. Losos. 2001. Natural restoration of the species–area relation for a widespread lizard following a hurricane. *Science* 294:1525–1528.

Schoener, T. W., D. A. Spiller, and J. B. Losos. 2002. Predation on a common *Anolis* lizard: Can the food-web effects of a devastating predator be reversed? *Ecological Monographs* 72:383–407.

Schoener, T. W., D. A. Spiller, and L. W. Morrison. 1995. Variation in the hymenopteran parasitoid fraction on Bahamian islands. *Acta Oecologica* 16:103–121.

Schoener, T. W., and C. A. Toft. 1983. Spider populations: Extraordinarily high densities on islands without top predators. *Science* 219:1353–1355.

Scholes, R. J., and S. R. Archer. 1997. Tree–grass interactions in savannas. *Annual Review of Ecology and Systematics* 28:517–544.

Scholes, R. J., W. J. Bond, and H. C. Eckhardt. 2003. Vegetation dynamics in the Kruger ecosystem. Pp. 242–262 in J. T. du Toit, K. H. Rogers, and H. C. Biggs, eds. *The Kruger experience. Ecology and management of savanna heterogeneity.* Island Press, Washington, DC.

Seidensticker, J. 1976. On the ecological separation between tigers and leopards. *Biotropica* 4:225–234.

Seigler, D. S. 1998. *Plant secondary metabolism.* Kluwer Academic Publishers, New York.

Sergio, F., and F. Hiraldo. 2008. Intraguild predation in raptor assemblages: A review. *Ibis* 150:132–145.

Sessions, L. A., and D. Kelly. 2001. Heterogeneity in vertebrate and invertebrate herbivory and its consequences for New Zealand mistletoes. *Austral Ecology* 26:571–581.

Setälä, H. 1995. Growth of birch and pine seedlings in relation to grazing by soil fauna on ectomycorrhizal fungi. *Ecology* 76:1844–1851.

Shachak, M., S. Brand, and Y. Gutterman. 1991. Porcupine disturbances and vegetation pattern along a resource gradient in a desert. *Oecologia* 88:141–147.

Shackell, N. L., and K. T. Frank. 2007. Compensation in exploited marine fish communities on the Scotian Shelf, Canada. *Marine Ecology Progress Series* 336:235–247.

Shahabuddin, G., and J. W. Terborgh. 1999. Frugivorous butterflies in Venezuelan forest fragments: Abundance, diversity and the effects of isolation. *Journal of Tropical Ecology* 15:703–722.

Shapiro, J., V. Lamarra, and M. Lynch. 1975. *Biomanipulation: An ecosystem approach to lake restoration.* Paper presented at Symposium on Water Quality Management Through Biological Control, Gainesville, FL.

Shapiro, J., and D. I. Wright. 1984. Lake restoration by biomanipulation: Round Lake, Minnesota, the first 2 years. *Freshwater Biology* 14:371–383.

Sharam, G. 2005. *The decline and restoration of riparian and hilltop forests in the Serengeti National Park, Tanzania.* University of British Columbia, Vancouver.

Sharam, G., A. R. E. Sinclair, and R. Turkington. 2009. Serengeti birds maintain forests by inhibiting seed predators. *Science* 325:51.

Shead, J. 2005. *Deer hunter's almanac 2006.* F+W Publications, Iola, WI.

Shears, N. T., and R. C. Babcock. 2003. Continuing trophic cascade effects after 25 years of no-take marine reserve protection. *Marine Ecology Progress Series* 246:1–16.

Sherman, P. M. 2002. Effects of land crabs on seedling densities and distributions in a mainland Neotropical rain forest. *Journal of Tropical Ecology* 18:67–89.

Shiomoto, A., K. Tadokoro, K. Nagasawa, and Y. Ishida. 1997. Trophic relations in the subarctic North Pacific ecosystem: Possible feeding effect from pink salmon. *Marine Ecology: Progress Series* 150:75–85.

Short, F. T., and S. Wyllie-Echeverria. 1996. Natural and human-induced disturbance of seagrasses. *Environmental Conservation* 23:17–27.

Shulman, M. J., and J. C. Ogden. 1987. What controls tropical reef fish populations: Recruitment or benthic mortality? An example in the Caribbean reef fish *Haemulon flavolineatum. Marine Ecology Progress Series* 39:233–242.

Shurin, J. B., E. T. Borer, E. W. Seabloom, K. Anderson, C. A. Blanchette, B. Broitman, S. D. Cooper, and B. S. Halpern. 2002. A cross-ecosystem comparison of the strength of trophic cascades. *Ecology Letters* 5:785–791.

Shurin, J. B., D. S. Gruner, and H. Hillebrand. 2006. All wet or dried up? Real differences between aquatic and terrestrial food webs. *Proceedings of the Royal Society B: Biological Sciences* 273:1–9.

Shurin, J. B., and E. W. Seabloom. 2005. The strength of trophic cascades across ecosystems: Predictions from allometry and energetics. *Journal of Animal Ecology* 74:1029–1038.

Sieving, K. E. 1992. Nest predation and differential insular extinction among selected forest birds of central Panama. *Ecology* 73:2310–2328.

Sih, A. 1991. Reflections on the power of a grand paradigm. *Bulletin of the Ecological Society of America* 72:174–178.

Sih, A. 1992. Prey uncertainty and the balancing of antipredator and feeding needs. *American Naturalist* 139:1052–1069.

Silliman, B. R., and M. D. Bertness. 2002. A trophic cascade regulates salt marsh primary production. *Proceedings of the National Academy of Sciences* 99:10500–10505.

Silliman, B. R., J. Van de Koppel, M. D. Bertness, L. E. Stanton, and I. R. Mendelssohn. 2005. Drought, snails, and large-scale die-off of southern U.S. salt marshes. *Science* 310:1803–1806.

Silman, M. R., J. W. Terborgh, and R. A. Kiltie. 2003. Population regulation of a dominant rain forest tree by a major seed predator. *Ecology* 84:431–438.

Sinclair, A. R. E. 1975. The resource limitation of trophic levels in tropical grassland ecosystems. *Journal of Animal Ecology* 44:497–520.

Sinclair, A. R. E. 1977. *The African buffalo: A study of resource limitation of populations.* University of Chicago Press, Chicago.

Sinclair, A. R. E. 1979. The eruption of the ruminants. Pp. 82–103 in A. R. E. Sinclair and M. Nortin-Griffiths, eds. *Serengeti: Dynamics of an ecosystem.* University of Chicago Press, Chicago.

Sinclair, A. R. E. 1989. Population regulation in animals. Pp. 197–241 in J. M. Cherrett, ed. *Ecological concepts.* Blackwell, Oxford.

Sinclair, A. R. E. 1995. Equilibria in plant–herbivore interactions. Pp. 91–114 in A. R. E. Sinclair and P. Arecese, eds. *Serengeti II: Dynamics, management, and conservation of an ecosystem.* University of Chicago Press, Chicago.

Sinclair, A. R. E. 2003. Mammal population regulation, keystone processes, and ecosystem dynamics. *Philosophical Transactions of the Royal Society of London. Series B, Biological Sciences* 358:1729–1740.

Sinclair, A. R. E., and P. Arcese. 1995. Population consequences of predation-sensitive foraging: The Serengeti wildebeest. *Ecology* 76:882–891.

Sinclair, A. R. E., H. Dublin, and M. Borner. 1985. Population regulation of Serengeti Wildebeest: A test of the food hypothesis. *Oecologia* 65:266–268.

Sinclair, A. R. E., J. G. C. Hopcraft, H. Olff, S. A. R. Mduma, K. A. Galvin, and G. J. Sharam. 2008. Historical and future changes to the Serengeti ecosystem. Pp. 7–46 in A. R. E. Sinclair, C. Packer, S. A. R. Mduma, and J. M. Fryxell, eds. *Serengeti III: Human impacts on ecosystem dynamics.* University of Chicago Press, Chicago.

Sinclair, A. R. E., and C. J. Krebs. 2002. Complex numerical responses to top-down and bottom-up processes in vertebrate populations. *Philosophical Transactions of the Royal Society of London. Series B, Biological Sciences* 357:1221–1231.

Sinclair, A. R. E., C. J. Krebs, J. M. Fryxell, R. Turkington, S. Boutin, R. Boonstra, P. Seccombe-Hest, P. Lundberg, and L. Oksanen. 2000. Testing hypotheses on trophic level interactions: A boreal forest ecosystem. *Oikos* 89:313–328.

Sinclair, A. R. E., S. Mduma, and J. S. Brashares. 2003. Patterns of predation in a diverse predator–prey system. *Nature* 425:288–290.

Sinclair, A. R. E., S. A. R. Mduma, J. G. C. Hopcraft, J. M. Fryxell, R. Hilborn, and S. Thirgood. 2007. Long-term ecosystem dynamics in the Serengeti: Lessons for conservation. *Conservation Biology* 21:580–590.

Sinclair, A. R. E., and K. Metzger. 2009. Advances in wildlife ecology and the influence of Graeme Caughley. *Wildlife Research* 36:8–15.

Sinclair, A. R. E., P. O. Olsen, and T. D. Redhead. 1990. Can predators regulate small mammal populations? Evidence from house mouse outbreaks in Australia. *Oikos* 59:382–392.

Sinclair, A. R. E., R. P. Pech, C. R. Dickman, D. Hik, P. Mahon, and A. E. Newsome. 1998. Predicting effects of predation on conservation of endangered prey. *Conservation Biology* 12:564–575.

Singer, F. J. 1996. Differences between willow communities browsed by elk and communities protected for 32 years in Yellowstone National Park. Pp. 279–290 in *Effects of grazing by wild ungulates in Yellowstone National Park*. Technical Report NPS/NRYELL/NRTR/96-01. USDI, National Park Service, Natural Resource Information Division, Yellowstone National Park, WY.

Singer, M. S., D. Rodrigues, J. O. Stireman III, and Y. Carrière. 2004. Roles of food quality and enemy-free space in host use by a generalist insect herbivore. *Ecology* 85:2747–2753.

Skarpe, C., P. A. Aarrestad, H. P. Andreassen, S. S. Dhillion, T. Dimakatso, J. T. du Toit, D. J. Halley, et al. 2004. The return of the giants: Ecological effects of an increasing elephant population. *Ambio* 33:276–282.

Skogland, T. 1990. Density dependence in a fluctuating wild reindeer herd: Maternal vs. offspring effects. *Oecologia* 84:442–450.

Skottsberg, C. 1922. The phanerogams of the Juan Fernández Islands. Pp. 95–240 in C. Skottsberg, ed. *The natural history of Juan Fernández and Easter Island. Vol. II. Botany.* Almquist and Wiksell, Uppsala, Sweden.

Slobodkin, L. B., F. E. Smith, and N. C. Hairston. 1967. Regulation in terrestrial ecosystems and the implied balance of nature. *American Naturalist* 101:109–124.

Slusarczyk, M. 2005. Food threshold for diapause in *Daphnia* under the threat of fish predation. *Ecology* 82:1089–1096.

Smith, B. L., and R. L. Robbins. 1994. *Migrations and management of the Jackson elk herd*. Resource Publication 199. U.S. Department of the Interior, National Biological Survey, Washington, DC.

Smith, D. W., R. O. Peterson, and D. B. Houston. 2003a. Yellowstone after wolves. *BioScience* 53:330–340.

Smith, F. A., S. K. Lyons, and S. K. M. Ernest. 2003b. Extinctions of herbivorous mammals in the late Pleistocene of Australia in relation to their feeding ecology: No evidence for environment change as cause of extinction. *Austral Ecology* 29:553–557.

Smith, J. E., M. Shaw, R. A. Edwards, D. Obura, O. Pantos, E. Sala, S. A. Sandin, S. Smriga, M. Hatay, and F. Rohwer. 2006. Indirect effects of algae on coral: Algae-mediated, microbe-induced coral mortality. *Ecology Letters* 9:835–845.

Smith, J. E., C. M. Smith, and C. L. Hunter. 2001. An experimental analysis of the effects of herbivory and nutrient enrichment on benthic community dynamics on a Hawaiian reef. *Coral Reefs* 19:332–342.

Smith, S. V. 1978. Coral-reef area and the contributions of reefs to processes and resources of the world's oceans. *Nature* 273:225–226.

Smuts, G. L. 1978. Interrelations between predators, prey, and their environment. *BioScience* 28:316–320.

Smythe, N. 1982. The seasonal abundance of night-flying insects in a Neotropical forest. Pp. 309–318 in E. G. Leigh Jr., A. S. Rand, and D. M. Windsor, eds. *The ecology of a tropical forest: Seasonal rhythms and long-term changes.* Smithsonian Institution Press, Washington, DC.

Soler, R., T. M. Bezemer, A. M. Cortesero, W. H. Van der Putten, L. E. M. Vet, and J. A. Harvey. 2007. Impact of foliar herbivory on the development of a root-feeding insect and its parasitoid. *Oecologia* 152:257–264.

Sommer, U., and F. Sommer. 2006. Cladocerans versus copepods: The cause of contrasting top-down controls on freshwater and marine phytoplankton. *Oecologia* 147:183–194.

Sorokin, Y. I. 1995. *Coral reef ecology.* Springer, Berlin.

Soulé, M., D. T. Bolger, A. C. Alberts, J. Wright, M. Sorice, and S. Hill. 1988. Reconstructing dynamics of rapid extinctions of chaparral-requiring birds in urban habitat islands. *Conservation Biology* 2:75–92.

Soulé, M. E., J. A. Estes, B. Miller, and D. L. Honnold. 2005. Strongly interacting species: Conservation policy, management and ethics. *BioScience* 55:168–176.

Soulé, M. E., and J. Terborgh. 1999. *Continental conservation: Scientific foundations for regional reserve networks.* Island Press, Washington, DC.

Sovada, M. A., A. B. Sargeant, and J. W. Grier. 1995. Differential effects of coyotes and red foxes on duck nest success. *Journal of Wildlife Management* 59:1–9.

Spalding, M. D., H. E. Fox, G. R. Allen, N. Davidson, Z. A. Ferdana, M. Finlayson, B. S. Halpern, et al. 2007. Marine ecoregions of the world: A bioregionalization of coastal and shelf areas. *BioScience* 57:573–583.

Spiller, D. A., and A. A. Agrawal. 2003. Intense disturbance enhances plant susceptibility to herbivory: Natural and experimental evidence. *Ecology* 84:890–897.

Spiller, D. A., J. B. Losos, and T. W. Schoener. 1998. Impact of a catastrophic hurricane on island populations. *Science* 281:695–697.

Spiller, D. A., and T. W. Schoener. 1988. An experimental study of the effect of lizards on web-spider communities. *Ecological Monographs* 58:57–77.

Spiller, D. A., and T. W. Schoener. 1990a. Lizards reduce food consumption by spiders: Mechanisms and consequences. *Oecologia* 83:150–161.

Spiller, D. A., and T. W. Schoener. 1990b. A terrestrial field experiment showing impact of eliminating top predators on foliage damage. *Nature* 347:469–472.

Spiller, D. A., and T. W. Schoener. 1994. Effects of top and intermediate predators in a terrestrial food web. *Ecology* 75:182–196.

Spiller, D. A., and T. W. Schoener. 1995. Long-term variation in effect of lizards on spider density is linked to rainfall. *Oecologia* 103:133–139.

Spiller, D. A., and T. W. Schoener. 1996. Food-web dynamics on some small subtropical islands: Effects of top and intermediate predators. Pp. 160–169 in G. A. Polis and K. O. Winemiller, eds. *Food webs: Integration of patterns and dynamics.* Chapman & Hall, New York.

Spiller, D. A., and T. W. Schoener. 1997. Folivory on islands with and without insectivorous lizards: An eight-year study. *Oikos* 78:15–22.

Spiller, D. A., and T. W. Schoener. 1998. Lizards reduce spider species richness by excluding rare species. *Ecology* 79:503–516.

Spiller, D. A., and T. W. Schoener. 2001. An experimental test for predator-mediated interactions among spider species. *Ecology* 82:1560–1570.

Spiller, D. A., and T. W. Schoener. 2007. Alteration of island food-web dynamics following major disturbance by hurricanes. *Ecology* 88:37–41.

Spiller, D. A., and T. W. Schoener. 2008. Climatic control of trophic interaction strength: The effect of lizards on spiders. *Oecologia* 154:763–771.

Spiller, D. A., and T. W. Schoener. 2009. Species area. Pp. 857–861 in R. Gillespie and D. Clague, eds. *Encyclopedia of islands*. University of California Press, Berkeley.

Springer, A. M., J. A. Estes, G. B. van Vliet, T. M. Williams, D. F. Doak, E. M. Danner, K. A. Forney, and B. Pfister. 2003. Sequential megafaunal collapse in the North Pacific Ocean: An ongoing legacy of industrial whaling? *Proceedings of the National Academy of Sciences* 100:12223–12228.

Stachowicz, J. J., J. F. Bruno, and J. E. Duffy. 2007. Understanding the effects of marine biodiversity on communities and ecosystems. *Annual Review of Ecology, Evolution, and Systematics* 38:739–766.

Stahler, D., B. Heinrich, and D. Smith. 2002. Common ravens, *Corvus corax*, preferentially associate with gray wolves, *Canis lupus*, as a foraging strategy in winter. *Animal Behaviour* 64:283–290.

Stallings, C. D. 2008. Indirect effects of an exploited predator on recruitment of coral-reef fishes. *Ecology* 89:2090–2095.

Steadman, D. W. 2006. *Extinction and biogeography of tropical Pacific birds*. University of Chicago Press, Chicago.

Steinberg, P. D., J. A. Estes, and F. C. Winter. 1995. Evolutionary consequences of food chain length in kelp forest communities. *Proceedings of the National Academy of Sciences* 92:8145–8148.

Steinfeld, H. P., T. Gerber, T. Wassenaar, V. Castel, M. Rosales, and C. de Haan. 2006. *Livestock's long shadow: Environmental issues and options*. Retrieved November 15, 2009 from www.fao.org/docrep/010/a0701e/a0701e00.htm.

Steneck, R. S. 1983. Escalating herbivory and resulting adaptive trends in calcareous algal crusts. *Paleobiology* 9:44–61.

Steneck, R. S. 1997. *Fisheries-induced biological changes to the structure and function of the Gulf of Maine ecosystem*. Paper presented at the Gulf of Maine Ecosystem Dynamics Scientific Symposium and Workshop. Regional Association for Research in the Gulf of Maine (RARGOM), Hanover, NH.

Steneck, R. S., M. H. Graham, B. J. Bourque, D. Corbett, J. M. Erlandson, J. A. Estes, and M. J. Tegner. 2002. Kelp forest ecosystem: Biodiversity, stability, resilience and their future. *Environmental Conservation* 29:436–459.

Steneck, R. S., and E. Sala. 2005. Large marine carnivores: Trophic cascades and top-down controls in coastal ecosystems past and present. Pp. 110–137 in J. C. Ray, K. H. Redford, R. S. Steneck, and J. Berger, eds. *Large carnivores and the conservation of biodiversity*. Island Press, Washington, DC.

Steneck, R. S., J. Vavrinec, and A. V. Leland. 2004. Accelerating trophic level dysfunction in kelp forest ecosystems of the western North Atlantic. *Ecosystems* 7:323–331.

Steneck, R. S., and C. J. Wilson. 2001. Long-term and large scale spatial and temporal patterns in demography and landings of the American lobster, *Homarus americanus*, in Maine. *Journal of Marine and Freshwater Research* 52:1302–1319.

Stephens, D. W., J. S. Brown, and R. C. Ydenberg, eds. 2007. *Foraging: Behavior and ecology.* University of Chicago Press, Chicago.

Stewart, G. H., and L. E. Burrows. 1989. The impact of white-tailed deer *Odocoileus virginianus* on regeneration in the coastal forests of Stewart Island, New Zealand. *Biological Conservation* 49:275–293.

Stibor, H., O. Vadstein, S. Diehl, A. Gelzleichter, T. Hansen, F. Hantzsche, A. Katechakis, et al. 2004. Copepods act as a switch between alternative trophic cascades in marine pelagic food webs. *Ecology Letters* 7:321–328.

Stireman, J. O. III, L. A. Dyer, D. H. Janzen, M. S. Singer, J. T. Lill, R. J. Marquis, R. E. Ricklefs, et al. 2005. Climate unpredictability and parasitism of caterpillars: implications of global warming. *Proceedings of the National Academy of Sciences* 102:17384–17387.

Stockton, S. A., S. Allombert, A. J. Gaston, and J. L. Martin. 2005. A natural experiment on the effects of high deer densities on the native flora of coastal temperate rain forests. *Biological Conservation* 126:118–128.

Stoddart, D. R. 1968. The conservation of Aldabra. *Geographical Journal* 134:418–482.

Stoddart, D. R. 1971. Settlement, development and conservation of Aldabra. *Philosophical Transactions of the Royal Society of London, Series B. Biological Sciences* 260:611–628.

Stoddart, D. R., and J. F. Peake. 1979. Historical records of Indian-Ocean giant tortoise populations. *Philosophical Transactions of the Royal Society of London. Series B, Biological Sciences* 286:147–161.

Stokstad, E. 2004. Loss of dung beetles puts ecosystems in deep doo-doo. *Science* 305:1230.

Stolzenburg, W. 2008. *Where the wild things were.* Bloomsbury, New York.

Stow, C. A., S. R. Carpenter, C. P. Madenjian, L. A. Eby, and L. J. Jackson. 1995. Fisheries management to reduce contaminant consumption. *BioScience* 46:752–758.

Straub, C. S., D. L. Finke, and W. E. Snyder. 2008. Are the conservation of natural enemy biodiversity and biological control compatible goals? *Biological Control* 45:225–237.

Straub, C. S., and W. E. Snyder. 2008. Increasing enemy biodiversity strengthens herbivore suppression on two plant species. *Ecology* 89:1605–1615.

Strauss, S. Y., and A. R. Zangerl. 2002. Plant–insect interactions in terrestrial ecosystems. Pp. 77–106 in C. M. Herrera and O. Pellmyr, eds. *Plant–animal interactions, an evolutionary approach.* Blackwell Science, Oxford.

Stromayer, K. A. K., and R. J. Warren. 1997. Are overabundant deer herds in the eastern United States creating alternate stable states in forest plant communities? *Wildlife Society Bulletin* 25:227–234.

Strong, A. M., T. W. Sherry, and R. T. Holmes. 2000. Bird predation on herbivorous insects: Indirect effects on sugar maple saplings. *Oecologia* 125:370–379.

Strong, D. R. 1992. Are trophic cascades all wet? Differentiation and donor-control in speciose ecosystems. *Ecology* 73:747–754.

Strong, D. R., D. Simberloff, L. G. Abele, and A. B. Thistle. 1984. *Ecological communities: Conceptual issues and the evidence.* Princeton University Press, Princeton, NJ.

Strong, D. R., A. V. Whipple, A. L. Child, and B. Dennis. 1999. Model selection for a subterranean trophic cascade: Root feeding nematodes and entomopathogenic nematodes. *Ecology* 80:2750–2761.

Strum, S. C. 1975. Primate predation: Interim report on the development of a tradition in a troop of olive baboons. *Science* 187:755–757.

Sukumar, R. 1988. *The Asian elephant: Ecology and management.* Cambridge University Press, Cambridge.

Summerson, H. C., and C. H. Peterson. 1984. Role of predation in organizing benthic communities of a temperate-zone seagrass bed. *Marine Ecology Progress Series* 15:63–77.

Sundell, J. 2006. Experimental tests of the role of predation in the population dynamics of voles and lemmings. *Mammalogy Review* 36:197–141.

Suominen, O. 1999. Impact of cervid browsing and grazing on the terrestrial gastropod fauna in the boreal forests of Fennoscandia. *Ecography* 22:651–658.

Sutherland, J. P. 1974. Multiple stable points in natural communities. *American Naturalist* 108:859–873.

Swift, E. 1948. *Wisconsin's deer damage forest reproduction survey: Final report.* Wisconsin Conservation Department Publication 347.

Swingland, I. R., and M. J. Coe. 1979. The natural regulation of giant tortoise populations on Aldabra Atoll: Recruitment. *Philosophical Transactions of the Royal Society of London. Series B, Biological Sciences* 286:177–188.

Switalski, T. A. 2003. Coyote foraging ecology and vigilance in response to gray wolf reintroduction in Yellowstone National Park. *Canadian Journal of Zoology* 81:985–993.

Taillon, J., D. G. Sauvé, and S. D. Côté. 2006. The effects of decreasing winter diet quality on foraging behavior and life-history traits of white-tailed deer fawns. *Journal of Wildlife Management* 70:1445–1454.

Takatsuki, S., and T. Gorai. 1994. Effects of sika deer on the regeneration of a *Fagus crenata* forest on Kinkazan Island, northern Japan. *Ecological Research* 9:115–120.

Tansley, A. G. 1926. Studies of the vegetation of the English chalk. II: Early stages of redevelopment of woody vegetation on chalk grassland. *Journal of Ecology* 14:1–32.

Tansley, A. G. 1939. *The British islands and their vegetation.* Cambridge University Press, Cambridge.

Taylor, J. D., C. J. R. Braithwaite, J. F. Peake, and E. N. Arnold. 1979. Terrestrial faunas and habitats on Aldabra during the late Pleistocene. *Philosophical Transactions of the Royal Society of London, Series B, Biological Sciences* 286:47–66.

Teer, J. G. 1990. Review of game ranch management. *Journal of Wildlife Management* 54:369.

Tegner, M. J., and P. K. Dayton. 1991. Sea urchins, El Niños, and the long term stability of southern California kelp forest communities. *Marine Ecology Progress Series* 77:49–63.

Tegner, M. J., and P. K. Dayton. 2000. Ecosystem effects of fishing in kelp forest communities. *ICES Journal of Marine Science* 57:579–589.

Telford, S. R. III. 2002. Deer tick–transmitted zoonoses in the eastern United States. Pp. 310–324 in A. A. Aguirre, R. F. Ostfeld, G. M. Tabor, C. House, and M. C. Pearl, eds. *Conservation medicine: Ecological health in practice.* Oxford University Press, Oxford.

Terborgh, J. 1988. The big things that run the world: A sequel to E. O. Wilson. *Conservation Biology* 2:402–403.

Terborgh, J. 1989. *Where have all the birds gone?* Princeton University Press, Princeton, NJ.

Terborgh, J. 1992. Maintenance of diversity in tropical forests. *Biotropica* 24:283–292.

Terborgh, J. 1999. *Requiem for nature.* Island Press, Washington, DC.

Terborgh, J. 2009. The trophic cascade on islands. Pp. 116–142 in J. B. Losos and R. E. Ricklefs, eds. *The theory of island biogeography revisited.* Princeton University Press, Princeton, NJ.

Terborgh, J., J. A. Estes, P. Paquet, K. Ralls, D. Boyd-Heigher, B. J. Miller, and R. F. Noss. 1999. The role of top carnivores in regulating terrestrial ecosystems. Pp. 39–54 in M. Soulé and J. Terborgh, eds. *Continental conservation: Scientific foundations of regional reserve networks.* Island Press, Washington, DC.

Terborgh, J., and J. Faaborg. 1973. Turnover and ecological release in the avifauna of Mona Island, Puerto Rico. *The Auk* 90:759–779.

Terborgh, J., K. Feeley, M. Silman, P. Nuñez, and B. Balukjan. 2006. Vegetation dynamics on predator-free land-bridge islands. *Journal of Ecology* 94:253–263.

Terborgh, J., L. Lopez, P. Nuñez, M. Rao, G. Shahabuddin, G. Orihuela, M. Riveros, et al. 2001. Ecological meltdown in predator-free forest fragments. *Science* 294:1923–1926.

Terborgh, J., L. Lopez, and J. Tello. 1997a. Bird communities in transition: The Lago Guri Islands. *Ecology* 78:1494–1501.

Terborgh, J., L. Lopez, J. Tello, D. Yu, and A. R. Bruni. 1997b. Transitory states in relaxing ecosystems of land bridge islands. Pp. 256–274 in W. F. Laurance and R. O. Bierregaard Jr., eds. *Tropical forest remnants: Ecology, management, and conservation of fragmented communities.* University of Chicago Press, Chicago.

Terborgh, J., G. Nuñez-Ituri, N. Pitman, F. Cornejo, P. Alvarez, B. Pringle, V. Swamy, and T. Paine. 2008. Tree recruitment in an "empty" forest. *Ecology* 89:1757–1768.

Terborgh, J., and G. W. Winter. 1980. Some causes of extinction. Pp. 119–131 in M. E. Soulé and B. A. Wilcox, eds. *Conservation biology: An ecological perspective.* Sinauer, Sunderland, MA.

Terborgh, J. W. 2000. In the company of humans. *Natural History* 109:54–62.

Tessier, A. J., and P. Woodruff. 2002. Cryptic trophic cascade along a gradient of lake size. *Ecology* 83:1263–1270.

Testa, J. W., E. F. Becker, and G. R. Lee. 2000. Movements of female moose in relation to birth and death of calves. *Alces* 36:155–162.

Thacker, R. W., D. W. Ginsburg, and V. J. Paul. 2001. Effects of herbivore exclusion and nutrient enrichment on coral reef macroalgae and cyanobacteria. *Coral Reefs* 19:318–329.

Thayer, G. W., K. A. Bjorndal, J. C. Ogden, S. L. Williams, and J. C. Zieman. 1982. Role of large herbivores in seagrass communities. *Estuaries* 7:351–376.

Thom, R. 1993. *Structural stability and morphogenesis: An outline of a general theory of models.* Addison-Wesley, Reading, MA.

Thompson, J. N. 1994. *The coevolutionary process.* University of Chicago Press, Chicago.

Thompson, J. N., O. J. Reichman, P. J. Morin, G. A. Polis, M. E. Power, R. W. Sterner, C. A. Couch, et al. 2001. Frontiers of ecology. *BioScience* 51:15–24.

Thompson, R. M., M. Hemberg, B. M. Starzomski, and J. B. Shurin. 2007. Trophic levels and trophic tangles: The prevalence of omnivory in real food webs. *Ecology* 88:612–617.

Tilghman, N. G. 1989. Impacts of white-tailed deer on forest regeneration in northwestern Pennsylvania. *Journal of Wildlife Management* 53:524–532.

Tilman, D. 1982. *Resource competition and community structure*. Princeton University Press, Princeton, NJ.

Tilman, D. 1997. Mechanisms of plant competition. Pp. 239–261 in M. J. Crawley, ed. *Plant ecology*. Blackwell Science, Oxford.

Tilman, D., P. Reich, H. Phillips, M. Menton, A. Patel, E. Vos, D. Peterson, and J. Knops. 2000. Fire suppression and ecosystem carbon storage. *Ecology* 81:2680–2685.

Toft, C. A., and T. W. Schoener. 1983. Abundance and diversity of orb spiders on 106 Bahamian islands: Biogeography of an intermediate trophic level. *Oikos* 41:411–426.

Towns, D. R., D. A. Wardle, C. P. H. Mulder, G. W. Yeates, B. M. Fitzgerald, G. R. Parrish, P. J. Bellingham, and K. I. Bonner. 2009. Predation of seabirds by invasive rats: Multiple indirect consequences for invertebrate communities. *Oikos* 118:420–430.

Tremblay, J.-P., I. Thibault, C. Dussault, J. Hout, and S. D. Côté. 2005. Long-term decline in white-tailed deer browse supply: Can lichens and litterfall act as alternative food sources that preclude density dependent feedbacks. *Canadian Journal of Zoology* 83:1087–1096.

Tscharntke, T. 1997. Vertebrate effects on plant–invertebrate food webs. Pp. 277–297 in A. C. Gange and V. K. Brown, eds. *Multitrophic interactions in terrestrial systems*. Blackwell Science, Oxford.

Tsou, T. S., and J. S. Collie. 2001. Estimating predation mortality in the Georges Bank fish community. *Canadian Journal of Fisheries and Aquatic Sciences* 58:908–922.

Turchin, P., L. Oksanen, P. Ekerholm, T. Oksanen, and H. Henttonen. 2000. Lemmings: Prey or predators. *Nature* 405:562–564.

U.S. Fish and Wildlife Service. 2006. *2006 National survey of fishing, hunting, and wildlife-associated recreation*. Retrieved November 15, 2009 from wsfrprograms.fws.gov/Subpages/NationalSurvey/National_Survey.htm.

Utah Division of Wildlife Resources. 2005. *Utah wolf management plan*. Utah Division of Wildlife Resources, Salt Lake City.

Vadas, R. L., and R. S. Steneck. 1995. Overfishing and inferences in kelp–sea urchin interactions. Pp. 509–524 in H. R. Skjoldal, C. Hopkins, K. J. Erikstad, and H. P. Leinaas, eds. *Ecology of fjords and coastal waters*. Elsevier Science, Amsterdam.

Valiela, I. 1995. *Marine ecological processes*. Springer-Verlag, New York.

Van Bael, S. A., and J. D. Brawn. 2005. The direct and indirect effects of insectivory by birds in two contrasting Neotropical forests. *Oecologia* 143:106–116.

Van de Koppel, J., R. D. Bardgett, J. Bengtsson, C. Rodriguez-Barrueco, M. Rietkerk, M. J. Wassen, and V. Wolters. 2005. The effects of spatial scale on trophic interactions. *Ecosystems* 8:801–807.

Van de Koppel, J., P. M. J. Herman, P. Thoolen, and C. H. R. Heip. 2001. Do multiple stable states occur in natural ecosystems? Evidence from tidal flats. *Ecology* 82:3449–3461.

Van Deelen, T. R., and D. R. Etter. 2003. Effort and functional response of deer hunters. *Human Dimensions of Wildlife* 8:97–108.

Van der Putten, W. H., L. E. M. Vet, J. A. Harvey, and F. L. Wackers. 2001. Linking above- and belowground multitrophic interactions of plants, herbivores, pathogens and their antagonists. *Trends in Ecology and Evolution* 16:547–554.

Van der Stap, I., M. Vos, A. M. Verschoor, N. R. Helmsing, and W. M. Mooli. 2007. Induced defenses in herbivores and plants differentially modulate a trophic cascade. *Ecology* 88:2474–2481.

van der Wal, R. 2006. Do herbivores cause habitat degradation or habitat state transition? Evidence from the tundra. *Oikos* 114:177–186.

Van Leeuwen, A., A. M. De Roos, and L. Persson. 2008. How cod shapes its world. *Journal of Sea Research* 60:89–104.

Van Nes, E. H., and M. Scheffer. 2004. Large species shifts triggered by small forces. *American Naturalist* 164:255–266.

Van Nes, E. H., and M. Scheffer. 2007. Slow recovery from perturbations as a generic indicator of a nearby catastrophic shift. *American Naturalist* 169:738–747.

Van Orsdol, K. G., J. P. Hanby, and J. D. Bygott. 1985. Ecological correlates of lion social organization. *Journal Zoology London* 206:97–112.

van Rijn, P. C. J., and M. W. Sabelis. 2005. Impact of plant-provided food on herbivore–carnivore dynamics. Pp. 223–266 in F. L. Wackers, P. C. J. van Rijn, and J. Bruin, eds. *Plant-provided food for carnivorous insects*. Cambridge University Press, Cambridge.

van Schaik, C. P., and M. A. van Noordjwick. 1985. Evolutionary effect of the absence of felids on the social organization of the macaques on the island of Simeulue. *Folia Primatologica* 44:138–147.

van Veen, F. J. F., A. Rajkumar, C. B. Müller, and H. C. J. Godfray. 2001. Increased reproduction by pea aphids in the presence of secondary parasitoids. *Ecological Entomology* 26:425–429.

Vander Zanden, M. J., T. E. Essington, and Y. Vadeboncoeur. 2005. Is pelagic top-down control in lakes augmented by benthic energy pathways? *Canadian Journal of Fisheries and Aquatic Sciences* 62:1422–1431.

Vander Zanden, M. J., and J. B. Rasmussen. 1996. A trophic position model of pelagic food webs: Impact on contaminant bioaccumulation in lake trout. *Ecological Monographs* 66:451–477.

Vanni, M. J., C. Luecke, J. F. Kitchell, Y. Allen, J. Temte, and J. J. Magnuson. 1990. Effects on lower trophic levels of massive fish mortality. *Nature* 344:333–335.

VanNimwegen, R. E., J. Kretzer, and J. F. Cully Jr. 2008. Ecosystem engineering by a colonial mammal: How prairie dogs structure rodent communities. *Ecology* 89:3298–3305.

Vavrinec, J. 2003. *Resilience of green sea urchin (*Strongylocentrotus droebachiensis*) populations following fishing mortality: Marine protected areas, larval ecology and post-settlement survival*. Ph.D. dissertation, University of Maine, School of Marine Sciences.

Veblen, T. T., M. Mermoz, C. Martin, and E. Ramilo. 1989. Effects of exotic deer on forest regeneration and composition in northern Patagonia. *Journal of Applied Ecology* 26:711–724.

Veblen, T. T., and G. H. Stewart. 1982. The effects of introduced wild animals on New Zealand forests. *Annals of the Association of American Geographers* 72:372–397.

Vera, F. W. M. 2000. *Grazing ecology and forest history*. CABI, Oxford.

Verity, P. G., and V. Smetacek. 1996. Organism life cycles, predation, and the structure of marine pelagic ecosystems. *Marine Ecology: Progress Series* 130:277–293.

Vermeij, G. J. 1977. Mesozoic marine revolution: Evidence from snails, predators and grazers. *Paleobiology* 3:245–258.

Vermeij, G. J. 1987. *Evolution and escalation: An ecological history of life*. Princeton University Press, Princeton.

Vermeij, M. J. A., and S. A. Sandin. 2008. Density-dependent settlement and mortality structure the earliest life phases of a coral population. *Ecology* 89:1994–2004.

Vermeij, M. J. A., J. E. Smith, C. M. Smith, R. Vega Thurber, and S. A. Sandin. 2009. Survival and settlement success of coral planulae: Independent and synergistic effects of macroalgae and microbes. *Oecologia* 159:325–336.

Viljoen, P. C. 1993. The effects of changes in prey availability on lion predation in a large, natural ecosystem in northern Botswana. *Symposia of the Zoological Society of London* 65:193–213.

Vos, M., S. M. Berrocal, F. Karamaouna, L. Hemerik, and L. E. M. Vet. 2001. Plant-mediated indirect effects and the persistence of parasitoid–herbivore communities. *Ecology Letters* 4:38–45.

Vucetich, J. A., R. O. Peterson, and T. A. Waite. 2004. Raven scavenging favors group foraging in wolves. *Animal Behaviour* 67:1117–1126.

Waldram, M. S., W. J. Bond, and W. D. Stock. 2008. Ecological engineering by a megagrazer: White rhino impacts on a South African savanna. *Ecosystems* 11:101–112.

Walker, B. H., R. H. Emslie, R. N. Owen-Smith, and R. J. Scholes. 1987. To cull or not to cull: Lessons from a southern African drought. *Journal of Applied Ecology* 24:381–401.

Wallace, A. R. 1860. On the zoological geography of the Malay Archipelago. *Journal of the Linnean Society* 4:172–184.

Wallace, J. B., S. L. Eggert, J. L. Meyer, and J. R. Webster. 1997. Multiple trophic levels of a forest stream linked to terrestrial litter inputs. *Science* 277:102–104.

Waller, D. M., and T. P. Rooney. 2008. *The vanishing present, Wisconsin's changing lands, waters, and wildlife*. University of Chicago Press, Chicago.

Walter, H. 1964. *Die Vegetation der Erde in öko-physiologischer Betrachtung. II Die tropischen und subtropischen Zonen*. Gustav Fischer Verlag, Jena, Germany.

Walter, H. 1968. *Die Vegetation der Erde in öko-physiologischer Betrachtung. II Die gemäßigten und arktischen Zonen*. Gustav Fischer Verlag, Jena, Germany.

Walter, H. 1985. *Vegetation of the earth and ecological systems of the geo-biosphere*, 3rd ed. Springer-Verlag, New York.

Walter, K. S. 1983. Orchidaceae. Pp. 282–292 in D. H. Janzen, ed. *Costa Rican natural history*. University of Chicago Press, Chicago.

Walters, C. J., and F. Juanes. 1993. Recruitment limitation as a consequence of natural selection for use of restricted feeding habitats and predation risk taking by juvenile fishes. *Canadian Journal of Fisheries and Aquatic Sciences* 50:2058–2070.

Walters, C. J., and J. F. Kitchell. 2001. Cultivation/depensation effects on juvenile survival and recruitment. *Canadian Journal of Fisheries and Aquatic Sciences* 58:39–50.

Walters, C. J., and S. J. Martell. 2004. *Fisheries ecology and management.* Princeton University Press, Princeton, NJ.

Wang, B. C., and T. B. Smith. 2002. Closing the seed dispersal loop. *Trends in Ecology and Evolution* 17:379–385.

Wangchuk, T. 2004. Predator–prey dynamics: The role of predators in the control of problem species. *Journal of Bhutan Studies* 10:68–89.

Ward, P., and R. A. Myers. 2005. Shifts in open-ocean fish communities coinciding with the commencement of commercial fishing. *Ecology* 86:836–847.

Wardle, D. A. 2002. *Communities and ecosystems: Linking the aboveground and belowground components.* Princeton University Press, Princeton, NJ.

Wardle, D. A., and R. D. Bardgett. 2004. Human induced changes in large herbivorous mammal density: The consequences for decomposers. *Frontiers in Ecology and the Environment* 2:145–153.

Wardle, D. A., R. D. Bardgett, J. N. Klironomos, H. Setälä, W. H. Van der Putten, and D. H. Wall. 2004. Ecological linkages between aboveground and belowground biota. *Science* 304:1629–1633.

Wardle, D. A., G. M. Barker, G. W. Yeates, K. I. Bonner, and A. Ghani. 2001. Introduced browsing mammals in natural New Zealand forests: Aboveground and belowground consequences. *Ecological Monographs* 71:587–614.

Wardle, D. A., P. J. Bellingham, K. I. Bonner, and C. P. H. Mulder. 2009. Indirect effects of invasive predators on plant litter quality, decomposition and nutrient resorption on seabird-dominated islands. *Ecology* 90:452–464.

Wardle, D. A., P. J. Bellingham, C. P. H. Mulder, and T. Fukami. 2007. Promotion of ecosystem carbon sequestration by invasive predators. *Biology Letters* 3:479–482.

Wardle, D. A., W. M. Williamson, G. W. Yeates, and K. I. Bonner. 2005. Trickle-down effects of aboveground trophic cascades on the soil food web. *Oikos* 111:348–358.

Wardle, D. A., and G. W. Yeates. 1993. The dual importance of competition and predation as regulatory forces in terrestrial ecosystems: Evidence from decomposer food-webs. *Oecologia* 93:303–306.

Ware, D. M., and R. E. Thomson. 2005. Bottom-up ecosystem trophic dynamics determine fish production in the northeastern Pacific. *Science* 308:1280–1284.

Warfe, D. M., and L. A. Barmuta. 2006. Habitat structural complexity mediates food web dynamics in a freshwater macrophyte community. *Oecologia* 150:141–154.

Warren, R. J. 1991. Ecological justification for controlling deer populations in eastern national parks. *Transactions of the North American Wildlife and Natural Resource Conference* 56:56–66.

Watson, A. 1983. Eighteenth-century deer numbers and pine regeneration near Braemar, Scotland. *Biological Conservation* 25:289–305.

Watters, G. M., R. J. Olson, R. C. Francis, P. C. Fiedler, J. J. Polovina, S. B. Reilly, K. R. Aydin, et al. 2003. Physical forcing and the dynamics of the pelagic ecosystem in the eastern tropical Pacific: Simulations with ENSO-scale and global-warming climate drivers. *Canadian Journal of Fisheries and Aquatic Sciences* 60:1161–1175.

Weis, A. E., and M. R. Berenbaum. 1989. Herbivorous insects and green plants. Pp. 123–162 in W. G. Abrahamson, ed. *Plant–animal interactions*. McGraw-Hill, New York.

Welch, D., B. W. Staines, D. Scott, D. D. French, and D. C. Catt. 1991. Leader browsing by red and roe deer on young Sitka spruce trees in western Scotland. 1. Damage rates and the influence of habitat factors. *Forestry* 64:61–82.

Wendell, F. E., R. A. Hardy, J. A. Ames, and R. T. Burge. 1986. Temporal and spatial patterns in sea otter, *Enhydra lutris*, range expansion and in the loss of Pismo clam fisheries. *California Fish and Game* 72:197–212.

Werner, E. E. 1991. Non-lethal effects of a predator on competitive interactions between two anuran larvae. *Ecology* 72:1709–1720.

Werner, E. E., and J. F. Gilliam. 1984. The ontogenetic niche and species interactions in size-structured populations. *Annual Review of Ecology and Systematics* 15:393–425.

Werner, E. E., J. F. Gilliam, D. J. Hall, and G. G. Mittelbach. 1983. An experimental test of the effects of predation risk on habitat use in fish. *Ecology* 64:1540–1548.

Werner, E. E., and S. D. Peacor. 2003. A review of trait-mediated indirect interactions in ecological communities. *Ecology* 84:1083–1110.

Werner, E. E., and S. D. Peacor. 2006. Lethal and non-lethal predator effects on an herbivore guild mediated by system productivity. *Ecology* 87:347–361.

White, R. W., and R. Wiseman. 2002. The success of a soft-release reintroduction of the flightless Aldabra rail (*Dryolimnas* [*cuvieri*] *aldabranus*) on Aldabra Atoll, Seychelles. *Biological Conservation* 107:203–210.

White, T. C. R. 2005. *Why does the world stay green? Nutrition and survival of plant-eaters*. CSIRO, Melbourne.

Whitham, T. G., and S. Mopper. 1985. Chronic herbivory: Impacts on architecture and sex expression in pinyon pine. *Science* 228:1089–1091.

Whitney, G. G. 1984. Fifty years of change in the arboreal vegetation of Heart's Content, an old-growth hemlock–white pine–northern hardwood stand. *Ecology* 65:403–408.

Whitney, G. G. 1994. *From coastal wilderness to fruited plain*. Cambridge University Press, Cambridge.

Whittaker, R. H. 1975. *Communities and ecosystems*. Collier MacMillan, London.

Whyte, I. J., R. van Aarde, and S. L. Pimm. 2003. Kruger's elephant population: Its size and consequences for ecosystem heterogeneity. Pp. 332–348 in J. T. du Toit, K. H. Rogers, and H. C. Biggs, eds. *The Kruger experience. Ecology and management of savanna heterogeneity*. Island Press, Washington, DC.

Wiegmann, S. M., and D. M. Waller. 2006. Fifty years of change in northern upland forest understories: Identity and traits of "winners" and "loser" plant species. *Biological Conservation* 129:109–123.

Wilby, A., M. Shachak, and B. Boecken. 2001. Integration of ecosystem engineering and trophic effects of herbivores. *Oikos* 92:436–444.

Wilcove, D. S. 1985. Nest predation in forest tracts and the decline of migratory songbirds. *Ecology* 66:1211–1214.

Williams, I. D., and N. V. C. Polunin. 2001. Large-scale associations between macroalgal cover and grazer biomass on mid-depth reefs in the Caribbean. *Coral Reefs* 19:358–366.

Williams, S. L., and K. L. Heck Jr. 2001. Seagrass community ecology. Pp. 317–337 in M. D. Bertness, S. D. Gaines, and M. E. Hay, eds. *Marine community ecology*. Sinauer, Sunderland, MA.

Williams, T. M., J. A. Estes, D. F. Doak, and A. M. Springer. 2004. Killer appetites: Assessing the role of predators in ecological communities. *Ecology* 85:3373–3384.

Wilmers, C. C., R. L. Crabtree, D. W. Smith, K. M. Murphy, and W. M. Getz. 2003. Trophic facilitation by introduced top predators: Grey wolf subsidies to scavengers in Yellowstone National Park. *Journal of Animal Ecology* 72:909–916.

Wilson, J. B., and A. D. Q. Agnew. 1992. Positive feedback switches in plant communities. *Advances in Ecological Research* 23:263–336.

Winemiller, K. O. 1990. Spatial and temporal variation in tropical fish trophic networks. *Ecological Monographs* 60:331–367.

Winemiller, K. O., and K. A. Rose. 1992. Patterns of life-history diversification in North-American fishes: Implications for population regulation. *Canadian Journal of Fisheries and Aquatic Sciences* 49:2196–2218.

Winemiller, K. O., and K. A. Rose. 1993. Why do most fish produce so many tiny offspring? *American Naturalist* 142:585–603.

Winnie, J. Jr., and S. Creel. 2007. Sex-specific behavioural responses of elk to spatial and temporal variation in the threat of wolf predation. *Animal Behaviour* 73:215–225.

Winter, F. C., and J. A. Estes. 1992. Experimental evidence for the effects of polyphenolic compounds from *Dictyoneurum californicum* (Phaeophyta; Laminariales) on feeding rate and growth in the red abalone (*Haliotus rufescens*). *Journal of Experimental Marine Biology and Ecology* 155:263–277.

Wirsing, A. A., M. R. Heithaus, A. Frid, and L. M. Dill. 2008. Seascapes of fear: Evaluating sublethal predator effects experienced and generated by marine mammals. *Marine Mammal Science* 24:1–15.

Wirsing, A. J., M. R. Heithaus, and L. M. Dill. 2007. Living on the edge: Dugongs prefer to forage in microhabitats that allow escape from rather than avoidance of predators. *Animal Behaviour* 74:93–101.

Witman, J. D., and K. P. Sebens. 1992. Regional variation in fish predation intensity: A historical perspective in the Gulf of Maine. *Oecologia* 90:305–315.

Wollkind, D. J. 1976. Exploitation in three trophic levels: An extension allowing intraspecies carnivore interaction. *American Naturalist* 110:431–447.

Woodley, S., J. Kay, and G. Francis, eds. 1993. *Ecological integrity and the management of ecosystems*. St. Lucie Press, Delray Beach. FL.

Woodroffe, R., and J. R. Ginsberg. 1998. Edge effects and extinction of populations inside protected areas. *Science* 280:2126–2128.

Woodroffe, R., and J. R. Ginsberg. 2005. King of the beasts? Evidence for guild redundancy among large mammalian carnivores. Pp. 154–176 in J. C. Ray, K. H. Redford, R. S. Steneck, and J. Berger, eds. *Large carnivores and the conservation of biodiversity*. Island Press, Washington, DC.

Woodward, F. I., and M. R. Lomas. 2004. Vegetation dynamics: Simulating responses to climatic change. *Biological Reviews* 79:643–670.

Woodward, F. I., T. M. Smith, and W. R. Emanuel. 1995. A global land primary productivity and phytogeography model. *Global Biogeochemical Cycles* 9:471–490.

Wootton, J. T. 1993. Size-dependent competition: Effects on the dynamics vs. the endpoint of mussel bed succession. *Ecology* 74:195–206.

Wootton, J. T. 1997. Estimates and tests of per capita interaction strength: Diet, abundance, and impact of intertidally foraging birds. *Ecological Monographs* 67:45–64.

Wootton, J. T., and M. E. Power. 1993. Productivity, consumers and the structure of a river food chain. *Proceedings of the National Academy of Sciences* 90:1384–1387.

Worm, B., and H. K. Lotze. 2006. Effects of eutrophication, grazing, and algal blooms on rocky shores. *Limnology and Oceanography* 51:569–579.

Worm, B., and R. A. Myers. 2003. Meta-analysis of cod–shrimp interactions reveals top-down control in oceanic food webs. *Ecology* 84:162–173.

Worthy, T. H., and R. N. Holdaway. 2002. *The lost world of the moa*. University of Canterbury Press, Christchurch, NZ.

Wright, J. P., A. S. Flecker, and C. G. Jones. 2003. Local vs. landscape controls on plant species richness in beaver meadows. *Ecology* 84:3162–3173.

Wright, S. J. 2003. The myriad effects of hunting for vertebrates and plants in tropical forests. *Perspectives in Plant Ecology, Evolution and Systematics* 6:73–86.

Wright, S. J., M. E. Gompper, and B. DeLeon. 1994. Are large predators keystone species in Neotropical forests? The evidence from Barro Colorado Island. *Oikos* 71:279–294.

Wright, S. J., A. Hernandez, and R. Condit. 2007. The bushmeat harvest alters seedling banks by favoring lianas, large seeds, and seeds dispersed by bats, birds and wind. *Biotropica* 39:363–371.

Wright, S. J., H. Zeballos, I. Domínguez, M. M. Gallardo, M. C. Moreno, and R. Ibáñez. 2000. Poachers alter mammal abundance, seed dispersal, and seed predation in a Neotropical forest. *Conservation Biology* 14:227–239.

Wuerthner, G. 1997. *Are cows just domestic bison? Behavioral and habitat use differences between cattle and bison*. Paper presented at International Symposium on Bison Ecology and Management in North America, Montana State University, Bozeman.

Wurst, S., R. Langel, A. Reineking, M. Bonkowski, and S. Scheu. 2003. Effects of earthworms and organic litter distribution on plant performance and aphid reproduction. *Oecologia* 137:90–96.

Wyatt, J. L., and M. R. Silman. 2004. Distance-dependence in two Amazonian palms: Effects of spatial and temporal variation in seed predator communities. *Oecologia* 140:26–35.

Wyman, R. L. 1998. Experimental assessment of salamanders as predators of detrital food webs: Effects on invertebrates, decomposition and the carbon cycle. *Biodiversity and Conservation* 7:641–650.

Yamaguchi, A., I. Kawahara, and S. Ito. 2005. Occurrence, growth and food of long-headed eagle ray, *Aetobatus flagellum*, in Ariake Sound, Kyushu, Japan. *Environmental Biology of Fishes* 74:229–238.

Young, T. P., M. L. Stanton, and C. E. Christian. 1987. Effects of natural and simulated herbivory on spine lengths of *Acacia drepanolobium* in Kenya. *Oikos* 101:171–179.

Zangerl, A. R., and M. R. Berenbaum. 2005. Increase in toxicity of an invasive weed after reassociation with its coevolved herbivore. *Proceedings of the National Academy of Sciences* 102:15529–15532.

Zimov, S. A. 2005. Pleistocene park: Return of the mammoth's ecosystem. *Science* 308:796–798.

Zimov, S. A., V. I. Chuprynin, A. P. Oreshko, F. S. Chapin III, J. F. Reynolds, and M. C. Chapin. 1995. Steppe–tundra transition: A herbivore-driven biome shift at the end of the Pleistocene. *American Naturalist* 146:765–794.

Zobel, M., and A. Kont. 1992. Formation and succession of alvar communities in the Baltic land uplift area. *Nordic Journal of Botany* 12:249–256.

Contributors

Joel Berger
Wildlife Biology Program
Division of Biological Sciences
Department of Organismic Biology and Ecology
University of Montana
Missoula, MT 59812

Robert L. Beschta
Department of Forest Ecosystems and Society
231 Peavy Hall
Oregon State University
Corvallis, OR 97331

William J. Bond
Botany Department
University of Cape Town
Private Bag
Rondebosch 7701, South Africa

Justin S. Brashares
Department of Environmental Science, Policy and Management
13 Mulford Hall
University of California
Berkeley, CA 94720

Stephen R. Carpenter
Center for Limnology
680 North Park Street
University of Wisconsin
Madison, WI 53706

Jonathan J. Cole
Cary Institute for Ecosystem Studies
Box AB, Route 44A
Millbrook, NY 12545

Jonas Dahlgren
Department of Forest Resource Management
Swedish University of Agricultural Sciences
SE-901 83, Umeå, Sweden

Per Ekerholm
Section for Environment
Administration of Kronoberg County
SE-351 86, Bäxjü, Sweden

Clinton W. Epps
Department of Fisheries and Wildlife
Nash Hall, Room #104
Oregon State University
Corvallis, OR 97331

Tim Essington
School of Aquatic and Fisheries Science
University of Washington
P.O. Box 355020
Seattle, WA 98195

James A. Estes
Department of Ecology and Evolutionary Biology
University of California
Center for Ocean Health
100 Shaffer Road
Santa Cruz, CA 95060

Kenneth Feeley
Department of Biological Sciences
Florida International University
Miami, FL 33199

John M. Fryxell
Department of Integrative Biology
University of Guelph
50 Stone Road E, Guelph
Ontario, N1G 2W1, Canada

Peter Hambäck
Department of Botany
Stockholm University
SE-106 91, Stockholm, Sweden

Ricardo M. Holdo
Department of Zoology
P.O. Box 118525
University of Florida
Gainesville, FL 32611

Robert D. Holt
Department of Zoology
P.O. Box 118525
University of Florida
Gainesville, FL 32611

Jeremy B. C. Jackson
Center for Marine Biodiversity and Conservation
Scripps Institution of Oceanography
9500 Gilman Drive
La Jolla, CA 92093-0202

James F. Kitchell
Center for Limnology
680 North Park Street
University of Wisconsin
Madison, WI 53706

Åsa Lindgren
Department of Botany
Stockholm University
S-106 91, Stockholm, Sweden

Russell W. Markel
Department of Zoology
University of British Columbia
6270 University Boulevard
Vancouver, BC V6T 1Z4, Canada

Robert J. Marquis
Department of Biology
University of Missouri–St. Louis
One University Boulevard
St. Louis, MO 63121

Blake Matthews
Department of Zoology
University of British Columbia
6270 University Boulevard
Vancouver, BC V6T 1Z4, Canada

Kristine Metzger
Centre for Biodiversity Research
6270 University Boulevard
University of British Columbia
Vancouver, V6T 1Z4, Canada

Ally Nkwabe
Serengeti Biodiversity Program
Tanzania Wildlife Research Institute
P.O. Box 661
Arusha, Tanzania

Lauri Oksanen
Section of Ecology
Department of Biology
University of Turku
FI-20014, Turku, Finland and
Department of Natural Sciences
Finnmark University College
N-9509, Alta, Norway

Tarja Oksanen
Department of Ecology and Environmental Science
Umeå University
SE-901 78, Umeå, Sweden

Johan Olofsson
Department of Ecology and Environmental Science
Umeå University
SE-901 78, Umeå, Sweden

Michael L. Pace
Environmental Sciences Department
Clark Hall
291 McCormick Road
P.O. Box 400123
University of Virginia
Charlottesville, VA 22904

Robert T. Paine
Department of Biology
P.O. Box 351800
University of Washington
Seattle, WA 98195-1800

Charles H. Peterson
Institute of Marine Sciences
University of North Carolina at Chapel Hill
3431 Arendell Street
Morehead City, NC 28557

Laura R. Prugh
Department of Environmental Science, Policy
    and Management
13 Mulford Hall
University of California
Berkeley, CA 94720

William J. Ripple
Department of Forest Ecosystems and Society
314 Richardson Hall
Oregon State University
Corvallis, OR 97331

Thomas P. Rooney
Department of Biological Sciences
Wright State University
3640 Colonel Glenn Hwy
Dayton, OH 45435

Stuart A. Sandin
Center for Marine Biodiversity and Conservation
Scripps Institution of Oceanography
9500 Gilman Drive
La Jolla, CA 92093-0202

Marten Scheffer
Aquatic Ecology and Water Quality Management Group
Department of Environmental Sciences
Wageningen University
P.O. Box 8080
6700 DD, Wageningen, The Netherlands

Thomas W. Schoener
College of Biological Sciences
6328 Storer Hall
University of California
Davis, CA 95616

Gregor Sharam
Centre for Biodiversity Research
6270 University Boulevard
University of British Columbia
Vancouver, V6T 1Z4, Canada

Jonathan B. Shurin
Department of Zoology
University of British Columbia
6270 University Boulevard
Vancouver, BC V6T 1Z4, Canada

A. R. E. Sinclair
Centre for Biodiversity Research
6270 University Boulevard
University of British Columbia
Vancouver, V6T 1Z4, Canada

Michael E. Soulé
P.O. Box 1808
Paonia, CO 81428

David A. Spiller
Department of Evolution and Ecology
4342 Storer Hall
University of California
Davis, CA 95616

Robert S. Steneck
School of Marine Sciences
University of Maine
Darling Marine Center
193 Clarks Cove Road
Wapole, ME 04573

Chantal J. Stoner
Department of Environmental Science, Policy and Management
13 Mulford Hall
University of California
Berkeley, CA 94720

John Terborgh
Center for Tropical Conservation
P.O. Box 90381
Duke University
Durham, NC 27708

F. J. Frank van Veen
Centre for Ecology & Conservation
School of Biosciences
University of Exeter, Cornwall Campus
Penryn, Cornwall
TR10 9EZ
United Kingdom

Sheila M. Walsh
Center for Marine Biodiversity and Conservation
Scripps Institution of Oceanography
9500 Gilman Drive
La Jolla, CA 92093-0202

David A. Wardle
Department of Forest Ecology and Management
Faculty of Forestry
Swedish University of Agricultural Sciences
SE901-83, Umeå, Sweden